Biological Electron Microscopy

Theory, Techniques, and Troubleshooting

Biological Electron Microscopy

Theory, Techniques, and Troubleshooting

MICHAEL J. DYKSTRA

College of Veterinary Medicine
North Carolina State University
Raleigh, North Carolina

PLENUM PRESS • NEW YORK AND LONDON

Library of Congress Cataloging in Publication Data

Dykstra, Michael J.
  Biological electron microscopy: theory, techniques, and troubleshooting / Michael
J. Dykstra.
      p.   cm.
  Includes bibliographical references (p.   ) and index.
  ISBN 0-306-44277-9
  1. Electron microscopy. 2. Scanning electron microscopy. 3. Transmission electron
microscopes. I. Title.
QH212.E4D95   1992   92-34102
578'.45 – dc20   CIP

ISBN 0-306-44277-9

© 1992 Michael J. Dykstra
Plenum Press is a division of Plenum Publishing Corporation
233 Spring Street, New York, N.Y. 10013

To the two Susans, for patience

Preface

Electron microscopy is frequently portrayed as a technically demanding discipline operating largely in the sphere of "black boxes" and governed by many absolute laws of procedure. At the introductory level, this description does the discipline and the student a disservice. The instrumentation we use is complex but ultimately understandable and, more important, reparable. The chemical and physical processes underlying the procedures we employ for preparing tissues and cells are not totally elucidated, but enough information is available to allow investigators to make reasonable choices concerning the best techniques to apply to their particular problems. There are countless specialized techniques in the field of electron microscopy that require the acquisition of specialized knowledge, particularly for the interpretation of results (energy-dispersive spectroscopy comes immediately to mind), but most laboratories with equipment to utilize these approaches have specialists to help the casual user.

There are several books available that deal with biological electron microscopy. Some (e.g., Meek, 1976) are extremely good in regard to instrumentation, whereas others are heavily skewed toward specimen preparation (e.g., Hayat, 1989). Some, such as that by Wischnitzer (1981), cover both specimen preparation and instrumentation in almost sufficient detail for an introductory course. Unfortunately, most single-volume electron microscopy texts were written prior to 1983 and are thus quite dated. There are also several series of books (see the Hayat series and the Glauert series in Appendix B) containing quite complete and up-to-date information on many subjects, but these are too expensive and too extensive to be considered as textbooks. There are also a number of laboratory procedure manuals available, but these consistently state that the limited techniques they list are the only ones that can be used. This is clearly not the case.

Most of my contemporaries were taught fear, rather than respect, for the instrumentation of electron microscopy. In addition, we were generally taught that there was only one right way to fix and embed a given class of organisms, only one way to properly break a glass knife, only one side of a grid to use for section retrieval, and only one way to properly post-stain grids. After going to various meetings for 20 years and hearing all the different types of approaches utilized to obtain publishable ultra-structural work, I see clearly that the system is much more flexible than our training had suggested.

This textbook is offered to help a beginning student see how things work in electron microscopy. It presents the principles of fixation and specimen preparation, showing the breadth of approaches that are practical. It gives students sufficient information about basic specimen preparation to make cogent decisions for their own projects after consulting the relevant literature for their own group of organisms or tissues. There are still black boxes in the discipline, but I will try to minimize them.

A major component of this book is suggested by the subtitle: Theory, Techniques, and Troubleshooting. Too many of my generation were given in their training little theory beyond optical theory; too little information about techniques, except the ones used in their specific laboratory; and almost nothing about troubleshooting problems, particularly in regard to instrumentation. Clearly, things do go wrong, and with alarming regularity.

In a discipline with so many varied approaches from which to choose, teaching how to approach a problem logically and how to correct a problem that presents itself probably represents the highest aim of a course in electron microscopy.

This book is intended for a one-semester course that covers all the basic approaches utilized in transmission and scanning electron microscopy. Its arrangement is somewhat unusual in being more or less chronological, so that the student can be reading about the techniques as they are employed in class.

Finally, one of the major purposes of this text is to provide more up-to-date coverage of the two areas that have revolutionized electron microscopy over the last 8–10 years: cryotechniques and immunolabeling. When glutaraldehyde was introduced by Sabatini *et al.* in 1963, chemical fixation more or less reached its apex. Until the advent of easily reproducible cryotechniques, no further advances took place in specimen preparation for transmission electron microscopy. By the late 1970s, the discipline was perceived by many as a fairly passé descriptive science with little innovation except in the area of high-voltage electron microscopy. Fortunately, in the 1980s there was improved instrumentation for cryo work, and the coupling of immunology and electron microscopy through immunogold and immunoperoxidase techniques has brought our discipline back on line. The power of current techniques to utilize molecular approaches to probe the location of various cellular products and activities has rekindled interest in the truly dynamic aspects of cellular behavior.

A student who masters the concepts in this text will be capable, with continued practice in technical skills, to utilize electron microscopy techniques productively in his or her research. It is my hope that this text will give a firm foundation on which the student can build knowledge of cellular structure and behavior far beyond the scope of this work.

Hayat, M.A. 1989. *Fixation for electron microscopy.* Academic Press, New York.
Meek, G.A. 1976. *Practical electron microscopy for biologists*, 2nd ed. John Wiley & Sons, New York.
Wischnitzer, S. 1981. *Introduction to electron microscopy*, 3rd ed. Pergamon Press, New York.

Contents

Introduction **1**

Chapter 1

Specimen Preparation for Transmission Electron Microscopy **5**
    I. Fixatives 5
        A. Purpose; Killing versus Fixing 5
        B. Methods 6
    II. Buffers 24
        A. Molarity/Molality/Osmolarity/Tonicity 24
        B. Purpose 29
        C. Types, Characteristics, and Uses of Buffers 30
    III. Dehydration 32
        A. Purpose 32
        B. Agents 32
    IV. Embedding Media 34
        A. Ideal Qualities 34
        B. Classes and Characteristics of Resins 34
        C. Embedding Mold Types 38
        D. Agar Embedment 38
    V. Examination of Four Tissues Prepared with a Variety of Fixatives
        and Buffers 39
    VI. A Quasi-Universal Fixation, Dehydration, and Embedment Schedule
        Successfully Used on Organisms from All Five Kingdoms of Life .. 68
    References 77

Chapter 2

Ultramicrotomy **79**
    I. Ultramicrotomes 79

A. Purpose   79
B. Design   79
C. History   81
II. Knives   82
A. History   82
B. Glass Knife Manufacturing   83
C. Section Handling   85
D. Knife Storage   86
E. Diamond Knives   87
III. Block Trimming   87
A. Trimming Procedures   88
B. Block-Trimming Tools   90
IV. Sectioning Procedures   91
A. Working Area   91
B. Sectioning Procedures   93
V. Eight Commonly Encountered Sectioning Problems   97
References 101

Chapter 3

Support Films **103**

I. Purpose 103
II. Types 103
A. Nitrocellulose 103
B. Formvar 103
C. Carbon 104
III. Methods 104
A. Droplet 105
B. Slide Stripping 105
C. Holey Films 111
D. Carbon Films 112
References 112

Chapter 4

Transmission Electron Microscopy **113**

I. Historical Review of Microscopy (1590–1990s) 113
II. Theory of Electron Optics 115
A. Light Microscopes versus the TEM 115
B. Resolution 115
C. Electron Lenses 117
D. Properties of Electron Lenses 118

III. Four Aspects of Image Formation 127
    A. Absorption 127
    B. Interference 127
    C. Diffraction 127
    D. Scattering 128
IV. General TEM Features 128
    A. Operating Voltage 128
    B. Resolution 128
    C. Magnification 129
    D. High Vacuum, Electronic, Magnetic, and Physical Stability 129
V. Parts of the Electron Microscope: Functional Aspects 129
    A. The Electron Gun 131
    B. Condenser Lens System 134
    C. Deflector Coils 135
    D. Objective Lens 135
    E. Diffraction Lens 138
    F. Projector System 138
    G. Camera Systems 138
    H. Specimen Holders 139
    I. Viewing System 139
    J. Detectors 141
VI. Operation of the TEM: Decision Making 141
    A. Accelerating Voltage 141
    B. Choice of Beam Current and Bias 141
    C. Condenser Settings 142
    D. Objective Settings 142
    E. Alignment 142
    F. Taking a Photograph 148
    G. Specimen Radiation Dose 148
    H. Microscope Calibration 149
References 150

Chapter 5

Vacuum Systems **151**

I. Types of Gauges 151
    A. Direct Reading 151
    B. Indirect Gauges 152
II. Vacuum Pumps 155
    A. Types of Pumping Systems 155
    B. Mechanical/Rotary Vane Pumps 155
    C. Diffusion Pumps 157
    D. Turbomolecular Pumps 159

E. Sputter Ion (Ion Getter) Pumps 161
F. Cryopumps and "Cold Fingers" 162
III. Sequential Operation of a Complete Vacuum System to Achieve
High Vacuum 162
IV. Lubrication of Vacuum Seals and Leak Detection 164
References 165

Chapter 6

Staining Methods for Semithins and Ultrathins **167**

I. Semithin Section Staining 167
   A. Toluidine Blue-O 168
   B. Toluidine Blue-O and Acid Fuchsin 169
   C. Basic Fuchsin/Methylene Blue 170
   D. Methylene Blue 171
   E. Periodic Acid/Schiff's Reagent 171
II. Ultrathin Section Staining 171
   A. Purpose 171
   B. *En Bloc* versus Post-Staining 172
   C. Commonly Used Post-Stains 173
   D. Microwave Staining 180
   E. Dark-Field Imaging without Staining 180
References 181

Chapter 7

Photography **183**

I. Emulsion Composition 184
II. Film Types 185
III. Producing a Latent Image 187
IV. Film Processing 187
   A. Developer 188
   B. Stop Baths 189
   C. Fixer 190
V. Development Controls 190
   A. Time 190
   B. Temperature 190
   C. Agitation 191
   D. Developer Choice 191
VI. Paper Types 191
VII. Keeping Properties of Chemicals and Precautions 193
VIII. Sharpness 193

   IX. Films Used in the Electron Microscopy Laboratory 194
      A. Negative-Release Films 194
      B. Positive-Release Films 195
    X. Copy Work 198
      A. Films 198
      B. Improving Copy-Stand Images 199
   XI. Types of Enlargers 204
      A. Diffusion Enlargers 204
      B. Condenser Enlargers 205
      C. Point Light Source Enlargers 205
      D. LogEtronics Enlargers 206
  XII. Viewing a Print in Perspective 206
  References 207

Chapter 8

Replicas, Shadowing, and Negative Staining **209**

    I. Shadowing Casting 209
      A. Mechanism 210
      B. Metals Used 211
      C. Vacuum Evaporators 211
      D. Electrodes 214
      E. Factors Leading to Fine Grains of Shadowed Metal 215
      F. Shadowing Techniques 216
      G. Sputter Coating 216
   II. Negative Staining 218
      A. Mechanism 218
      B. Methods 219
  References 221

Chapter 9

Scanning Electron Microscopy **223**

    I. History 223
   II. The Use of SEM in Biological Research and Medicine 224
  III. Principles of the SEM 228
  IV. Operation of the SEM 229
   V. Interaction of the Electron Beam and Specimen 230
      A. Resolution 230
      B. Secondary Electrons 232
      C. Backscattered Electrons 234
      D. Auger Electrons 235

        E. Energy-Dispersive Spectroscopy (EDS) 236
        F. Cathodoluminescence 237
    VI. Specimen Preparation 237
        A. Fixation 237
        B. Dehydration and Transition Fluids 238
        C. Drying 238
        D. Mounting Specimens 240
        E. Coating Specimens 241
    VII. Artifacts and Their Correction 242
    VIII. Specialty SEMS: FEG, LV, and ESEM Devices 243
    References 245

Chapter 10

Cryotechniques **247**

    I. History 248
        A. Organismal Period 248
        B. Mechanistic Period 248
        C. Cytological Period 249
    II. Purpose 249
    III. Cryogens 251
    IV. Safety Precautions 251
    V. Freezing Methods 252
        A. High-Speed Plunging/Immersion 252
        B. Spray Freezing 254
        C. Jet Freezing 254
        D. High-Pressure Freezing 255
        E. Metal Mirror/Slam Freezing 257
    VI. Uses of Frozen Specimens 258
        A. Cryoultramicrotomy 260
        B. Cryosubstitution 262
        C. Freeze-Fracture 264
        D. Freeze-Drying as Typified by Molecular Distillation 267
    VII. Artifacts and Their Correction 270
    References 272

Chapter 11

High-Voltage Electron Microscopy **275**

    I. History 275
    II. Purpose 276
    III. Functional Aspects of HVEMS 276
        A. Resolution 276

B. Radiation Damage 277
C. Contrast 278
IV. Microscope Construction 279
V. Sample Preparation 280
VI. Applications 280
VII. Intermediate-Voltage Electron Microscopy (IVEM) 281
References 283

Chapter 12

Microanalysis **285**

I. Microanalysis Techniques 285
A. Energy-Dispersive Spectroscopy (EDS) 285
B. Electron Energy Loss Spectroscopy (EELS) 291
References 293

Chapter 13

Cytochemistry **295**

I. Problems 295
II. Specific Reaction Products 297
A. Peroxidase Procedures 298
B. Lead Capture 298
C. Ferritin 298
D. Colloidal Gold 299
E. Ruthenium Red, Alcian Blue, Pyroantimonate 299
III. Examples of Enzyme Cytochemistry 299
A. Peroxidase Methods 299
B. Hatchett's Brown Methods 300
C. Lead-Capture Methods 300
IV. Examples of Nonenzymatic Cytochemistry 302
A. Cationic Dyes 302
B. Polysaccharide Stains 302
C. Monosaccharide and Disaccharide Stains 305
D. Calcium Staining 305
References 307

Chapter 14

Immunocytochemistry **309**

I. Purpose 309
II. Preparative Techniques 310

A. Pre-Embedding Labeling 310
B. Post-Embedding Procedures 311
C. Cryoultramicrotomy Technique 313
D. Negative Staining Procedures 314
III. Immunoglobulins 315
A. Protein A and Protein G Techniques 315
B. Polyclonal and Monoclonal Antibodies 315
IV. Common Immunolabeling Techniques for Electron Microscopy 315
A. Immunoferritin 315
B. Immunoperoxidase Techniques 318
C. Immunogold Techniques 318
References 319

Chapter 15

Autoradiography **321**

I. History 321
II. Purpose 322
III. Theoretical Aspects 322
A. Detection of Radioactivity 322
B. Types of Particles 322
C. Nuclear Emulsions 324
D. Determination of Isotope Dose Level 325
E. Rules for Autoradiography 326
F. Light Microscopy Autoradiography 330
References 331

Chapter 16

Computer-Assisted Imaging **333**

I. Purpose 333
II. Resolution and Discrimination 334
III. Image Processing 335
IV. Morphometric Analysis 337
V. Stereology 337
VI. Computer-Assisted Analysis of Movement 339
References 340

Chapter 17

Scanning Tunneling Microscopy and Its Derivatives **341**

Appendix A

Laboratory Safety **345**

Appendix B

Literature Sources for Electron Microscopy **347**

    I. Atlases 347
    II. Journals 347
    III. Society Publications 348
    IV. Book Series 348
    V. NIH Resources for Intermediate Voltage Electron Microscopy
       (IVEM) 348

Appendix C

Electron Microscopy Equipment and Supplies **349**

    I. Expendable Supplies and Small Equipment 349
    II. Electron Microscopes 350
    III. Diamond Knives 351
    IV. High-Vacuum Pumps 352
    V. Ultramicrotomes 352
    VI. Equipment for Cryotechniques 352
    VII. Sputter Coaters and Vacuum Evaporators 353

Index **355**

Biological Electron Microscopy

Theory, Techniques, and Troubleshooting

Introduction

Although electron microscopy is usually thought of as a fairly recent development dating from Knoll and Ruska's report on the first transmission electron microscope in 1932, the discipline actually has much older antecedents. The science of light microscope optics is directly connected with the beginnings of electron microscopy because many of the shared principles of optics were first defined by developers of early light microscopes. The roots of our discipline can be ultimately traced back to the Janssen brothers, who developed the first compound light microscope. In the 1600s, the field expanded to include offerings by Malpighi, Hooke, and Leeuwenhoek, who worked with light microscope optics to describe the living organisms in water and the microstructure of various tissues. By the 1800s, the essential cellular nature of life was confirmed independently by Schleiden and Schwann, and by 1840 the fact that most cells contained a nucleus was established. By 1886, the maximum resolution of glass optics had been achieved by Ernst Abbe through his development of apochromatic objectives and matched oculars that were highly corrected for spherical aberrations and were color corrected for three wavelengths of light. Thus, in approximately 300 years light microscopy was born, expanded, and perfected.

Like the current "information age," electron microscopy steamrolled forward from its birth in 1932. The first instrument had no focusing system for illumination and had less resolution than a light microscope. By 1934, a condenser lens for focusing the electron beam had been added by Ruska. The resolving power surpassed that of a light microscope for the first time in 1935. During the late 1930s, the gathering storms of war caused the small electron microscopy community to begin working with decreased intergroup communication. Siemens and Halske produced the first commercial electron microscope in Germany in 1939. Burton, Hillier, and Prebus constructed an electron microscope in Canada, which led to the production of the 1941 version, known as the RCA type B, which could resolve 25 Å. By 1946, Hillier had a microscope that could resolve 10 Å, which exceeds the level of resolution normally possible with chemically fixed biological tissue (about one third of a plasmalemma, or 20 Å, is usually resolvable).

During the 1950s and 1960s, power supplies were refined and lens manufacture, vacuum systems, and both mechanical and electronic stability were improved. By the 1970s, high-voltage electron microscopes, scanning transmission electron microscopes, and microanalytical instruments were commercially available. At the end of the 1980s, computer-assisted instruments capable of 3.44-Å resolution and from 200,000 to 500,000× maximum magnification were standard, with some instruments capable of resolving 1.27 Å. In addition, the development of intermediate voltage electron microscopy (IVEM), turbomolecular pumps, fiberoptic plates, and computer storage, manipulation, and analysis of images caused an expansion in the field of electron microscopy.

Scanning electron microscopy (SEM) underwent a more or less parallel development to transmission electron microscopy into the 1940s from the first instrument built in 1938 by Von Ardenne. The first commercially available scanning electron microscope, the Cambridge Stereoscan, became available from Oatley's group in Cambridge, England, in 1965 after a development period that began in 1948. By the late 1970s, 50- to 60-Å resolution and better than 100,000× magnification was considered a standard in the SEM trade.

When Ernst Ruska died in 1988, the optical device that he invented with Knoll in 1932 had achieved the practical limits of resolution imposed by electron optics. In one-fifth the time that it took to perfect light microscope optics, electron optics had peaked out.

Unfortunately for biologists, the techniques for handling cells and tissues have not had quite as dizzying a record of success. In virtually all cases, the instrumentation of electron microscopy was first used by materials scientists, due to the inherent stability of their specimens. Biologists were beset with difficulties from the beginning, in that their samples needed to be stabilized by some means and dehydrated so that they could be put into a vacuum. In addition, they needed some surrounding medium to hold them while they were cut into thin enough sections for the electron beam to penetrate. Finally, biological investigators had no good instrumentation for cutting ultrathin sections, nor any knives capable of doing the job in the early days of electron microscopy.

Thus, even though TEM instruments with 25-Å resolution capabilities were available to examine biological specimens by the early 1940s, it was not until the late 1940s that a picture of a sectioned tissue first appeared. Advances in microtome modifications and the introduction of glass knives, and then shortly after, diamond knives, along with the introduction of improved resins, led to the burgeoning publications on biological sections seen in the early 1950s. By the 1960s, all the current commonly used fixatives and classes of resins had been discovered. Little substantive change in tissue preparation for TEM occurred from the early 1960s up until the late 1970s, when cryotechniques moved tissue preservation to a different plane and the introduction of improved acrylic resins allowed a vast expansion of cellular probing techniques that continues to this day.

The preparation of samples for biological scanning electron microscopy lagged considerably behind TEM techniques, largely due to the fact that SEMs were not commercially available until 1965. As was true for TEM, the more conductive and stable specimens from the materials sciences investigators were examined by SEM prior to biological samples.

By the late 1970s, the basic instrumentation for biological electron microscopy (TEMs, SEMs, ultramicrotomes, critical point dryers, vacuum evaporators, sputter coaters, freeze-fracture devices, and other equipment for cryotechniques) was in place and functioning adequately. Since then, most of the instrumentation and techniques of preparation that have been developed are largely modifications of previous techniques.

We will now begin a systematic examination of all the approaches leading to the 8 × 10 inch glossy prints characteristic of our work. We will start with fixation, dehydration, and embedding, and then move to a discussion of the tools and functional characteristics of microtomy. Then we will fully explore the workings of a transmission electron microscope. Methods to impart contrast to specimens will be provided, as well as information about producing photographic images. The book will then cover a number of useful ancillary techniques, such as replica production and negative staining of TEM samples.

Next, we will discuss aspects of scanning electron microscopes that differ from TEMs and the fine points of specimen preparation for such examination.

Finally, we will end with brief discussions of specialized techniques, such as various cryotechniques, high-voltage EM, X-ray microanalysis, EELS, cytochemistry, immunocytochemistry, and autoradiography.

By the end of the text, you should have a good idea of what is available and what kinds of questions to ask the personnel in laboratories set up to do these techniques.

CHAPTER 1

Specimen Preparation for Transmission Electron Microscopy

I. FIXATIVES

Before beginning this section, I want to recommend the book *Fixation for Electron Microscopy* (Hayat, 1981), which is an invaluable source and contains information concerning all aspects of biological specimen preparation in much greater depth than can be provided in this text.

A. Purpose; Killing versus Fixing

The purpose of fixation is two-fold: to bring about the rapid cessation of biological activity and to preserve the structure of the cell. Ideally, the colloidal suspension of cytoplasm and organelles within a cell are turned into a gel that maintains the spatial relationship of the components while providing sufficient stability for them to survive the solvent action of aqueous buffers, dehydration agents, and plastic resins. The aim is to be able to process the tissue without significant changes in size, shape, and positional relationships of the cellular components and to preserve the chemical reactivity of cellular constituents, such as enzymes and antigenic proteins.

One of the first concepts to consider is the difference between killing and fixing. Most of the time, the term *fixing* is used to include both processes, which sometimes leads us to forget that they are not one and the same thing. Most primary fixatives used today are aldehyde fixatives, and it is well documented that cells have significant periods of response to these solutions before cell death (Table 1). Slime mold amoebae

TABLE 1. Rates of Fixation

I. Chemical Fixation
    A. 3% glutaraldehyde/3% acrolein: *Pyrsonympha* axoneme moves
       for 2 sec
    B. 3% glutaraldehyde: Chicken embryo fibroblast cytoplasmic in-
       clusions move for 30–45 sec
    C. Tomato petiolar hair cell cytoplasmic streaming
       1. 15 min with 0.5–5.3% glutaraldehyde
       2. 9 min with 5% acrolein
       3. 6 min with 2% glutaraldehyde + 5% acrolein
       4. 15 min with 2% glutaraldehyde + 1% osmium
II. Cryofixation: Movement ceases in 10 msec

From Gilkey and Staehelin, 1986, with permission.

grown to confluence on an agar surface can be seen to round up and develop intercell spaces during the first 30 sec after a glutaraldehyde mixture is poured onto them. If, instead, the Petri dish containing the cells is first inverted over a drop of 1% osmium tetroxide for 3 min, the addition of glutaraldehyde causes no obvious morphological change in the cells. Thus, the highly volatile and toxic osmium tetroxide vapor *kills* the cells and the later addition of glutaraldehyde *fixes* the cells. It is important not to forget that these processes are not one and the same thing when evaluating the success or failure of your particular fixation regimen.

B. Methods

1. Physical Fixation

Physical fixation involves the application of heat or cold to stabilize cellular components.

a. Heat

Heat fixation can be as simple as the changes that take place in an egg when it is fried: The translucent, colloidal proteinaceous material of the egg white becomes gelled into an opaque, rubbery solid. This material is truly fixed; however, it has been massively altered in structure. Heat fixing has also been used for decades to preserve blood smears in a clinical setting. Blood samples are smeared onto a glass slide and then passed through heat, such as a Bunsen burner flame, to semipermanently affix the cells to the slide and to stabilize their structure enough for rudimentary staining for further identification of cell populations. Neither of these techniques avoids serious precipitation of proteins and major changes in their tertiary structures. In addition, there is a rapid change in the states of hydration of cellular components. Various components, such as lipids, would not be expected to be particularly well preserved by such methods. Thus, these techniques are grossly inadequate for ultrastructural preservation.

In recent years, *microwave* use (Leong *et al.*, 1985; Leong and Gove, 1990; Login *et al.*, 1990) has resulted in some novel, extremely rapid fixation techniques. They are not, however, solely heat fixation techniques. Cells or tissues may be fixed directly by microwaves (after suspension in a suitable fluid such as phosphate-buffered saline) or may be put into a fixative (aldehydes or osmium tetroxide), which is then subjected to microwaves for a matter of seconds. In the latter case, this allows a much shorter time of fixation and is reported to result in better fixative penetration than conventional chilled or room-temperature fixation for an hour or two.

b. Cold

Cold, defined as the rapid freezing of cells or tissues, has been effectively used for preservation. The subject of cryopreservation will be thoroughly treated in the section on cryotechniques, but at this point certain aspects should be considered. Cryofixation, in general, is restricted to much smaller sample sizes than conventional chemical fixation. This is due to the fact that heat can be removed from the specimen quickly enough to prevent ice crystal damage only from a relatively superficial aspect of the tissue. Thus, with most techniques, a freezing depth of about 10–15 μm is achieved. The more basal layers of the specimen have ice crystals large enough that ultrastructural detail is unacceptably distorted. There are three major variations on the theme of cryopreservation that will be elaborated: (1) cryofixation followed by cryoultramicrotomy to produce ultrathin frozen sections, primarily for microanalysis and immunolabeling; (2) cryosubstitution, wherein a specimen is quickly frozen and then dehydrated at reduced temperatures ($-80°C$), followed by infiltration of chemical fixatives that are inactive at the low temperature at which the sample is maintained. After diffusion of fixative throughout the specimen, the temperature is raised in a controlled fashion. As the temperature increases, the fixative components become chemically active and stabilize all the cellular components with which they react simultaneously (as opposed to normal chemical fixation, which involves diffusion of the fixatives from outside the tissues/cells to the interior, resulting in different areas of the sample becoming fixed at different times); (3) freeze-drying, which involves cryofixation followed by the removal of water by sublimation while the specimen is maintained at a low temperature under vacuum.

2. Chemical Fixation

Chemical fixation has been the standard of the biological electron microscopy trade since its inception and, despite the advances in cryopreservation techniques, continues to dominate the field. The fixatives commonly in use have been derived to a large extent from the leather-tanning industry. The advantages of chemical fixation are that the chemicals are generally fairly stable, the specimen-handling techniques are simple, materials can be stored in the solutions for some period of time, no significant equipment is needed, and the fixatives are relatively inexpensive. In addition, although the rate of penetration is not particularly fast and specimens must be made small enough for rapid penetration of fixative to prevent autolysis of structures deep within the sample, a relatively large sample can be adequately preserved. The general rule of thumb is that most fixatives will penetrate at least 0.5 mm into a sample within 1 hour such that if you could produce a 1-mm-thin slice through the center of a watermelon, it would be expected to fix adequately. Needless to say, various softer tissues, such as the brain, are resistant to being neatly sliced into 1-mm-thick portions without some sort of fixation; so other techniques, such as vascular perfusion of fixatives, are often employed.

Evidence of good fixation is based on a number of different criteria, including (1) evidence of ground substance (a background density greater than that of non-tissue plastic areas outside the cells) in the cytoplasm and nucleoplasm, and membranes of the nuclear envelope virtually parallel without significant swelling, (2) mitochondria are not distended, and inner and outer membranes are more or less parallel, (3) the endoplasmic reticulum is not swollen, and most membranes and cellular components are intact. If these three criteria are met, fixation is probably adequate.

In a general contemporary fixation schedule, tissues are fixed in a primary aldehyde fixative followed by rinses in an appropriate buffer. The samples are then post-fixed in osmium tetroxide, rinsed to remove the osmium, dehydrated, infiltrated with a plastic resin, and then the resin containing the sample is polymerized by heat, catalysts, or ultraviolet radiation.

We will now examine individual chemical fixatives and evaluate their effects on various cellular components. In addition, we will discuss their formulation, storage, and parameters of use.

a. Osmium Tetroxide

Osmium tetroxide (OsO_4) has been used for over a century for the preservation of cellular detail. Botanists of the late 1800s used to put a drop of osmium solution on a section of tissue on a glass slide and examine it avidly until sufficient darkening had occurred. These individuals often suffered from a certain haziness of vision, but it eventually resolved (as their previously fixed corneal epithelium sloughed off and new tissue was formed). Osmium was the first fixative used with success for the ultrastructural preservation of animal tissues (Glauert, 1975).

This fixative has a high vapor pressure, is highly toxic, and should be used under a fume hood at all times. An excellent rule to follow is "if you can smell it, you are too close." If osmium contacts your skin, it will turn it black, and if you manage to cut yourself with an osmium-coated surface, the wound will not heal quickly because all of the adjacent tissue will be killed. Tissues generally turn black after osmium treatment, but this is not the best judge of its effectiveness, since tissues low in lipids (such as highly proteinaceous scar tissues) will not blacken. In addition, lipids with different degrees of saturation would be expected to reduce different amounts of osmium. If the unused aqueous osmium stock is straw-colored, it can be assumed to be chemically reactive. When the solution begins to gray or blacken, it should be discarded.

As a primary fixative, osmium causes rapid gross permeabilization of membranes with cessation of cytoplasmic movement within seconds to minutes in most cases. It has been used as a vapor fixative to quickly stop biological processes in cell monolayers and also to stabilize materials without having to go through an aqueous phase. In the latter case, samples containing potentially soluble products have been prepared for X-ray microanalysis by exposing freeze-dried tissues to osmium crystals in a dessicator for 8 hr (Hayat, 1981).

Osmium is most economically purchased in crystalline form in sealed ampules. About 10 years ago, the price per gram skyrocketed. At first, the price increased from about $10 per gram to $45 per gram, making most researchers in the field more careful about the potential of contaminating our osmium stocks. Cleanliness is extremely important when dealing with osmium because it is a strong oxidant and can react readily with virtually any buffer or speck of biological material. It is also photoreactive and is best stored in the refrigerator to increase its shelf life. Some people even fix tissues in the dark to prevent photoreactivity, though this seems to be unnecessary in most cases.

i. DISADVANTAGES

Osmium is one of the most slowly penetrating fixatives we use and has no cross-linking capabilites for cellular components. When used as a primary fixative, it typically makes the nucleoplasm and cytoplasm look somewhat extracted, giving the material more contrast than when using better fixation techniques. As it is a strong oxidant, osmium profoundly inactivates virtually all enzymes and can be expected to damage many antigens.

ii. ADVANTAGES

Osmium has nine different oxidation states (Hayat, 1981), and at least five of them are relatively stable and have different reactivity with cellular components. Osmium is soluble in both polar and nonpolar media, and thus can penetrate and fix both hydrophobic and hydrophilic domains in cells. Finally, when osmium reacts with cellular components in an oxidative fashion, it is concomitantly reduced. The reduced form is largely unextractable and also has the added benefit of exhibiting electron density under an electron beam, since it is a heavy metal.

iii. SPECIFIC ACTIVITY

Lipids. Osmium works directly on unsaturated lipids by oxidizing double bonds, leading to the formation of monoesters, diesters, and dimeric monoesters (Fig. 1). Saturated fatty acids are normally chemically unreactive with osmium, since they lack double bonds but often are preserved by an indirect route. As mentioned above, osmium is soluble in nonpolar materials such as lipids. Thus, osmium can be taken up in an unreduced form by these saturated lipids, and then is reduced by organic solvents during the ensuing dehydration steps. Thus, all classes of lipids are potentially preserved by osmium treatment, either by direct chemical reaction or indirectly in the case of saturated lipids.

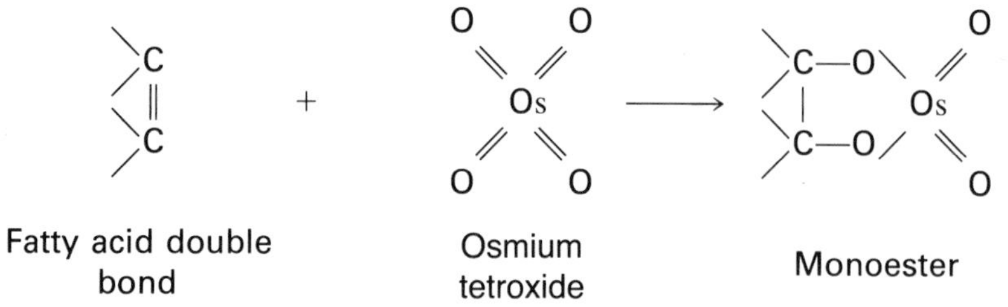

Fatty acid double bond   Osmium tetroxide   Monoester

FIGURE 1.

Proteins. Prolonged osmium fixation results in the progressive denaturation of proteins, leading to the eventual extraction of the smaller peptides during buffer washes and the dehydration series. According to Hayat (1981), a weak osmium solution can produce gels with proteins such as albumin, globulin, and fibrinogen, possibly due to the presence of tryptophan residues with reactive double bonds. Osmium also reacts with a number of other amino acids, such as cysteine and methionine (at alkaline pH), but only slightly with lysine, asparagine, and glutamine. Some reactivity also exists with sulfhydryl, phenolic, hydroxyl, carboxyl, amino, and certain heterocyclic groups and with disulfide linkages.

Nucleic Acids. Osmium can cause the formation of coarse aggregates of DNA. Some workers add calcium to osmium to aid in the stabilization of bacterial DNA. It is thought (Hayat, 1981) that chromosomal reactivity with osmium may be due to the presence of reactive amino acids in histones.

Carbohydrates. Osmium does not react with most pentoses and hexoses, though prolonged treatment of tissues with osmium at 50°C will cause glycogen blackening (Hayat, 1981). Most carbohydrates in osmium-fixed tissue are extracted during rinsing and dehydration.

iv. Physical Changes

Osmium causes some swelling (30% in liver after 4 hr, 15% after 15 min, according to Hayat, 1981). This is normally counteracted to some extent by the consequent shrinkage encountered during dehydration. Some workers add calcium chloride or sodium chloride to osmium when using it as a primary fixative to help reduce the swelling. Nonelectrolytes, such as sucrose or glucose, have also been used in an attempt to alleviate the problem. Some workers report that calcium added to osmium may cause a granular precipitate in tissue, while others claim that the calcium helps prevent this artifact (Millonig and Marinozzi, 1968).

Hardening. Hardening of tissues takes place as a result of osmium exposure. In most cases this is a trivial problem, but in cases in which structures are fragile and in which cells need to be centrifuged, this can cause severe problems. As discussed in the section on embedding media, cell suspensions are often embedded in molten water agar for subsequent ease of handling. If this is done after osmication, there is a great tendency for the cells to be sheared during centrifugation due to the hardening effects of osmium. This is not a significant problem with small cells such as bacteria, but with larger cells such as protozoa it can be a serious problem.

v. PARAMETERS

Concentration. Most workers utilize 1–2% osmium in buffers at slightly al-kaline pH (7.2–7.4). It has been suggested that higher concentrations of osmium result in greater conversion of polypeptides to soluble peptides, which can lead to cells with evident extraction of cytoplasmic components.

Temperature. Some workers insist on osmicating on ice or at 4°C, but other laboratories osmicate at room temperature with consistently good results.

Rate of Penetration. The addition of electrolytes or nonelectrolytes to osmium, as mentioned above, is said to decrease the rate of penetration (nonelectrolytes are said to reduce the rate of penetration more than electrolytes). At room temperature the rate of penetration is no more than 0.5 mm/hr. Thus, a 1-mm-thick piece of tissue will be fixed to the center in 1 hr. After 1 hr, fixation slows down because the fixed surface of the sample impedes the further incursion of fixative. In addition, fixative that penetrates to the center of a tissue is more dilute than fixative at the surface because osmium is being removed from the fixative solution at reactive sites in the tissue as the fixative diffuses to the center of the sample.

Duration. Fixation for 1–2 hr is usually sufficient for properly prepared samples (1 mm thin in at least one dimension) due to the penetration rate cited above, though some workers have utilized up to 12 hr with embryonic tissues or algae (Hayat, 1981).

Formulation and Storage. Osmium crystals can be melted by hot tap water. This allows the preparation of diluted osmium stocks from crystals while avoiding the possibility of contamination. All the following steps are done underneath a fume hood. The procedure is to carefully unwrap an ampule containing 1 g of osmium and to expose the ampule to hot tap water until the crystals melt and run down to one end of the ampule. In the meantime, 50 ml of distilled water is heated to barely steaming in an acid-cleaned 100-ml beaker on a hot plate. The vial is then scored and rolled up in several thicknesses of paper towels so that it can be snapped in two with no possibility of injury to the worker. The ampule half that contains the crystals is held carefully in one hand, while a clean Pasteur pipet is used to add a small volume of the heated water to the ampule. The crystals liquefy and are then pipetted into a previously acid-cleaned, glass 100-ml bottle with a secure screw cap (such as a growth-medium bottle). Care must be taken not to run the osmium down the side of the bottle, since it will cool and solidify instantly on the side of the container. Repeat the process until all the osmium is transferred to the bottom of the bottle. Pour the remaining volume of the heated water into the 100-ml bottle. Cap the bottle, put Parafilm® around the neck, and place the bottle into a larger capped container wrapped in aluminum foil to exclude light. Leave this package at room temperature under the fume hood for 2 days or until it is obvious that the crystals are all dissolved and the solution has a pale straw color. After 2 days, place the container at 4°C for storage. Unless the osmium gets contaminated, this 2% solution will remain usable for 6 months or more. If the solution becomes gray or black, it is contaminated and should be replaced. The best way to ensure the longest shelf life for osmium stocks is to always remove osmium with a fresh Pasteur pipet and to keep the container in the refrigerator whenever it is not being used.

b. Permanganates

Permanganates were initially introduced to overcome some of the fixation problems encountered with walled organisms such as plants as workers realized that osmium penetration was not adequate. Permanganates have now almost passed into history as primary fixatives, though papers utilizing potassium permanganate still appear periodically. They were introduced by Luft (1956) and were noted for their ability to penetrate tissue rapidly (1 mm/hr). Potassium, lithium, lanthanum, and sodium permanganates have all been used. They are all strong oxidants, like osmium, but they do not confer electron density to the cellular components with which they react. They preserve membranes well, but the rest of the cellular constituents are severely extracted. Ribosomes are lost, mitochondria and chloroplasts swell, and many lipids are lost, as are microtubules and microfilaments. Workers have generally used solutions containing 0.6–3.0% potassium permanganate buffered to pH 7.2–7.4. Fixation is usually on ice for about 1 hr, since higher temperatures and longer fixation times result in further cytoplasmic extractions.

c. Formaldehyde

Formaldehyde (Fig. 2) was used as a standard histological fixative component long before we began using it in electron microscopy. It is the smallest and simplest aldehyde used, with the formula CH_2O. It is available as commercial formalin (37–40% formaldehyde), which contains formic acid (less than 0.05%) and methanol (6–15%), the latter being added to prevent polymerization. When diluted to 2%, formaldehyde is primarily in monomeric form. Most electron microscopists prefer to make up formaldehyde fresh from the polymer, paraformaldehyde powder. Unfortunately, after paraformaldehyde is depolymerized to make formaldehyde, it immediately begins polymerizing again, leading to a limited shelf life.

Formaldehyde is usually regarded as an inferior primary fixative for electron microscopy, because even though it penetrates rapidly due to the small size of the molecule, it does not cross-link strongly, and the fixation it achieves is reversible with sufficient washing in aqueous solvents (the formaldehyde can be washed out of the tissue). The pH of buffered neutral formalin (BNF) solutions (4% formaldehyde in a phosphate buffer) made from 37% commercial formalin stocks and used in routine

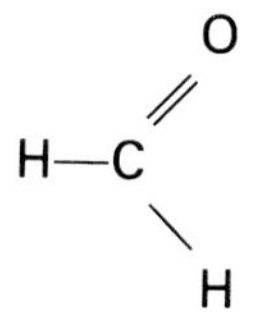

$$\overset{\displaystyle O}{\underset{\displaystyle H}{\overset{\displaystyle \|}{H-C}}}$$

FIGURE 2. Formaldehyde.

histological work frequently drops to 6.0–6.5 in a relatively short period of time on the shelf and causes poor ultrastructural fixation if not corrected. Carson *et al.* (1973) developed a formulation for phosphate-buffered 4% formaldehyde that produces acceptable ultrastructural preservation in many cases. It is generally accepted that the addition of methanol and other stabilizers to commercial-grade 37% formaldehyde makes it unsuitable for enzyme cytochemistry, but many procedures for light-level immunohistochemistry utilize BNF with great success, so this objection appears to be questionable. In addition, by the time the 37% stocks are diluted to 4% for fixation, the concentration of the stabilizers is not very significant.

We still make up fresh formaldehyde from paraformaldehyde when doing cytochemical procedures but utilize 37% formaldehyde stocks in fixative mixtures for conventional structural work. Examination of numerous clinical samples submitted to our service laboratory in formaldehyde for ultrastructural study over the last 13 years has demonstrated that some look as good as glutaraldehyde-fixed tissues, while others are so extracted and poorly fixed as to be virtually unusable. Clearly, the care with which formaldehyde is buffered and the handling of the tissue prior to its fixation (i.e., not letting it sit on a counter at room temperature for 30 min prior to being put into the fixative) probably accounts for this observed variation. Poor fixation with formalin is characterized by numerous membrane discontinuities; very "open"-appearing cytoplasm and nucleoplasm, due to massive extraction of ground substance; and swollen endoplasmic reticulum and mitochondria. Nuclear envelopes usually look fairly well preserved.

i. Disadvantages

As mentioned above, formaldehyde cross-links more weakly than other aldehydes and can be washed out of tissues upon sufficient aqueous rinses. As with all aldehydes, it provides no electron density to tissues.

ii. Advantages

The formaldehyde molecule is the smallest aldehyde and thus penetrates rapidly into tissue. Since it is not a strong cross-linker, it tends to leave most protein structures relatively unchanged and thus is a preferred fixative for various enzyme cytochemical procedures, as well as for immunocytochemical techniques.

iii. SPECIFIC ACTIVITY

Lipids. Formaldehyde can react with the double bonds of unsaturated lipids, but the lipids are extracted by various solvents in the processing series unless post-fixation with osmium is added to stabilize the lipids.

Proteins. Peptide chains are cross-linked, but the vast majority of the cross-links can be reversed by aqueous solutions. Increased formaldehyde binding is caused by using higher concentrations of the fixative. Higher temperatures and pH will also increase binding to proteins. Maximum binding to proteins is reported to be at pH 7.5–8.0 (Hayat, 1981).

Nucleic Acids. Formaldehyde reacts with nucleic acids and proteins without breaking the linear structure of either. The preservation of nucleic acids is probably due mostly to the interaction of formaldehyde with the protein component of nucleoproteins. As with pure proteins, the reaction of formaldehyde with nucleoproteins is reversible in aqueous solutions.

Carbohydrates. Small carbohydrate molecules are not preserved and are washed out during processing. Carbohydrate moieties complexed with other cellular components may be preserved. Mucoproteins are fixed, but acid mucopolysaccharides are not. Glycogen survives formaldehyde fixation, provided that fixation times are not too long.

iv. PHYSICAL CHANGES

Properly formulated formaldehyde solutions, such as Carson's (1973), applied to tissues 1 mm thin in at least one dimension should result in little appreciable shrinkage or swelling of tissues. Formaldehyde will toughen tissues, but they remain pliable (unlike osmium-fixed tissues, which become brittle). Cells can be easily pelleted through molten water agar without significant damage.

v. PARAMETERS

Concentration. Most workers use 4% formaldehyde solutions, though some of the recent immunocytochemical procedures utilize as little as 2% formaldehyde made up from paraformaldehyde.

Temperature. Room-temperature fixation is successful, though some workers believe ice-bath temperatures provide better fixation.

Rate of Penetration. Formaldehyde penetrates more rapidly than either osmium or glutaraldehyde due to the small size of the monoaldehyde and is thus the fixative of choice for large tissues.

Duration. Many workers also warn that since the reaction of formaldehyde is slow and reversible, tissues should not be left in formaldehyde for very long. On the other hand, Ghadially has published two books (1975, 1985) illustrating numerous pathology samples that have been successfully fixed with buffered neutral formalin (BNF) from histology labs, many of which have been stored in these solutions for extended periods of time. Generally, it would appear prudent to fix tissues for no more than 1–2 hr in fixative solutions containing only formaldehyde and buffers before proceeding with the rest of the processing schedule.

Formulation and Storage. As mentioned previously, many electron microscopists believe that formalin (37–40% formaldehyde solutions containing stabilizers) is inappropiate for ultrastructural preservation. This is clearly not the case when one considers the numerous medical case reports illustrated with electron micrographs prepared from tissues fixed in BNF. Poorly formulated BNF, particularly if the pH is below 7.0, is not to be used, however. If making fresh formaldehyde from paraformaldehyde powder, recognize that the monoaldehydes that are produced by putting the powder into heated water and raising the pH will begin repolymerizing immediately. Thus, most workers recommend utilizing freshly made formaldehyde immediately or at least within a week or two. If the solution is being stored for a week or two, it should be refrigerated. On the other hand, standard formalin (37–40% formaldehyde) has a storage life of years at room temperature. Dilute stocks (4% in buffer) stored under refrigeration are presumed to be even more stable.

d. Acrolein

Acrolein (Fig. 3) was first introduced as a primary fixative for electron microscopy by Luft in 1959. It is a three-carbon monoaldehyde (C_3H_4O) noted for its ability to penetrate tissues more rapidly than osmium, glutaraldehyde, or formaldehyde. It is more reactive than the other aldehydes and inactivates enzymes profoundly.

i. DISADVANTAGES

Acrolein polymerizes rapidly upon exposure to air, light, and chemicals, and generates heat. Commercial preparations usually contain 0.1% hydroquinone to prevent oxidation. Acrolein is flammable and its vapors can be quite irritating to mucous membranes (it is a component of tear gas).

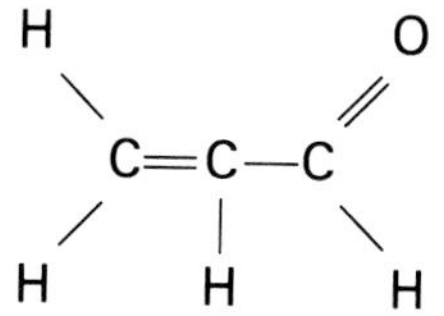

FIGURE 3. Acrolein.

ii. ADVANTAGES

Acrolein is more reactive with proteins than formaldehyde and penetrates tissues quite rapidly. It preserves microtubules better than either glutaraldehyde or formaldehyde alone.

iii. SPECIFIC ACTIVITY (GLAUERT, 1975)

Lipids. Acrolein is reported to solubilize lipids.

Proteins. This aldehyde is more reactive with proteins than any of the other aldehydes. For enzyme or antigen localization, it is contraindicated as a fixative because of possibly severe denaturation.

Nucleic Acids. Acrolein is thought to be somewhat reactive with charged groups, but the primary fixation of nucleic acids is probably due to their well-fixed associated proteins such as the histones.

Carbohydrates. These are generally regarded as unreactive with acrolein.

iv. PHYSICAL CHANGES

Acrolein causes a toughening of tissues without loss of pliability, a feature shared with other aldehyde fixatives. The loss of lipids reported by some authors may be associated with changes in membrane permeability.

v. PARAMETERS

Concentration. Some workers recommend 10% buffered acrolein, but others use 3–4% with success.

Temperature. Fixation at 4°C or at room temperature gives good results.

Rate of Penetration. Penetration of about 1 mm/hr has been reported for rat kidney. This fixative has been recommended for large tissue pieces because of its excellent penetration capabilities.

Duration. Due to its rapid rate of penetration, 1 hr at room temperature is quite adequate for the tissue sample sizes normally encountered in an electron microscopy lab.

Formulation and Storage. As previously mentioned, acrolein usually comes in concentrated form containing hydroquinone as a stabilizer. The presence of the stabilizer does not seem to adversely affect fixation. The shelf life of acrolein at room temperature is not long because of its propensity for auto-polymerization. Storage at 4°C is recommended. More dilute solutions will be less prone to polymerize. If a 10%

solution in water has a pH below 6.4, or if there is turbidity in the concentrated solution from which the dilute solution is made, discard the solution. This fixative has a high vapor pressure and is extremely toxic, so always work carefully with it under a fume hood.

e. Glutaraldehyde

When glutaraldehyde (Fig. 4) was introduced by Sabatini *et al.* (1963), it revolutionized chemical fixation because its capabilities were so much better than the preceding alternatives. The fact that it is a five-carbon dialdehyde ($C_5H_8O_2$) with two hydroxyl groups capable of binding proteins makes it an extremely good cross-linking fixative. This strong cross-linking is virtually irreversible.

i. DISADVANTAGES

Glutaraldehyde penetrates tissues more slowly than formaldehyde or acrolein and inactivates some enyzmes and antibodies. This is probably due to glutaraldehyde itself or some of the nearby cross-linked proteins sterically hindering access to reactive sites. Microtubules depolymerize and ribosomes move if they are exposed to cold fixative. No electron opacity is afforded the fixed tissue. Physiological activities of cells are not stopped immediately, resulting in cells changing conformation in response to the fixative under some conditions. In addition, membranes become more permeable after primary glutaraldehyde fixation, and cellular compartments can become leaky. This will result in some lytic enzymes escaping their normal cellular sequestration, which can cause some autolytic damage if the cells are not kept at depressed temperatures (4°C) during prolonged storage.

ii. ADVANTAGES

Even though the alpha helix of proteins is affected by this fixative, the tertiary structure of proteins is unaffected. As previously mentioned, reactive moieties of proteins may be occupied by glutaraldehyde and thus be inaccessible to cytochemical procedures, though most proteins maintain a fair amount of their activity, which is useful for cytochemical or immunocytochemical procedures.

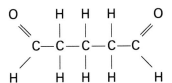

FIGURE 4. Glutaraldehyde.

iii. SPECIFIC ACTIVITY

Lipids. Glutaraldehyde probably reacts only with phospholipids containing free amino groups. If glutaraldehyde-fixed tissues are not post-fixed with osmium, most lipids are extracted during dehydration. Glutaraldehyde itself may make some phospholipids go into solution, which can result in the artifactual formation of myelin-like figures that are stabilized by consequent osmication. This appears to be responsible for the formation of "mesosomes" in bacteria, since they are not found in cryopreserved materials (Aldrich *et al.*, 1987). Myelin figures are usually more abundant in large tissue blocks and are more common when extended glutaraldehyde fixation times are used. Some investigators add 1–3 mM $CaCl_2$ to glutaraldehyde to diminish this phenomenon.

Proteins. These are the main target of glutaraldehyde, which is most reactive with many free amino groups and cross-links proteins through this agency.

Nucleic Acids. As with the other aldehydes, most of the preservation of nucleic acids is considered to be due to the interaction of this fixative with the proteins associated with nucleic acids more than by a direct interaction of nucleic acids and glutaraldehyde.

Carbohydrates. Most carbohydrates are not fixed well and are lost during subsequent rinses. Glycogen is a major exception; however, 40–65% of the total glycogen remains in fixed tissues, according to Hayat (1981).

iv. PHYSICAL CHANGES

Glutaraldehyde toughens tissues but does not impart electron density. It is known to cause shrinkage and so is generally made up in hypertonic solutions. Linear shrinkage in the realm of 5% is common for many tissues.

v. PARAMETERS

Concentration. A high concentration (37%) causes profound shrinkage, while a low concentration (0.5%, as used in a number of immunocytochemical procedures) can result in severe extraction. Typically, solutions of 1.5–4% glutaraldehyde in an appropriate buffer are used. With larger sample sizes, some workers use higher concentrations of glutaraldehyde, since the solution becomes more dilute as it passes to the center of the tissue block because the glutaraldehyde becomes bound and is no longer free in solution.

Temperature. Historically, it has been recommended that glutaraldehyde fixation be done on ice or at 4°C. At the same time, it was established early on that microtubules become depolymerized when held at such temperatures. Glutaraldehyde fixation at room temperature seems to give adequate results without the threat of microtubule depolymerization.

Rate of Penetration. Hayat (1981) reported that a 2% solution of glutaraldehyde at a tonicity of 200 mOsm penetrates liver to a depth of 0.7 mm after 3 hr
at room temperature. Periodic acid-Schiff's reagent staining to monitor the depth of
penetration is effective, since Schiff's reagent will react with the two hydroxyls of
glutaraldehyde.

Duration. Even though the fixative can reach depths in the tissue beyond
0.5 mm after 1 hr, most tissues will exhibit evidence of autolysis after an hour at room
temperature. Thus, if the fixative does not penetrate all areas of the tissue after 1 hr at
room temperature, most tissues will exhibit poorly fixed central areas.

Formulation and Storage. Glutaraldehyde is available in various forms, from a
70% aqueous solution stored under dry nitrogen gas in sealed ampules, to 50% or 25%
aqueous solutions stored in 500-ml brown glass bottles, to 8% solutions in sealed
ampules. Some users have redistilled their glutaraldehyde after the method of Smith
and Farquhar (1966), but this rarely is necessary. Glutaraldehyde at high concentrations
(over 25%) is subject to spontaneous polymerization if not stored under dry nitrogen.
Heat can also adversely affect such stocks, though glutaraldehyde is not photosensitive. If stored under dry nitrogen in sealed ampules, glutaraldehyde should be stable
indefinitely if stored at 4°C. Biological-grade glutaraldehyde (25%) stored in a refrigerator generally has a shelf life of several years. If glutaraldehyde stocks become
yellowed, polymerization may be taking place. A more reliable indicator is pH. If the
pH of 25% stocks is between 3 and 6, they are probably all right. If an ultraviolet
reading of the stock reveals a larger peak at 235 nm than at 280 nm, the stocks may be
bad. A 1:1 or 2:1 ratio of 280-nm to 235-nm peaks is associated with good glutaraldehyde. After distillation, only the 280 nm (monomeric) peak should remain.
Dimeric (235 nm) and higher polymeric forms are not considered good for fixation.
Glutaraldehyde solutions mixed with buffer at working concentrations (2–4%) are
good for at least a year at 4°C.

f. Fixative Supplements

Various agents have been previously mentioned, such as electrolytes, nonelectrolytes, calcium, and buffers that are added to fixatives to attempt to address the
problems of tonicity and pH, and to help stabilize various cellular constituents. These
agents usually deal broadly with cellular reactions rather than with a specific subset of
cellular components.

A variety of other supplements (Hayat, 1981) have been utilized to stabilize
specific compounds or structures. One of the most frequent problems addressed is that
of lipid stability. Since the most commonly used primary fixatives, aldehydes, do not
fix lipids, the lipids can undergo structural alterations and reconfigurations during the
primary fixation period (prior to osmium steps that will stabilize them). Thus, various
workers have added specific agents to aldehydes to help stabilize lipids until osmium
can react with them. Glutaraldehyde containing digitonin has been used to specifically
stabilize cholesterol and cholesterol esters, while glutaraldehyde containing malachite
green has been used to preserve some lipid-containing granules in mammalian sper-

matozoa. A combination of glutaraldehyde fixation followed by osmium with added potassium ferricyanide has been used to help preserve phospholipids in central nervous system samples as well as the surfactant material in type II cells of the lung (myelin-like figures in vacuoles of type II cells).

Glutaraldehyde containing lead acetate has been used to preserve soluble inorganic phosphate, which would otherwise be washed out during fixation and buffer washes. Glutaraldehyde containing phosphotungstic acid applied to tissues pretreated with polyethyleneimine has been used to demonstrate anionic sites in basement membranes and collagen. Biogenic amines, such as adrenaline, have been demonstrated with a combination of glutaraldehyde and potassium dichromate.

Borrowing from light microscopy techniques, trinitro compounds (most notably, picric acid and 2,4,6-trinitrocresol) have been added to primary fixatives such as glutaraldehyde to better preserve smooth endoplasmic reticulum in interstitial cells of testes and other steroid-secreting cells, although microtubules sometimes are sacrificed in the process. Another attempt to produce better preservation of male reproductive tissues has utilized glutaraldehyde primary fixation followed by post-fixation in osmium containing potassium ferrocyanide.

Tannic acid has been added to glutaraldehyde to help preserve various proteinaceous components of cells, such as microfilaments and microtubules, as well as cytomembranes. Its mode of action is purported to be reaction with peptide bonds, amine, and amide residues present in side chains of polar amino acids.

Glutaraldehyde containing uranyl acetate has been utilized to preserve structures with a tendency to lose DNA because the mixture has been shown to gel DNA in minutes. Uranyl acetate is also frequently used as a separate incubation step (usually a 0.5% aqueous mixture) after primary aldehyde fixation and osmium post-fixation. Tissues are fixed overnight at 4°C in the uranyl acetate solution to help preserve phospholipids and to stabilize nucleic acids. The need for a uranyl acetate post-staining step is then considered to be unnecessary by many workers. Uranyl acetate reacts with phosphate groups of phospholipids and nucleic acids and serves as an electron-dense stain after becoming complexed with these cellular constituents.

Finally, various workers have utilized alcian blue, ruthenium red, or cetylpyridinium chloride in aldehydes to help stabilize polysaccharide residues, particularly those on cell surfaces. In addition, ruthenium red has also been utilized for the same purpose as an additive in osmium post-fixations. This area will be considered again in the section on cytochemistry, since the compounds are semispecific stains as well as fixatives.

g. Procedural Comparisons

i. SEQUENTIAL FIXATION

Sequential aldehyde/osmium fixation with a primary fixation for 1–2 hr in 2% glutaraldehyde in a 100-mM phosphate or cacodylate buffer at pH 7.2–7.4, followed by a post-fixation in 1–2% osmium in the same buffer, is common in electron microscopy. As mentioned previously, lipids can sometimes migrate and/or change configuration in this type of fixation schedule.

ii. Simultaneous Fixation

Simultaneous aldehyde/osmium fixation can be used successfully with various single-celled protozoans, but can be expected to be most useful in a situation where lipid lability is of concern. It can exhibit penetration problems with blocks of tissue, resulting in only a superficial band of adequate fixation. Customarily, quadruple-strength glutaraldehyde (8%) is mixed in equal volume with quadruple-strength buffer (0.4 M phosphate buffer, pH 7.2–7.4), and this combined mixture is added to an equal volume of double-strength aqueous osmium (4%) just before adding tissue. This results in a fixative containing 2% glutaraldehyde, 2% osmium, and 0.1 M phosphate buffer at pH 7.2–7.4. The various components can easily be made up far in advance of use and stored separately in the refrigerator. However, once they are mixed together, they begin reacting, with the glutaraldehyde reducing the osmium tetroxide. At room temperature, the mixture will turn an opaque black within 1 hr after mixing. This is not of concern because there is such an excess of both fixative components. Some workers insist that this sort of procedure should be done at 4°C to minimize the fixative interactions, but it is unnecessary and theoretically would impede the interaction of the fixatives with the tissue a bit. The whole point of this regimen is to be able to stabilize both proteins and lipids in the early moments of the fixation schedule, rather than to stabilize the lipids with an osmium step an hour or so later. This technique is very effective for certain protozoans and seems to improve some neural samples.

iii. Aldehyde Mixtures

Aldehyde mixtures as primary fixatives were first used by Karnovsky (1965). His original formula contained 5% glutaraldehyde and 4% formaldehyde (made from paraformaldehyde) in 80 mM cacodylate buffer at pH 7.3 containing $CaCl_2$ (5 mM). Various workers almost immediately modified this highly hypertonic medium (2010 mOsm) by using it in a half-strength formulation. Over the years, some investigators have used phosphate buffer rather than cacodylate buffer, have used other percentages of the constituent aldehydes, or have omitted the calcium chloride. Thus, when reading the literature, beware of papers that state they used "Karnovsky's" fixative. Unless the authors specifically spell out the formulation they employed, all you know is that they used some combination of formaldehyde and glutaraldehyde in their primary fixation.

Another major advance came in 1976 when McDowell and Trump examined a variety of aldehyde fixative combinations for their suitability for kidney perfusions. They finally determined that 4F:1G (4% formaldehyde, 1% glutaraldehyde buffered with monobasic sodium phosphate to pH 7.2–7.4) fulfilled their needs most completely. They were looking for a fixative of moderate osmolarity (760 mOsm) that could be made from commonly stocked chemicals with long shelf lives (25% biological-grade glutaraldehyde and 37% biological-grade formaldehyde) and that could be made into the final fixative solution and stored for some months before use (up to 3 months at 4°C). In addition, they wanted a fast-penetrating fixative that was suitable for perfusing kidneys. The formaldehyde component penetrated the tissue rapidly and was then followed by the more slowly diffusing glutaraldehyde, which stabilized the tissue more thoroughly and more permanently than the formaldehyde.

This fixative was easier to formulate than Karnovsky's, since the formaldehyde was not made up from paraformaldehyde and had a longer storage life before application to tissues. They also determined that kidneys could be stored in the fixative for up to a year at 4°C without any serious structural changes. Finally, they noted that 4F:1G-fixed tissues could be paraffin embedded, sectioned easily, and subjected to common histological staining techniques with great success. This contrasts significantly with most electron microscopy fixatives. Almost all procedures call for the use of at least 2% glutaraldehyde for fixing tissues for ultrastructural examination. Tissues prepared in this fashion and then embedded in paraffin are typically brittle and hard to section. In addition, the excess of glutaraldehyde in tissues makes them nonspecifically Schiff's-reagent positive (as mentioned earlier, this is one way to monitor the depth of fixative penetration in tissues fixed with glutaraldehyde). By dropping the glutaraldehyde concentration to only 1%, most of the glutaraldehyde hydroxyl groups that could react with Schiff's reagent are cross-linked to proteins in the tissue and are thus unavailable to react with the stain.

iv. TEMPERATURES FOR FIXATION

Temperatures for fixation are variable in the literature. Many workers are convinced that fixing on ice or at 4°C is necessary. During fixation, various hydrolytic enzymes, such as acid phosphatases and esterases, begin leaking from their storage compartments (lysosomes) as membranes become partially solubilized during fixation. These enzymes can hydrolyze various cytoplasmic components, culminating in their removal during ensuing aqueous and other solvent washes. It is recognized that formaldehyde and glutaraldehyde, our most commonly utilized primary fixatives, do not completely inactivate a broad assemblage of enzymes (Hopwood, 1967, 1972). We make use of this knowledge when designing protocols for enzyme cytochemistry but frequently overlook it when questioning a partially unsuccessful fixation procedure. Maintaining a temperature considerably below the normal V_{max} range for these enzymes thus makes some sense. In addition, lower temperatures should also decrease the amount of cellular extractions possible during various fluid washes during tissue processing.

On the other hand, low temperatures will depolymerize microtubules and will cause vasoconstriction if the fixative is being perfused into live animal tissues.

Our laboratory has used room-temperature fixation for 1 hr in McDowell's and Trump's 4F:1G fixative followed by post-fixation for 1 hr in 1% osmium in a 0.1-M phosphate buffer, which is suitable for a variety of tissues from mammals, birds, fish, and plants, as well as for viruses, bacteria, and protozoans.

v. STORAGE

Storage of fixatives has been covered under the sections on individual fixatives, but a general recapping is in order. Most aldehyde fixatives diluted to working strength and appropriately buffered for use are stable for at least a year at 4°C. Aqueous osmium stocks are stable for months at the same temperature, provided that no dirt, buffer, or other reactive substances have been introduced and that they are stored in the dark.

The electron microscopy community differ a great deal on their attitudes about storing tissues in fixatives. It is generally agreed that long-term storage of tissues in osmium solutions is not a good idea because within hours polypeptides are cleaved into peptides, which are solubilized by the various washes, resulting in significant extraction of cellular contents. Storage in aldehydes, however, is a more contentious subject. One of the advantages cited by McDowell and Trump for their formulation 4F:1G was that tissues could be stored in it for up to 1 year. Other workers insist that tissues should never be stored in aldehydes because excessive protein precipitation will take place over time. This suggestion has not been thoroughly examined, to my knowledge, for the various aldehydes utilized.

vi. PERFUSION METHODS

Perfusion methods for a variety of tissues are thoroughly discussed by Hayat (1981, 1989), and thus the technical aspects will be omitted here. Suffice it to say that if your research necessitates perfusion, it is best to consult recent literature in your area to determine what your colleagues are doing. Perfusions are indicated when the tissue you are dealing with autolyzes quickly, changes structurally when vascular supplies are compromised (kidney), or is deep within a large organism, necessitating lengthy dissection procedures before you can excise the small pieces mandated for good fixation. Clearly, utilizing the vasculature of an organism for the delivery of fixative to all cells in a tissue is quicker than relying on diffusion of the fixative from the surface into a millimeter cube.

Most textbooks and manuals discount the need for any consideration of transport when working with plants or other nonvascular (in the circulating-fluid sense) organisms. There are exceptions to this concept, however. Dr. Gene Shih (personal communication) studied phloem in higher plants back in the early 1970s and discovered that sieve-tube plasmodesmata presented two different pictures depending on whether he cut off the stems while they were submerged in fixative or whether he cut off the stems and then immersed sections of the stems in fixative. He suggested that the image produced from stems sliced open under fixative was less artifactual.

vii. MARINE MICROORGANISMS

Marine microorganisms collected from nature or grown in complex artificial media can present special problems on occasion. If the organisms cannot be spun down easily and pelleted (which is true for many ciliates and flagellates, as they will continue swimming to the top of a tube while you are trying to spin them to the bottom) prior to fixative addition, it becomes necessary to mix a double-strength fixative in equal volume with the surrounding medium in the tube to kill them prior to centrifugation. If the medium contains significant amounts of reactive salts, such as calcium or magnesium, phosphate buffers will become coprecipitated with them, thus potentially putting precipitates into the tissue, as well as losing the buffering capacity of the medium and the fixative buffer. To avoid this problem, it is advisable to utilize cacodylate buffers, as will be detailed in the section on buffers.

Mixing double-strength fixatives with organisms contained in growth media is also necessary if the organisms have cellular projections, such as bacterial pili or flagella, which are shed by the organism if handled excessively. In many cases, once the cells are killed by the fixative, these structures will not be shed.

viii. ANESTHETICS

Anesthetics are another often overlooked aspect of fixation. The various common anesthetics utilized in animal research affect some tissues in deleterious ways that must be considered when designing fixation protocols. Thus, if you are concerned with a tissue (such as liver) that can respond to certain classes of anesthetics (e.g., barbiturates), other anesthetics should be considered. A more thorough discussion of this topic can be found in Hayat's works (1981, 1989).

ix. NUTRITIONAL STATES

Nutritional states of some organisms can affect the perceived fixation quality. A liver from an animal fed right up to the moment of sacrifice is structurally different in terms of glycogen reserves and mitochondrial configurations from an animal taken off food 12 hr before sacrifice. Semithick sections from the liver of a well-fed rat compared to the liver of an unfed animal (Figs. 5 and 6) have the appearance in the former of poorly fixed tissue with large, open areas throughout the cytoplasm. Ultrastructural comparison of the two specimens (Figs. 7 and 8) reveals the open areas in the unfed animals to be large pools of alpha glycogen. These kinds of changes are important to consider when designing protocols.

This whole section on fixation should make it clear that a variety of approaches can usually be taken to solve the same problem. If the initial simple attempts fail, then move on to more esoteric formulations to stabilize the tissues in question. The approach in our laboratory is to try something like McDowell's and Trump's fixative first; we rarely have to move on to the more complicated procedures.

II. BUFFERS

A. Molarity/Molality/Osmolarity/Tonicity

Before any serious consideration of buffering systems can be undertaken, it is useful to review the various ways utilized to discuss the concentration of solutions. When making up chemical solutions, one normally talks in terms of molarity (1.0 M means one molecular weight [MW] of solute per liter of solvent) or molality (1.0 molal means one molecular weight of solute per kilogram of solvent).

A related concept used with acidic and basic solutions is normality (N). If the acid has but one hydrogen (HCl) or the base has but one hydroxyl group (NaOH), normality and molarity are identical. On the other hand, if there is more than one hydrogen or hydroxy group in the formula, the normality and molarity are not identical, and the concept of equivalent weight must be considered. Equivalent weight is the molecular

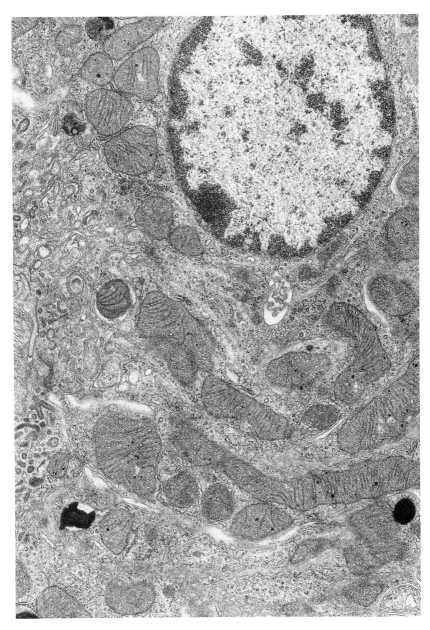

FIGURE 5. Mouse kidney fixed with Carson's fixative (4% formaldehyde in a phosphate buffer, pH 7.28). Post-fixed in 1% osmium in the same buffer (Fix #1). 14,200×.

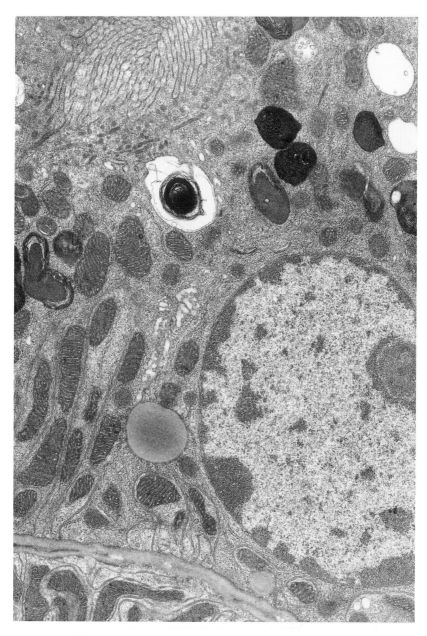

FIGURE 6. Mouse kidney fixed with McDowell's and Trump's 4F:1G in phosphate buffer, pH 7.22. Post-fixed in 1% osmium in phosphate buffer (Fix #2). 14,200×.

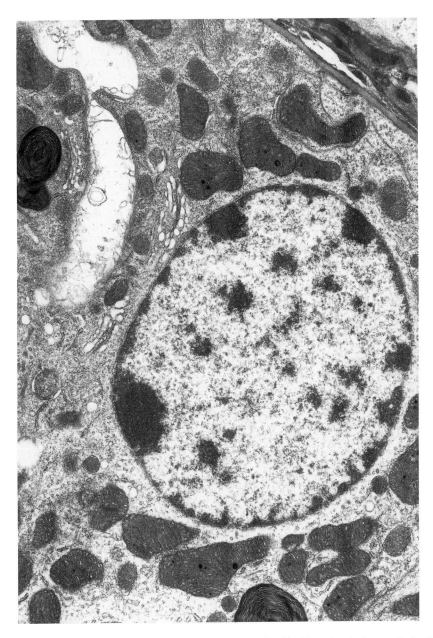

FIGURE 7. Mouse kidney fixed in Karnovsky's, pH 7.25. Post-fixed in 1% osmium in the same buffer (Fix #3). 14,200×.

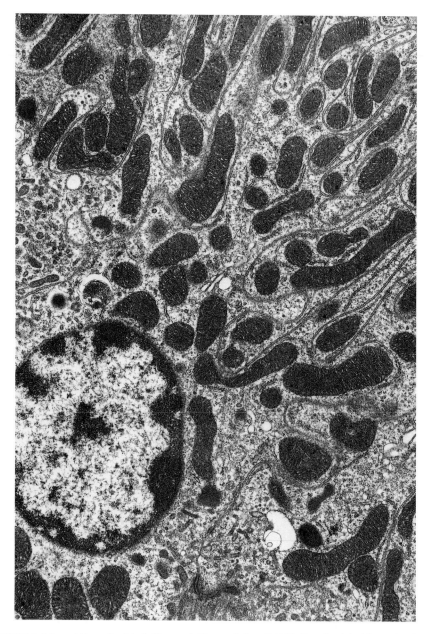

FIGURE 8. Mouse kidney fixed in half-strength Karnovsky's, pH 7.25. Post-fixed in 1% osmium in the same buffer (Fix #4). 14,200×.

weight divided by the number of unit charges. Hydrochloric acid has a molecular weight of approximately 36.45, which means that a 1.0-M solution would contain 36.45 g of pure HCl per liter of water. A 1.0-N solution would be formulated the same way, since the molecular weight divided by the number of unit charges (one hydrogen) is still 36.45 g/l. A 1.0-M solution of phosphoric acid (H_3PO_4), on the other hand, consists of 98 g/l, but only 32.67 g/l is needed to make a 1.0-N solution (98 MW/3 charges due to the three hydrogens = 32.67).

The term *osmolarity* (Osm/l) occurs frequently in discussions of buffers and fixative solutions. Tonicity and osmolarity are equivalent. An isotonic solution has the same osmotic potential as the cytoplasm of cells within it. If a nonelectrolytic solution is involved, molarity and osmolarity are the same. If a solution contains a dissociating electrolyte, then the osmolarity is greater than the molarity (the osmotic pressure on the cells is greater than it would be for a nonelectrolyte solution of the same molarity). Osmolarity is most frequently measured by freezing-point depression of a solution. Fixative solutions are typically either isotonic or slightly hypertonic (see Saito and Tanaka, 1980 for examples of the effect of different tonicities on cells).

Unfortunately, osmolarity of the cytoplasm of a living cell is not measureable. In practice, this means that fixatives with different tonicity have been tried on a variety of cell and tissue types. Some of those that have worked have had their osmolarity checked. Further studies can use an osmometer to ensure that all fixative solutions have the same osmolarity. Naturally, in a given organism (e.g., a rat) different tissues may have various osmolarities. It then becomes a task of empirically determining the best osmolarity for a fixative and buffering system. Once a suitable buffer strength is selected, an osmometer can be utilized to reproduce this osmolarity for ensuing studies. Thus, osmolarity of a fixative or buffer can be measured, but osmolarity of a cell is hard to determine. The best function of an osmometer is to measure the consistency of tonicity of solutions.

If you desire your primary aldehyde (glutaraldehyde) fixative solution and the osmium post-fixative solution to be truly at the same osmotic potential, it would be necessary to increase the tonicity of the vehicle (buffer) for the osmium because osmium has less osmotic potential than glutaraldehyde.

B. Purpose

The purpose of buffers is threefold. First, the buffer serves as a solvent for the fixative constituents. Secondly, the buffer helps maintain a specified pH. Third, the buffer maintains tonicity, as described above. In the primary fixative, all three factors are important. However, after primary fixation in aldehydes, tonicity becomes a trivial concern because the cell membranes are no longer subject to osmotic potentials due to a major increase in permeability for small molecules conferred by primary fixation.

This is best exemplified by one of the major techniques utilized for the preparation of samples for cryoultramicrotomy (Tokuyasu, 1983). In this technique, cells are lightly fixed with aldehydes prior to being rinsed and placed in 2.1–2.3 M sucrose in phosphate buffer. If the cells were placed in sucrose *prior* to the aldehyde fixation, they would plasmolyze badly, but they do not because of molecular changes in the plasmalemma induced by primary fixation.

Maintenance of pH is a significant activity of buffers, particularly if tissues are stored in the primary fixative or buffers for any period of time. It is of serious importance if cells are left in osmium for any length of time. As mentioned in the section on osmium as a fixative, it tends to cleave polypeptides into their constituent peptide molecules. As the process continues, more and more carboxy terminals are exposed, causing a tendency for a decrease in pH that the buffer must resist. Tissues left in unbuffered osmium go from pH 6.2 to 4.4 in 48 hr (Hayat, 1981). Cell necrosis itself can cause pH changes as the constituent proteins undergo autolysis, leading to the same potential for a drop in pH.

When designing buffering protocols, it is important to recognize that pH is somewhat dependent on temperature, so the buffer pH should be adjusted at the temperature at which the buffer is to be used. Another consideration is the effect of buffer concentration. A 0.2-M buffer diluted to 0.1 M by adding an equal volume of distilled water will still have the same pH, but its buffering capacity will be reduced.

C. Types, Characteristics, and Uses of Buffers

1. Phosphate Buffers

Phosphate buffers are probably the most widely used buffers in biological work. They are used in irrigation solutions for living tissues during some surgical procedures (phosphate-buffered saline [PBS]), are used to make up immunologically active solutions during immunochemical staining procedures, and are used widely to buffer fixatives for light and electron microscopy preparations. They can be used to wash living cells in culture without inducing injury. The pH is maintained more effectively than with the other most commonly used buffer, sodium cacodylate.

In electron microscopy, phosphate buffer may be formulated from monobasic sodium phosphate (NaH_2PO_4), as in Millonig's phosphate buffer, or from a combination of monobasic and dibasic (Na_2HPO_4) sodium phosphate, as in Sorenson's buffer. Phosphate buffers are preferred by many workers, partially because of their lack of toxicity compared to the second most commonly used buffer, sodium cacodylate. However, the reactivity of phosphate groups must be considered when formulating solutions containing positively charged compounds, such as calcium or magnesium salts. It has been reported that the dibasic form of phosphate may cause protein precipitation and can also precipitate lead, uranyl acetate, and other polyvalent cations. Glauert (1975) has shown dog lung cells covered with a black precipitate that is said to be due to mixing phosphate buffer with osmium. Similar work in our laboratory has never produced this artifact. Phosphate buffers are best used either at physiological pH or in a slightly alkaline formulation (pH 7.2–7.4). Their buffering capacity drops dramatically when significantly above or below this range.

2. Cacodylate Buffer

Cacodylate buffer is the second most commonly used buffering material used in electron microscopy. Like phosphate buffer, it is used primarily at physiological pH or at slightly alkaline conditions, though it buffers effectively at pH 6.4–7.4. It lacks the strongly reactive character of phosphate solutions and thus can be used in cytochemical reaction mixtures and with media containing various ions without fear of coprecipitation with other solution components. If, for example, you are faced with fixing a marine protist that cannot be centrifuged easily in a living state (the organism swims to the top of the tube), it is often advisable to mix double-strength buffered fixative in equal volume with seawater containing the organism. However, seawater contains numerous salts that can be precipitated by phosphate buffers with consequent drastic changes in pH. Cacodylate buffers will avoid this problem. On the other hand, it will precipitate uranyl acetate. Furthermore, sodium cacodylate contains arsenic and can produce arsenic gas with acids. It is toxic to the user and can cause dermatitis, so it must be handled with care. Sodium cacodylate is more toxic to cells than phosphate buffers. Finally, sodium cacodylate is also considerably more expensive than phosphate buffer components.

3. Collidine Buffer

Collidine buffer has mostly historical significance at this time. It was originally used because it had a effective pH range of 6.0–8.0, but it caused significant extraction of proteins in some tissues and was not compatible with formaldehyde (leading to membrane damage). It also is a relatively toxic compound.

4. Veronal Acetate

Veronal acetate buffer works most effectively at pH 4.2–5.2, is ineffective at physiological pH, and is reactive with aldehydes. It is thus contraindicated as a buffer for electron microscopy. Veronal acetate was first used as a buffer when osmium was being used as a primary fixative (Palade, 1952). Hayat (1989) states that it produces better membrane preservation than other buffering systems used with osmium.

5. Tris

Tris buffer is most effective above pH 7.5 and thus finds most use in cytochemical incubation solutions prepared for demonstrating enzymes with activity under alkaline conditions. It may react with glutaraldehyde, so washing after primary fixation must be thorough.

6. Sodium Bicarbonate

Sodium bicarbonate is frequently encountered in buffers prepared for physiological studies but is not commonly recommended for electron microscopy fixative preparation. Salema and Brandao (1973) concluded that it was superior to phosphate buffers for the preservation of plant tissues.

7. Zwitterionic Buffers

Zwitterionic buffers (PIPES, HEPES, MOPS) are a class of buffers little used in electron microscopy that are either amines or N-substituted amino acids. They have been used in tissue culture media, indicating their lack of toxicity and their effective buffering capabilities. Hayat (1981) suggests that they might find utility in micro-analytical studies because they do not contain ions that might compromise elemental analysis. He also suggests that they produce a relatively high cellular density because of increased retention of proteins and phospholipids.

8. Miscellaneous Buffers

Miscellaneous buffers are encountered in various enzymatic localization studies when maintenance of either high or low pH is necessary. It is helpful to recognize that many of these have incompatibilities with common components of fixative solutions, and appropriate washes need to be employed to prevent inappropriate interactions between the constituents of these different solutions.

III. DEHYDRATION

A. Purpose

The purpose of dehydration agents is to remove water from tissues so that they can be infiltrated with the most widely used embedding media that are typically immiscible with water. Dehydration is also important for the preparation of materials for scanning electron microscopy, removing water from the specimen, and typically replacing it with ethanol, which then can be exchanged with liquid carbon dioxide during the critical-point drying process.

B. Agents

Ethanol and acetone are the two most commonly used dehydration fluids, with propylene oxide frequently recommended as a final transition solvent used to make up diluted epoxide mixtures during the initial resin infiltration. Many of the electron microscopy texts and laboratory manuals suggest that propylene oxide is necessary for epoxide resin infiltration or that it is in some way superior to acetone as a transition solvent. This is not borne out in practice nor mandated by the resin manufacturers' and producers' instructions. The standard epoxide resins in use today (Spurr's resin, Poly-bed 812, LX-112, SPI-Pon 812) are easily mixed with acetone, and infiltration is excellent with this transition solvent. In addition, these resins are also miscible with ethanol and the resulting blocks section well, though the blocks are usually somewhat tacky on the surface. Despite its long record in the literature, because propylene oxide is the most toxic of the three solvents, has the highest vapor pressure, and confers no discernable advantage to infiltration processes for epoxide resins, it can no longer be recommended.

1. Ethanol

Ethanol continues to see the widest application as a dehydration agent. It is the most benign of the solvents used for dehydration because it is not as strong an organic solvent as acetone (and thus is potentially less injurious to the user) and exhibits less flammability than solvents such as acetone. On the other hand, Luft and Wood (1963) reported 4% of proteins were extracted during dehydration following primary fixation in osmium, with the majority being lost in the more dilute ethanol steps (none were lost in 95–100% ethanol steps or in propylene oxide). A study by Hanstede and Gerrits (1983) described linear shrinkage of about 9.3% in liver dehydrated with ethanol, mostly in ethanol concentrations above 95%.

2. Acetone

Acetone has been used by many workers as a dehydration medium and appears to cause less tissue shrinkage than an ethanolic series. An early study by Page and Huxley (1963) determined that while 10% linear shrinkage in skeletal muscle occurred with ethanol dehydration, none was noticeable with an acetone series. Phospholipids stabilized by *en bloc* uranyl acetate treatment should be relatively unextracted by acetone dehydration, while the more hydrophobic lipid entities (sterols and triglycerides) can be expected to be significantly extracted.

3. Dimethoxypropane

Dimethoxypropane (DMP) has been recommended as a superior dehydration agent compared to acetone or ethanol because more water-soluble entities remain in place (Thorpe and Harvey, 1979). Other workers (Beckmann and Dierichs, 1982) have suggested that DMP treatment extracts more lipids than ethanol or acetone. Due to the relatively limited application DMP for dehydration in comparison to the other two solvents listed, it is still difficult to make a strong case for the use of DMP.

4. Parameters

It should be evident from this brief discussion that shrinkage and extraction of various cellular components is to be expected with any dehydration series. It is thus important to consider ways to minimize these difficulties. Employing a protocol that provides the least amount of time in solvents (including the eventual infiltration of resins, which are also solvents) is clearly the best plan. A brief scanning of the literature will reveal that there are proponents of cold dehydrations to minimize extractions, while other workers routinely dehydrate at room temperature. If refrigerated dehydration series are used, it should be recognized that they will tend to develop increasing water contents after each exposure to room-temperature air (as they come out of the refrigerator) due to condensation of moisture from the air as it contacts the cold solvent. This can compromise the "dryness" of the final dehydration steps (100% ethanol, 100% acetone). As you will see in the section on resins, this is of extreme importance because any residual water in the specimen or transition solvent can ruin the polymerization capabilities of most epoxide resins.

There are various approaches to dehydration, some involving more incremental steps and some involving fewer steps. If speed is the most important consideration, as might be expected in certain clinical settings, extremely rapid dehydration can be accomplished if the tissue samples are kept small (0.25 mm^3). Coulter (1967) completed a successful dehydration using ethanol followed by propylene oxide in under 30 min utilizing such small pieces of tissue and constant agitation. Under most circumstances, however, it is more convenient to dehydrate at a more leisurely pace. In our laboratory we rinse out our post-fixative (osmium) and put it immediately into room-temperature 50% ethanol for 15 min followed by 75% ethanol for 15 min, 95% ethanol (two 15-min washes), 100% ethanol (two 30-min washes), and finish with two 10-min washes with 100% acetone prior to beginning our resin infiltration. We have never encountered any problems with insufficient dehydration or any appearance of excessive extraction. This procedure has been used with organisms from all five kingdoms of life with equally good results.

IV. EMBEDDING MEDIA

A. Ideal Qualities

Ideal qualities of embedding media include uniformity of batches of resin components, good solubility in dehydration agents, low viscosity as a monomer, along with minimal shrinkage, stability under the electron beam, minimal granularity, and good sectioning characteristics. Unfortunately, all these goals are not always achievable. For example, resins with low vicosity tend to shrink significantly. In addition, certain sacrifices are made when cytochemistry or immunocytochemistry is practiced on sections of resin-embedded materials.

B. Classes and Characteristics of Resins

1. Acrylic Resins

Acrylic resins were first utilized in the form of glycol methacrylates (Newman *et al.,* 1949). Glycol methacrylates have proved to be inadequate for electron microscopy but remain important in light microscopy in the form of JB-4 resin. Acrylics, in general, are prone to uneven polymerization, which can be due to impurities in the medium itself or the specimen catalyzing polymerization. They are strong lipid extractors and are more unstable under the electron beam than the epoxide resins. They must be polymerized in the absence of oxygen, so specimens are usually placed into resin-filled gelatin capsules, which are capped to exclude excess oxygen. Acetone also prevents polymerization, so the typical dehydration series is ethanolic, culminating in 95–100% ethanol for the water-miscible acrylics typically used (Lowicryl K4M, LR White). Osmium in tissues can lead to improper polymerization in some cases (notably glycol methacrylates). Osmium also renders tissues opaque, thus preventing proper polymerization if UV is used. Lowicryl resins will polymerize at room temperature, but the reaction is exothermic and produces an unacceptable temperature increase unless polymerized at lower temperatures. LR White resin sections can usually be picked up on uncoated grids, though they will respond to the beam by drifting initially. On the other hand, Lowicryl sections generally need a support film such as Formvar® to prevent excessive motion under the beam. See Table 2 for brief descriptions of some of the acrylic resins currently in use.

2. Polyester Resins

Polyester resins were first introduced by Kellenberger *et al.* (1956). The most commonly available resin is Vestopal W. It is more immune to beam damage than the methacrylates that it replaced, shrinks less, and polymerizes more evenly. The presence of air results in uneven polymerization. Polymerized blocks are harder than epoxy resin blocks. Two of the three components (benzoyl peroxide and cobalt napthenate) are unstable (even if held at 4°C for extended periods), and benzoyl peroxide must be mixed with Vestopal W prior to the addition of cobalt napthenate or a strongly exothermic reaction may take place. Heat may be used to polymerize the resin, eliminating the need for cobalt napthenate. The polymerized resin has excellent sectioning characteristics, low background, and consequent high contrast.

3. Epoxide Resins

Epoxide resins (see Table 2 for brief descriptions of readily available types) were introduced in 1956, with the Araldites (Glauert *et al.*, 1956) becoming widely used despite their relatively high viscosity. Luft (1961) published a seminal paper utilizing Epon 812 from Shell Oil Co., which has considerably lower viscosity than the Araldites. Several authors have utilized mixtures of Epon and Araldite (Mollenhauer, 1964) to produce blocks with the excellent sectioning qualities of Araldites, along with lesser viscosity, for improved specimen preparation.

Most of the epoxides are extremely intolerant of the presence of water and will fail to polymerize properly if the tissue is not totally dehydrated, resulting in extremely rubbery blocks from which tissues cannot be rescued. The epoxides are soluble in ethanol, acetone, and propylene oxide, though most schedules suggest passage through acetone or propylene oxide, in which the resins have the best solubility (though, as stated in the dehydration section, propylene oxide confers no advantages and is much more toxic than acetone and should not be used). A few of the more recently derived

TABLE 2. Selected Resin Embedding Media

Acrylic
  Hydrophilic (polar): Used primarily for immunocytochemistry
    Lowicryl K4M (polymerizes with UV at −35°C)
    Lowicryl K11M (polymerizes with UV at −60°C)
    LR White (polymerizes at room temperature with added accelerator; at 60°C overnight without
      accelerator)
    LR Gold (polymerizes at −25°C with UV)
  Hydrophobic (apolar): Good for lipid retention
    Lowicryl HM20 (polymerizes with UV at −70°C)
    Lowicryl HM23 (polymerizes with UV at −80°C)

Epoxy (viscosities in cp at 25°C)
  Water immiscible
    Epon 812 (150–210 cp; manufacture discontinued by Shell Oil Co. in 1979)
    Epon 812 replacements
      Medcast (Pelco)
      Eponate 12 (Pelco)
      SPI-pon 812 (SPI)
      Poly/Bed 812 (Polysciences; somewhat water soluble)
      EMbed 812 (EMS)
      LX-112
    Araldites
      502 (3000 cp)
      6005 (1300–1650 cp)
    Maraglas 655 (500 cp)
    Spurr (60 cp)
    HXSA-based low viscosity (21 cps; similar to Spurr, except NSA replaced with hexenyl succinic
      anhydride [HXSA]; Pelco)
  Water Miscible
    Durcupan
    Quetol 812 (140 cp; Kushida, H. 1983. *J. Electr. Microsc.* 32:65)
    Quetol 651 (15 cp; Fujita *et al.* 1977. *J. Electr. Microsc.* 23:165)
    Quetol 653 (60 cp; Kushida, H. 1980. *J. Electr. Microsc.* 29:193)

Melamine
  Nanoplast (water miscible, no dehydration necessary; good for LM and EM; 8–10 times more expensive
    than resins such as Spurr's)

Polyester
  Vestopal-W

resins are water miscible (see Table 2). Resins with viscosities higher than about 150 cP benefit from longer infiltration times, more gradual steps involving resin diluted with solvent, and rotation or tumbling during infiltration.

Mollenhauer (1986) suggested the addition to Spurr resin of 0.1–0.4 g lecithin dissolved in 0.1–0.4 g peanut oil to improve sectioning characteristics with glass knives. This technique may be applicable to other epoxide resins.

Spurr resin is an excellent all-purpose epoxide resin that penetrates most tissues easily, stores for up to 2 months at −20°C, and has a short infiltration series (30 min in

50% Spurr/50% acetone; 60 min in 100% Spurr; a second 60 min in Spurr; transfer tissues to new Spurr in molds and polymerize overnight at 70°C). If stored frozen, it must be warmed to room temperature before opening to prevent condensation of water, or the resin will not polymerize. The low viscosity of the monomer results in visible shrinkage in molds, and the resin chemically interacts with commonly used silicon embedding molds, unlike the other epoxides. This resin also does not stain effectively with aqueous uranyl acetate, so a 5% methanolic stain is usually employed, which can result in extensive wrinkling of sections (see section staining).

4. Polyethylene Glycol

Polyethylene glycol (PEG) was introduced as an embedding medium for producing sections for high-voltage electron microscopy (Wolosewick, 1980). The technique involves dehydrating normally fixed tissues in 100% ethanol and then placing the tissues in a 1:1 mixture of 100% ethanol and PEG 4000 overnight at 60°C in uncapped vials (to let the alchohol evaporate). The tissues are then removed, briefly blotted dry, and placed in predried gelatin capsules filled with fresh PEG 4000. When the specimen has sunk to the bottom of the PEG, the capsule is rapidly immersed in liquid nitrogen. Thin sections are then cut (though the blocks are quite hydrophilic and must be blotted dry frequently during sectioning) and placed on Formvar and poly-l-lysine-coated grids (the latter to keep the sections attached to the plastic film). They are next dehydrated again to remove the PEG and critical-point dried. This results in critical-point dried sections on grids without any surrounding medium, which thus have sufficient contrast that further staining is unnecessary. PEG-embedded tissues are relatively difficult to work with, and this technique thus is used almost exclusively in high-voltage electron microscopy.

5. Miscellaneous Water-Miscible Media

Miscellaneous water-miscible media, such as urea-aldehyde (Pease and Peterson, 1972) and Heckman's and Barrnett's (1973) glutaraldehyde-carbohydrazide (GACH), were introduced in the early 1970s to increase lipid retention and to produce sections that were accessible for enzyme histochemistry. These techniques have been supplanted by the modern acrylic resins introduced in the 1980s (Lowicryls, LR White resins).

6. Health Hazards

Health hazards must be considered with all resins and their associated solvents. As mentioned previously, many workers still utilize the epoxide, propylene oxide, as a transitional solvent for epoxide resins. This is unnecessary and contraindicated due to its toxicity. All epoxide resins should be treated with respect, but Spurr resin, in particular, is formulated with vinylcyclohexene dioxide, which is reported to be carcinogenic (Causton, 1981) even after polymerization. Care should be taken not to breathe resin vapors, to come into physical contact with the liquid resin, or to ingest plastic chips or dust produced during trimming procedures. Acrylic resins are reputed to be less toxic than epoxides. All resins and most of their solvents can cause dermatitis, and repeated contact can result in hypersensitivity to them. There are no readily available gloves guaranteed to be incapable of penetration by these substances. Double-gloving and short exposure times help, but the best policy is not to come into contact with these media at any time.

C. Embedding Mold Types

1. Choice of Molds

The choice of molds is determined, in part, by the characteristics of the embedding medium employed. If an acrylic or polyester medium is used, air must be excluded, so gelatin capsules are commonly used. Polyethylene molds, such as BEEM capsules, are incompatible with acrylic resins polymerized with heat because they will frequently partially dissolve. Another factor that must be considered is whether the specimen needs orientation or not. Flat embedding molds made of silicone rubber are available in a variety of designs to accomodate this need. When specimen orientation is not critical, cylindrical polyethylene molds, such as BEEM capsules, gelatin capsules, and silicon molds that yield cylindrical blocks with tapered tips are available.

Specialized molds have been developed to embed cells grown on coverslips (Chang Coverslip mold) or to serve as a substrate on which cells can be grown, fixed, dehydrated, and embedded, resulting in a 1 × 3″ slide that has excellent optical characteristics for selecting individual cells to section (Hanker-Giammara mold). Consult the catalogs from electron microscopy supplies vendors listed in the back of this book for details on different types of molds available.

Another approach to achieving the desired orientation is to flat-embed tissues, cut out areas of the blocks, and reorient them by gluing them to polymerized blocks devoid of tissues. Pieces of tissue can also be removed from polymerized resins with a jeweler's saw and oriented in empty, flat-embedding molds, which are then filled with unpolymerized resin and polymerized.

D. Agar Embedment

Agar embedment is a useful method for turning suspensions of cells or particulates (microsomes, mitochondria) into tissue-like blocks for ease of handling during processing. Cell suspensions or particulates should be centrifuged into pellets at the speeds normally used. After the suspensions have been fixed and then washed several times in an appropriate buffer, they are pelleted again. It is best to do this work in microfuge tubes. Next, molten (approximately 55–60°C) 3–4% water agar is introduced into the pellet by pulling the agar up into a preheated Pasteur pipet, putting the tip to the bottom of the pellet in the microfuge tube, and expelling the agar. The tube is quickly placed in a microfuge and run at the same speed used to pellet the original suspension for about 30 sec. A longer centrifugation time will allow the agar to begin solidifying, and shearing of the fixed materials will be likely. When utilizing this procedure, it is important not to pipet the cells up and down (this can cause shearing) and to do the procedure before osmication (osmium-treated materials are brittle enough to be sheared when centrifuged through the molten agar). After the agar has solidified, the tip of the centrifuge tube can be cut off with a razor blade, and the small pellet of agar-encased material can be removed and sliced into 1-mm-thick pieces for further processing.

Another use of agar for specimen handling is the agar-peel technique for cells *in situ*. In this procedure, adherent cells in flasks or Petri dishes are rinsed free of growth medium with PBS (to remove media proteins that otherwise would be fixed), and the culture is then covered with an appropriate primary fixative. After 30–60 min, the primary fixative is removed and the culture is rinsed two or three times with an appropriate buffer, osmicated for an appropriate period of time, and then rinsed in distilled water several times. Molten 3–4% water agar is then gently poured onto the surface of the culture and allowed to harden. Next, the culture is dehydrated in a normal fashion. During the 100% acetone steps, the agar should release from the plate, hopefully with the cultured cells attached to it. At that point, the agar with cells can be cut into strips for further processing in vials and then is finally flat-embedded. On some occasions, the agar lifts off without removing the cells from the plate. If this occurs, put the plate back into 100% ethanol (so the plate will not get totally dissolved by the acetone) and go directly into a mixture of plastic resin and ethanol after several rinses in 100% ethanol. After polymerization, pieces of the dish containing cells can be cut out with a jeweler's saw and glued to other blank blocks for sectioning.

V. EXAMINATION OF FOUR TISSUES PREPARED WITH A VARIETY OF FIXATIVES AND BUFFERS

We have already discussed various types of aldehyde primary fixatives and their combinations, as well as the two major types of buffers used (phosphate and cacodylate) in most structural studies. At this point, it would be useful to see what the effect of these protocols actually is on a set of tissues. We have chosen three mouse tissues (small intestine, kidney, and liver). All the tissues were collected at the same time and immediately placed in the various fixatives. Primary fixation was for 1 hr at room temperature, and in the cases where osmium was used as a post-fixative, that step also lasted 1 hr at room temperature. All tissues were rinsed in the same buffer used during primary fixation, rinsed in distilled water after osmium treatment, dehydrated, and embedded as described in the next section. The figure legends describe the fixation of each specimen.

Figures 5–14 illustrate mouse kidney fixed a variety of ways. No totally unacceptable images of fixation appear, except in two cases. Figure 11 from a kidney fixed with simultaneous glutaraldehyde and osmium shows mitochondria almost in negative contrast and an overall muddiness of preservation, while the kidney fixed in glutaraldehyde with no osmium post-fixation (Fig. 14) has indistinct membranous pro-

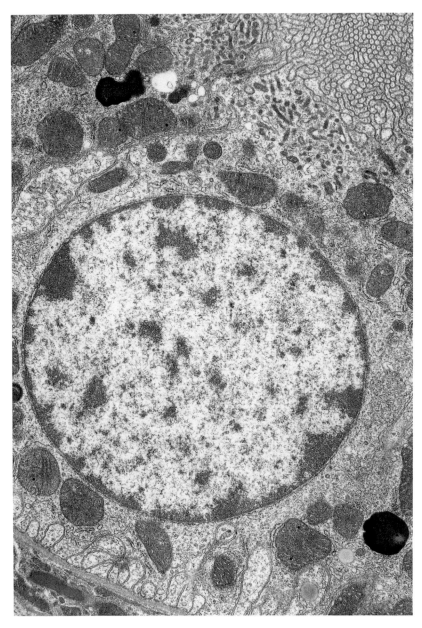

FIGURE 9. Mouse kidney fixed in 2% glutaraldehyde in 0.1 M phosphate buffer, pH 7.27. Post-fixed in 1% osmium in the same buffer (Fix #5). 14,200×.

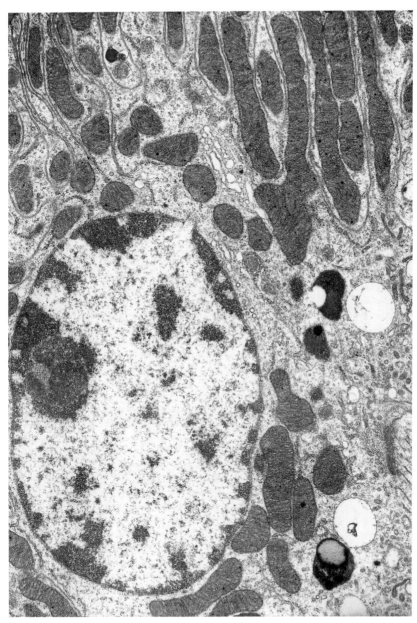

FIGURE 10. Mouse kidney fixed in 2% glutaraldehyde in 0.1 M cacodylate buffer, pH 7.26. Post-fixed in 1% osmium in the same buffer (Fix #6). 14,200×.

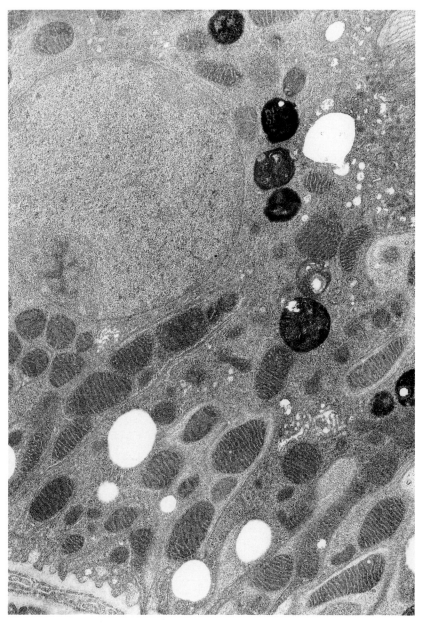

FIGURE 11. Mouse kidney fixed in 2% glutaraldehyde/0.5% osmium in 0.1 M phosphate buffer, pH 7.39. Simultaneous fixation with no post-fixation (Fix #7). 14,200×.

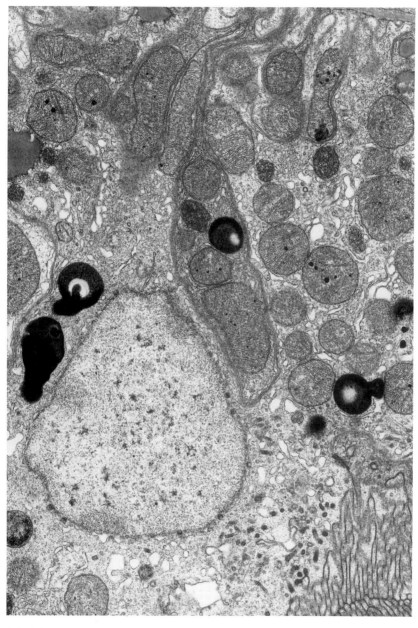

FIGURE 12. Mouse kidney fixed in 1% osmium in 0.1 M cacodylate buffer, pH 7.26. No post-fixation (Fix #8). 14,200×.

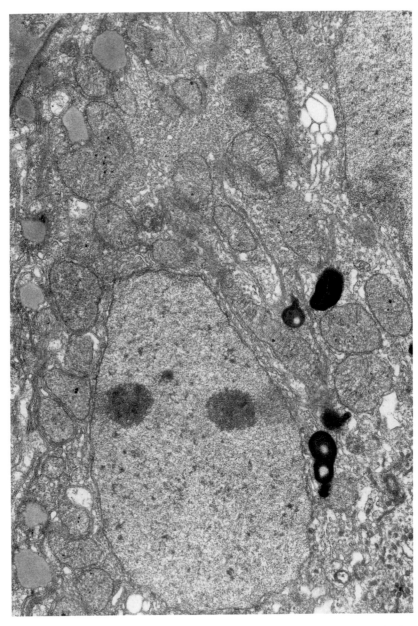

FIGURE 13. Mouse kidney fixed in 1% osmium in 0.1 M phosphate buffer, pH 7.39. No post-fixation (Fix #9). 14,200×.

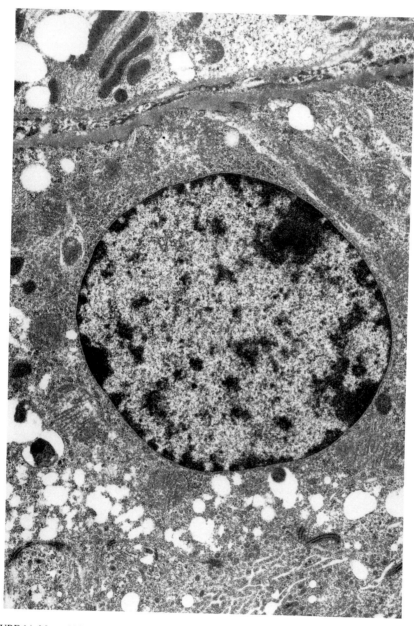

FIGURE 14. Mouse kidney fixed in 2% glutaraldehyde in 0.1 M phosphate buffer, pH 7.27. No post-fixation (Fix #10). 14,200×.

files, as well as open holes where materials were extracted, and a general granularity and muddiness to the image. There are variations between the other preparations from the standpoint of mitochondrial density and nuclear and cytoplasmic extraction, but there is nothing in the series that is clearly unacceptable. In normal practice, the variation between samples prepared exactly the same way with the same fixatives will show variation in the same range as seen in these samples.

Examining the liver samples (Figs. 15–24) reveals a different situation than found with the kidney samples. The Carson's fixed material (Fig. 15) showed very irregular nuclear outlines and poorly differentiated glycogen. The full-strength Karnovksy's sample (Fig. 17) showed marked distension of endoplasmic reticulum, along with an apparent leaching of cytoplasmic materials. The material exposed to half-strength Karnovsky's (Fig. 18) showed some dilation of endoplasmic reticulum, along with a swollen and ruffled nuclear envelope profile. Tissues treated with osmium as a primary fixative (Figs. 22 and 23) showed some cytoplasmic extraction and compacted-appearing mitochondria. Finally, the glutaraldehyde-fixed tissue (Fig. 24) that had no osmication had large extracted areas and indistinct membranous profiles.

The intestine samples (Figs. 25–34) looked well fixed except for the sample fixed with glutaraldehyde in phosphate buffer with osmium post-fixation (Fig. 29), the sample treated with glutaraldehyde with phosphate buffer and osmium post-fixation (Fig. 30), the simultaneous glutaraldehyde/osmium fixation (Fig. 31), and the glutaraldehyde without osmium post-fixation treatment (Fig. 34).

The glutaraldehyde-fixed materials that had an osmium post-fixation had mitochondrial abnormalities (swelling) in both cases, though this was most noticeable in the cacodylate-buffered glutaraldehyde (Fig. 30). The intestine fixed with glutaraldehyde with no osmication (Fig. 34) had empty spaces and indistinct membrane profiles, as found in the kidney and liver samples fixed in the same way.

This series of photographs is meant to illustrate that good fixation can be influenced by the type of tissue and the individual sample of tissue, as well as the constituents of the fixative. Most of the fixations in the series resulted in usable material, but an aldehyde fixation followed by osmium post-fixation seemed to offer the best pictures. The poor image quality for the intestine samples fixed in glutaraldehyde with an osmium post-fixation appears to be anomalous. Unfortunately, it represents the fact that good fixatives can yield bad samples, even with rapid collection of samples below 1 mm in thickness.

Evaluation of semithick sections (Figs. 35 and 36) also can deliver a wealth of information about the success of fixation techniques. Figure 35 shows a rat liver fixed in 4F:1G (McDowell and Trump, 1976), followed by osmication. The surface of the tissue block is at the lower left of the photograph. Large grayish glycogen pools are seen in the cytoplasm. As one moves deeper in the block (upper right), it becomes evident that these glycogen pools are progressively more extracted, indicating poor fixation. The nuclei also become more open as the nucleoplasm is extracted. Figure 36 shows a rat kidney semithick section from a block of tissue simultaneously fixed with glutaraldehyde/osmium in a phosphate buffer. The surface of the tissue (lower left) is well-fixed, but only two to three kidney tubules into the block, the tissue becomes grossly understained due to poor fixation. In this case, the combined action of both

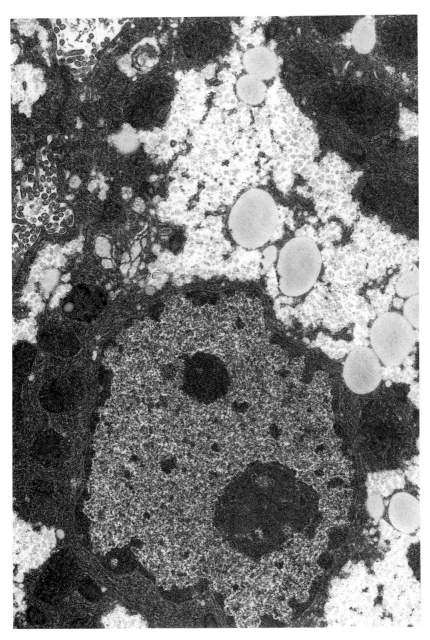

FIGURE 15. Mouse liver, Fix #1. 14,200×.

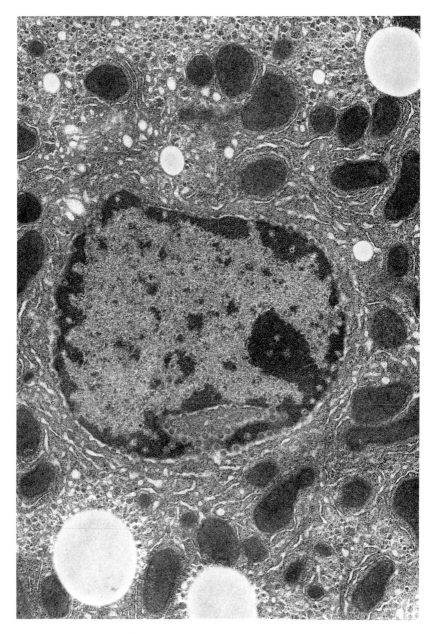

FIGURE 16. Mouse liver, Fix #2. 14,200×.

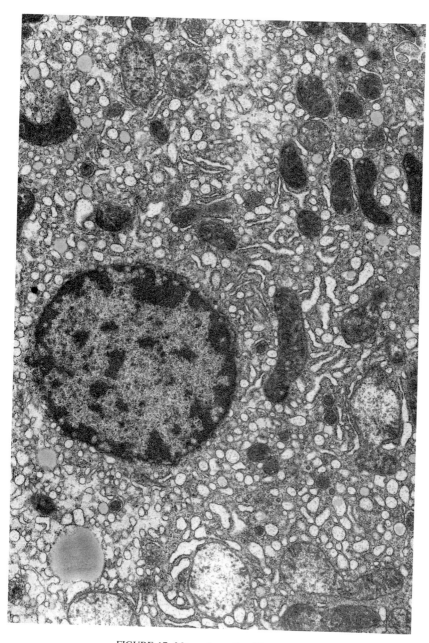

FIGURE 17. Mouse liver, Fix #3. 14,200×.

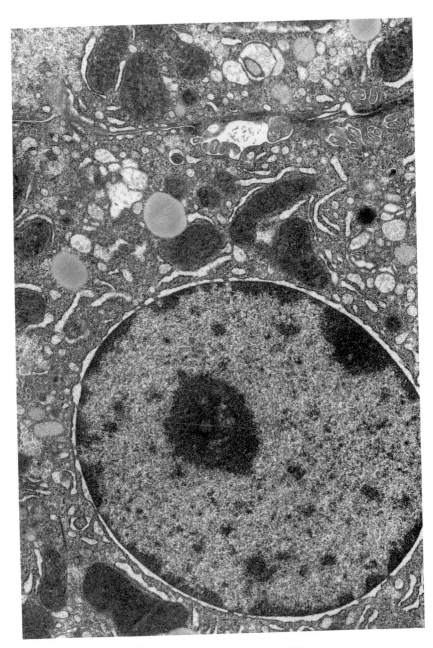

FIGURE 18. Mouse liver, Fix #4. 14,200×.

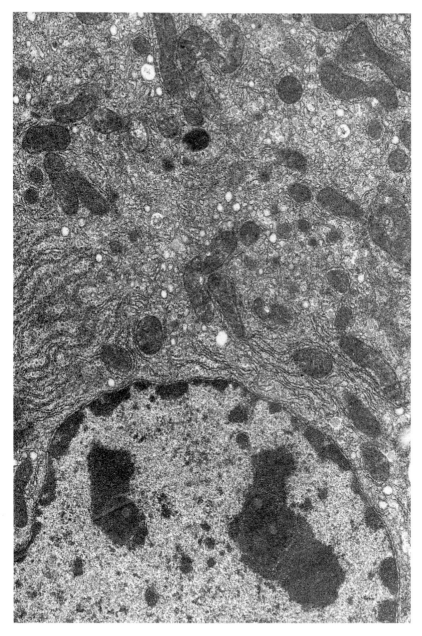

FIGURE 19. Mouse liver, Fix #5. 14,200×.

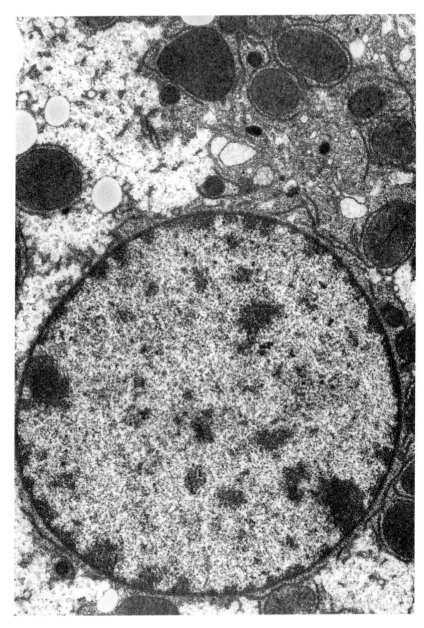

FIGURE 20. Mouse liver, Fix #6. 14,200×.

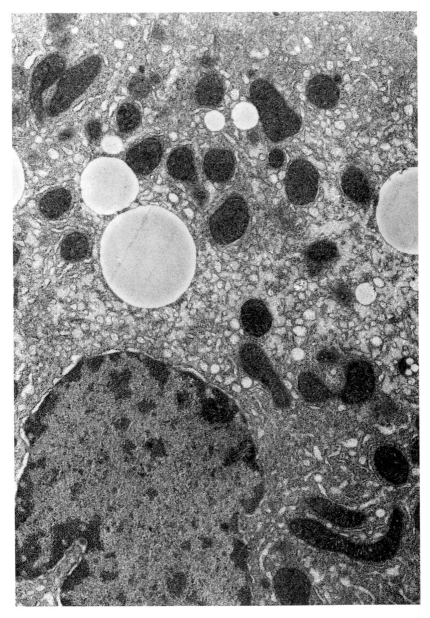

FIGURE 21. Mouse liver, Fix #7. 14,200×.

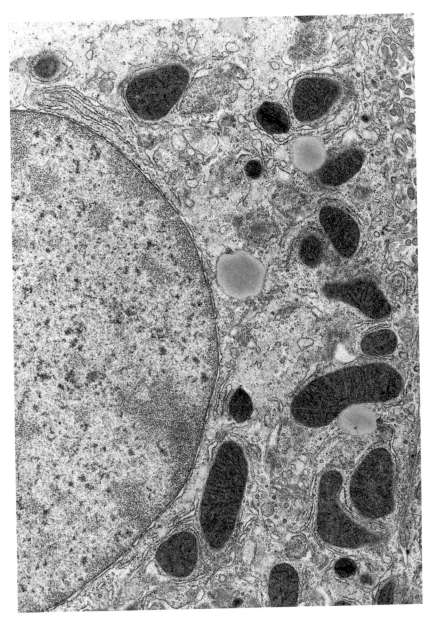

FIGURE 22. Mouse liver, Fix #8. 14,200×.

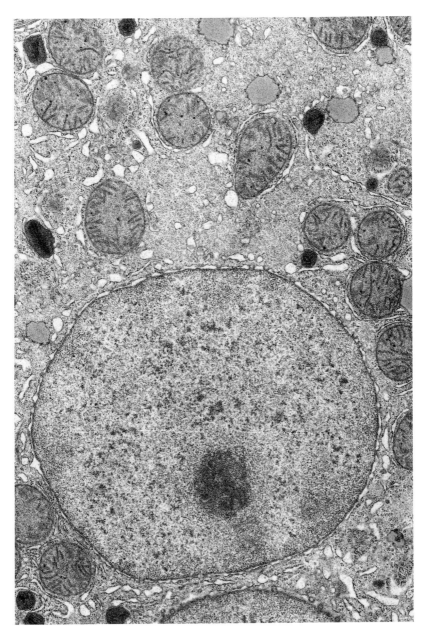

FIGURE 23. Mouse liver, Fix #9. 14,200×.

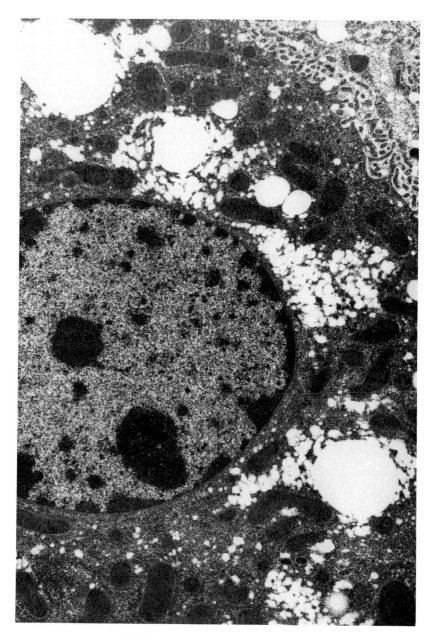

FIGURE 24. Mouse liver, Fix #10. 14,200×.

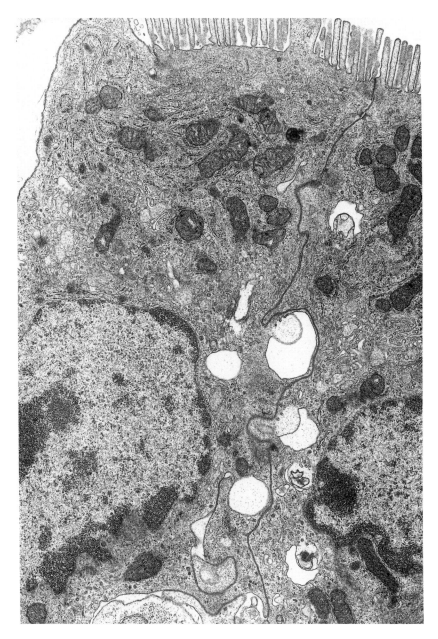

FIGURE 25. Mouse small intestine, Fix #1. 14,200×.

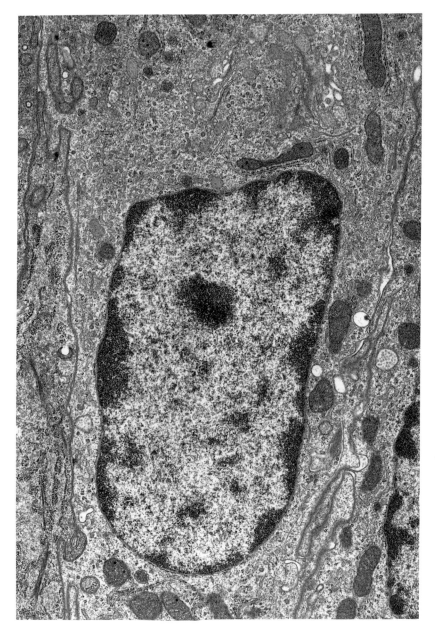

FIGURE 26. Mouse small intestine, Fix #2. 14,200×.

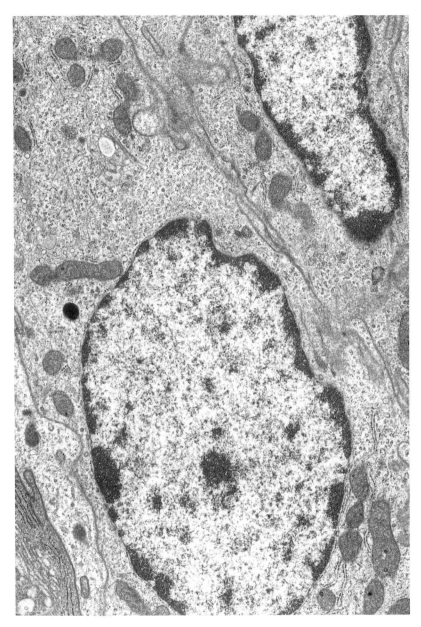

FIGURE 27. Mouse small intestine, Fix #3. 14,200×.

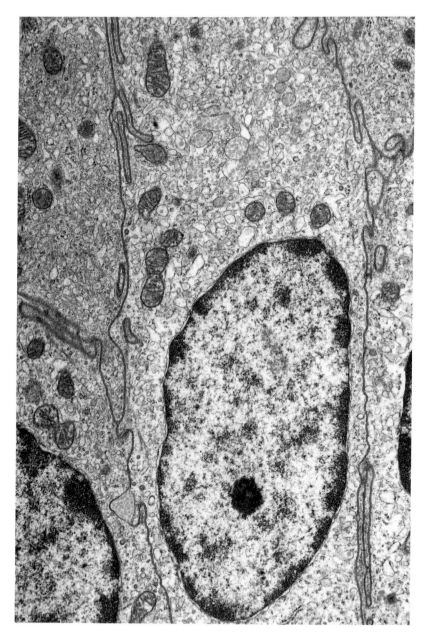

FIGURE 28. Mouse small intestine, Fix #4. 14,200×.

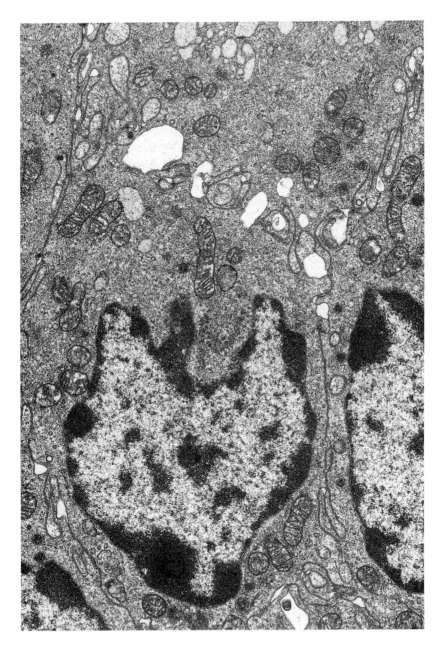

FIGURE 29. Mouse small intestine, Fix #5. 14,200×.

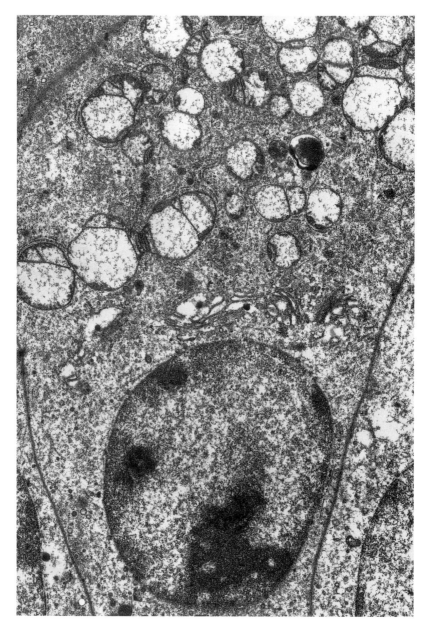

FIGURE 30. Mouse small intestine, Fix #6. 14,200×.

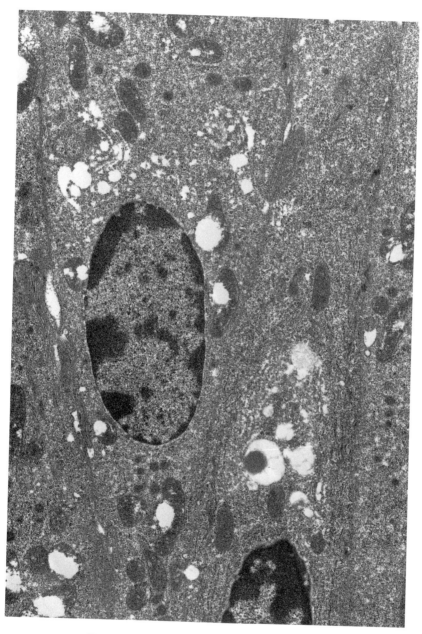

FIGURE 31. Mouse small intestine, Fix #7. 14,200×.

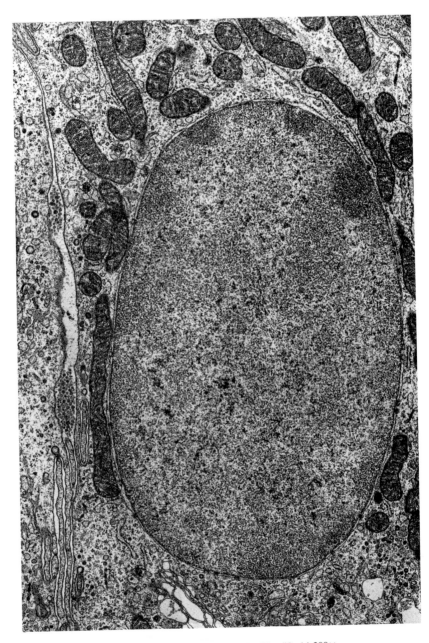

FIGURE 32. Mouse small intestine, Fix #8. 14,200×.

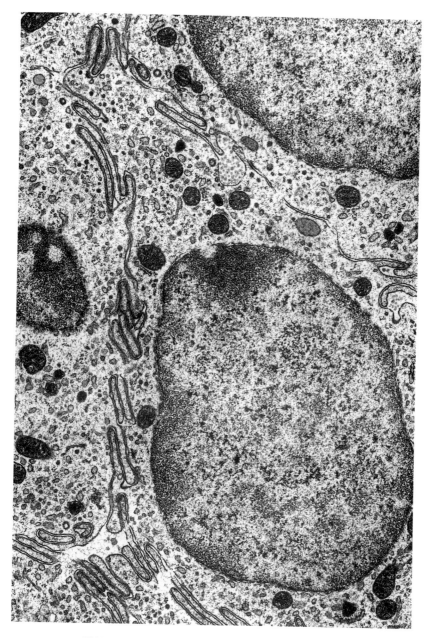

FIGURE 33. Mouse small intestine, Fix #9. 14,200×.

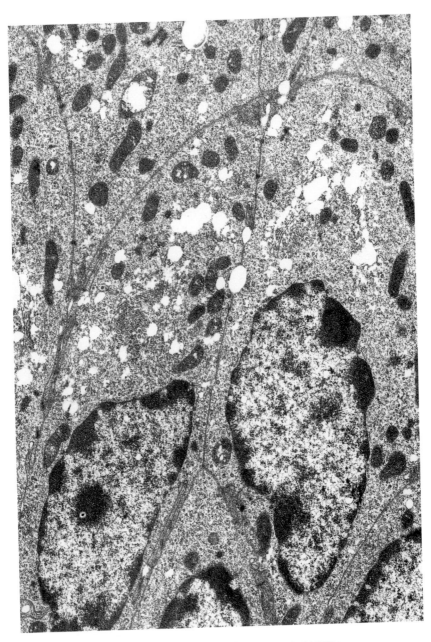

FIGURE 34. Mouse small intestine, Fix #10. 14,200×.

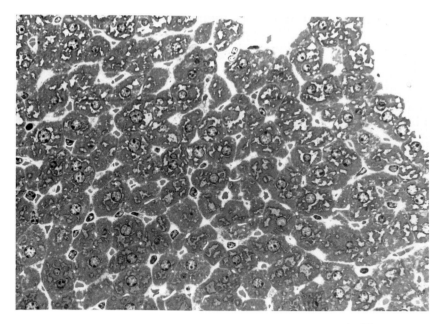

FIGURE 35. Rat liver, semithin section stained with toluidine blue O. 460×.

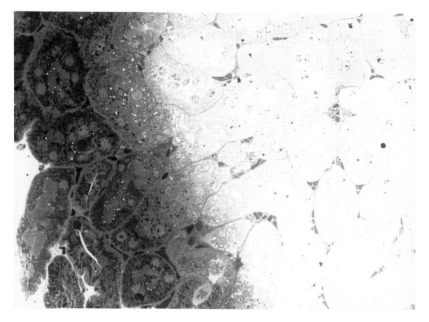

FIGURE 36. Mouse kidney, semithin section stained with toluidine blue O. 460×.

fixatives working on the tissue simultaneously apparently blocked the ingress of the fixative components deeper into the tissue. Similar-sized blocks fixed with sequential procedures would not have shown this artifact.

VI. A QUASI-UNIVERSAL FIXATION, DEHYDRATION, AND EMBEDMENT SCHEDULE SUCCESSFULLY USED ON ORGANISMS FROM ALL FIVE KINGDOMS OF LIFE

Routine TEM Fixation

1. 1 hr to months in McDowell's and Trump's (1976) 4F:1G fixative. Initial fixation can be done at room temperature, but longer storage should be at 4°C.
2. Rinse tissue 15 min two times in 0.1 M phosphate buffer, pH 7.2–7.4.
3. Post-fix tissue in 1% osmium tetroxide/0.1 M phosphate buffer pH 7.2–7.4 for 1 hr at room temperature.
4. Rinse tissue in distilled water two times (5 min each).
5. Dehydrate tissue:
     50% ethanol, 15 min
     75% ethanol, 15 min (Can leave tissues here overnight at 4°C.)
     95% ethanol, 15 min, two times
     100% ethanol, 30 min, two times
     100% acetone, 10 min, two times
6. Infiltrate with Spurr (1969) resin (recipe with 6.3 g DER for good sections from diamond or glass knives):
     50% Spurr in 100% acetone, 30 min
     100% Spurr resin, 60 min
     100% Spurr resin, 60 min
     NEW 100% Spurr resin; put in appropriate molds
7. Polymerize in 70°C oven overnight to 3 days.

Figures 37–39 illustrate distal convoluted tubule epithelial cells from a rat kidney that was perfused with 4F:1G in March 1985. Figure 37 was taken of a piece of the kidney that was processed immediately after perfusion, and Figure 38 was taken from the same kidney stored 4 years at 4°C before being further processed. Finally, Figure 39 illustrates the same material stored for 5 years before subsequent processing. All three figures illustrate cells containing mitochondria that have normal density and show no evidence of artifactual swelling. None of the samples exhibited any degree of nuclear envelope dilation. The 4- and 5-year samples have slight evidence of extraction of nucleoplasmic materials, and the outer nuclear envelope membrane appears less distinct than in the immediately processed sample. The plasmalemma appears normal in all three samples.

Figure 40 shows the small intestine of a bobwhite quail infected with *Cryptosporidium*. The enterocytes contain mitochondria, lysosomes, and nuclei, none of which show any signs of fixation damage. Part of an adjacent goblet cell also has normal ultrastructural conformation. The *Cryptosporidium* trophozoites contained nu-

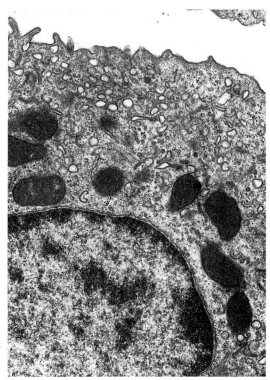

FIGURE 37. Distal convoluted tubule of rat kidney fixed and processed in 1985. 16,800×.

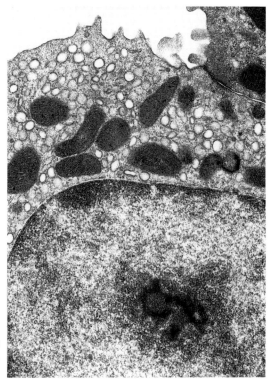

FIGURE 38. Distal convoluted tubule of rat kidney fixed in 1985 and processed in 1989. 16,800×.

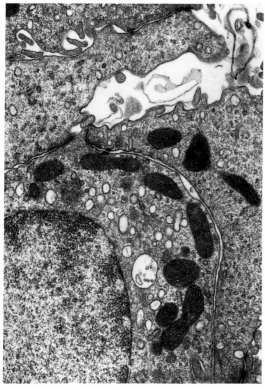

FIGURE 39. Distal convoluted tubule of rat kidney fixed in 1985 and processed in 1990. 16,800×.

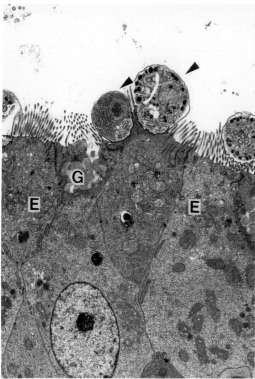

FIGURE 40. Bobwhite quail intestine infected with *Cryptosporidium* (arrowheads). Note the normal appearance of goblet cells (G) and enterocytes (E). 4,600×.

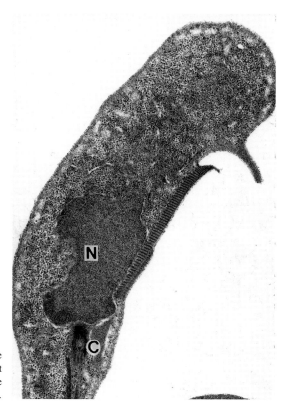

FIGURE 41. *Giardia* from liquid culture with centriole (C) and associated rootlet and nucleus (N) closely associated with the subpellicular microtubule array. 15,600×.

clei and characteristic electron-dense granules, as well as a network of tubules constituting the feeding apparatus along the border of the attachment zone on host cells. There is obvious separation of the protozoan pellicle from the cytoplasmic contents in the two more-mature cells. All of these features are consistent with images published of various coccidians, including *Cryptosporidium*. The density of the nucleoplasm and cytoplasm indicate good chemical fixation.

Figure 41 shows a cell of *Giardia* that was grown in liquid artificial medium. A centriole and associated rootlet for a flagellum are shown, as well as the complex subpellicular microtubule array characteristic of this organism. The nuclear envelope and endoplasmic reticulum reveal no artifactual swelling. No extraction of the cytoplasm or nucleoplasm is evident.

Mammalian cells infected with the herpesvirus HHV6 (Fig. 42) have slight mitochondrial abnormalities in the form of swelling and moth-eaten areas, which are consistent with profiles of a great number of cells grown in culture. Otherwise, the cytoplasmic and nuclear preservation looks normal for chemical fixation.

Healthy fish skin containing both cytoplasmically dense and relatively electron-lucent epidermal cells is shown in Figure 43. Desmosomes and their associated tonofilaments are well preserved. The nucleus and cytoplasm of the less electron-dense cell appear to be somewhat extracted, and the nuclear envelope profiles appear to be somewhat distended. Figure 44 is an illustration of hypertrophied fish epidermal cell

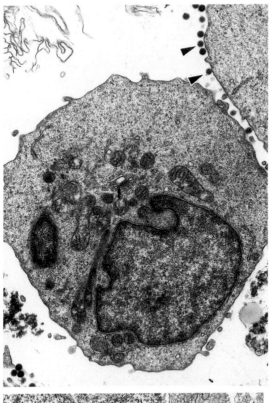

FIGURE 42. Human cord blood cells infected with HHV6 *Herpes* virus (arrowheads). 12,700×.

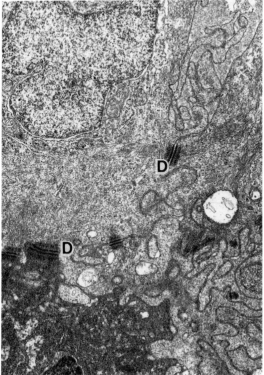

FIGURE 43. Healthy fish skin (*Tilapia*) showing desmosomes (D) with associated tonofilaments. 12,300×.

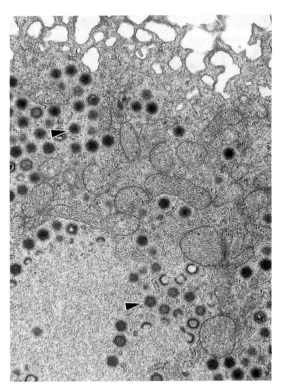

FIGURE 44. Field-collected fish skin cell infected with *Lymphocystis* (arrows). 12,000×.

from a field-collected sample infected with the iridovirus, *Lymphocystis*. The cell surface has a network of cytoplasmic extensions, and the cytoplasm contains numerous mitochondria in apparently good condition.

Rhodotorula (a yeast) and *Aspergillus* (a filamentous fungus) are shown in Figures 45 and 46, respectively. They both have well-fixed wall materials, mitochondria, and cytoplasmic ground substance. The *Aspergillus* cell contains two nuclei, which have well-preserved nuclear membranes, as well as one lipid droplet. An unevenly distributed extracellular matrix is revealed (Fig. 46).

Escherichia coli (Fig. 47) cells appear normal and contain the expected distribution of ribosomes and bacterial chromosome material typical for this organism after chemical fixation. The cell membranes appear well preserved, and the cytoplasmic ground substance appears unextracted.

Root meristem cells from *Zea mays* (Fig. 48) contain well-fixed nuclei, mitchondria, vacuoles, and plasmodesmata through the cell walls. No swelling of endoplasmic reticulum, nuclear, or mitochondrial profiles was noted. Starch grains are somewhat extracted, as is typical for chemically fixed plant cells. The nucleoplasm and cytoplasm appear unextracted.

McDowell and Trump designed their fixative 4F:1G to address a number of problems peculiar to their kidney research: (1) Kidneys need to be perfused with fixatives at physiological pressure to maintain proper structural integrity of their proximal convoluted tubules. (2) Kidney researchers customarily examine materials both by

FIGURE 45. Yeast cell (*Rhodotorula*) with mitochondria (M) and endoplasmic reticulum (arrowheads). 29,400×.

light microscopy of paraffin sections, as well as by electron microscopy of plastic sections. (3) Tissues fixed in glutaraldehyde concentrations of 2% or greater become difficult to section due to brittleness if embedded in paraffin; they also exhibit non-specific periodic acid-Schiff's staining throughout due to the free hydroxyl groups of glutaraldehyde, which are not totally bound to the fixed tissues. (4) Collection of large numbers of tissues from relatively complicated surgical procedures frequently results in sampling over long time periods and the consequent physical impossibility of processing tissues immediately.

Our experience with this fixation regimen over the last 13 years has illustrated that it accomplishes the goals listed above but, fortunately, has even greater capabilities than originally envisioned.

We have shown that good-quality, publishable electron micrographs can be produced from a tissue (kidney) stored 5 years at 4°C in 4F:1G. The slight apparent extraction of nucleoplasm and the degraded appearance of the nuclear envelopes shown in Figures 1B and 1C may be a function of storage or may simply reflect the kind of sample fixation variation encountered with any tissue and fixation regimen.

Probably the most significant aspect of this procedure is the potentially broad applicability of the 4F:1G formulation to organisms from all five kingdoms of life and from various environments. The examples illustrated are only a small sampling of the various samples we have processed over the years. There are still some tissues and cell

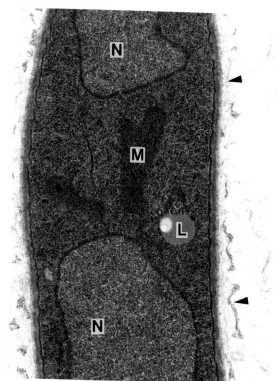

FIGURE 46. Filamentous fungus (*Aspergillus*) showing nuclei (N), mitochondrion (M), lipid droplet (L), and extracellular matrix (arrowheads). 29,100×.

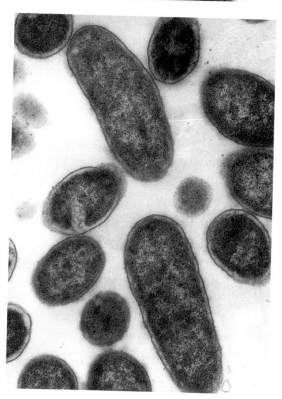

FIGURE 47. Bacteria (*Eschericia coli*). 23,900×.

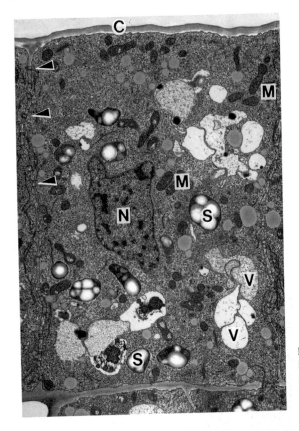

FIGURE 48. Corn root meristem showing nucleus (N), mitochondria (M), starch grains (S), vacuoles (V), cell wall (C), and plasmodesmata (arrowheads). 5,400×.

types that do not fix well with 4F:1G. An example would be mammalian leukocytes, which periodically show poor fixation with 4F:1G, illustrated by numerous discontinuities in plasmalemmal profiles. These samples are always collected in our clinics and then forwarded to our laboratory in the 4F:1G that we provide, so sample collection faults could easily explain our variable results. Some cells and tissues will show slightly improved images when prepared with other fixatives, but the 4F:1G procedure is still adequate for publication.

Protozoans from saline environments, protozoans from freshwater environments, parasitic protozoans, and free-living protozoans all seem to fix quite well with 4F:1G. Bacteria and fungi from the same range of environments also seem to fix comparably. Mammalian, reptilian, amphibian, and arthropod tissues all fix well (though arthropod cuticles must first be penetrated mechanically for good fixation). We have had little experience applying this fixative to plant tissues, but we have had good fixation with corn apices (root and shoot) as well as mung bean leaves.

We would like to suggest that many fixation regimens will work equally well with the variety of tissues and cells from the variety of environments that we have studied. We have chosen to stay with 4F:1G because of the variety of specimens that respond well to the fixative and the flexibility available in terms of being able to submit subsamples of fixed tissues for histological preparation for light microscopy. In addition, 4F:1G can be stored for at least 3 months before use at 4°C, and once tissues are immersed in the fixative, they can be stored for weeks or months (and in one case, at least 5 years) at the same temperature and still provide publishable results.

REFERENCES

Aldrich, H.C., Biemborn, D.B., and Schonheit, P. 1987. Creation of artifactual internal membranes during fixation of *Methanobacterium thermoautotrophicum*. *Can. J. Microbiol.* 33:844.

Beckman, H.-J., and Dierichs, R. 1982. Lipid extracting properties of 2,2-dimethoxypropane as revealed by electron microscopy and thin layer chromatography. *Histochemistry* 76:407.

Carson, F., Martin, J.H., and Lynn, J.A. 1973. Formalin fixation for electron microscopy: A reevaluation. *Am. J. Clin. Pathol.* 59:365.

Causton, B.E. 1981. Resins: Toxicity, hazards, and safe handling. *Proc. Ry. Microsc. Soc.* 16:265.

Coulter, H.D. 1967. Rapid and improved methods for embedding biological tissues in Epon-812 and Araldite 502. *J. Ultrastruct. Res.* 20:346.

Ghadially, F.N. 1975. *Ultrastructural pathology of the cell*. Butterworths, London.

Ghadially, F.N. 1985. *Diagnostic electron microscopy of tumors*, 2nd ed. Butterworths, London.

Gilkey, J.C., and Staehelin, L.A. 1986. Advances in ultrarapid freezing for the preservation of cellular ultrastructure. *J. Electron Microsc. Tech.* 3:177.

Glauert, A.M. 1975. *Fixation, dehydration and embedding of biological specimens*. Elsevier North-Holland, New York.

Glauert, A.M., Rogers, G.E., and Glauert, R.H. 1956. A new embedding medium for electron microscopy. *Nature* 178:803.

Hanstede, J.G., and Gerrits, P.O. 1983. The effects of embedding in water-soluble plastics on the final dimensions of liver sections. *J. Microsc.* 131:79.

Hayat, M.A. 1981. *Fixation for electron microscopy*. Academic Press, New York.

Hayat, M.A. 1989. *Principles and techniques of electron microscopy: Biological applications*, 3rd ed. CRC Press, Boca Raton, FL.

Heckman, C.A., and Barrnett, R.J. 1973. GACH: A water-miscible, lipid-retaining embedding polymer for electron microscopy. *J. Ultrastruct. Res.* 42:156.

Hopwood, D. 1967. Some aspects of fixation with glutaraldehyde: A biochemical and histochemical comparison of the effects of formaldehyde and glutaraldehyde fixation on various enzymes and glycogen with a note on penetration of glutaraldehyde into liver. *J. Anat.* 101:83.

Hopwood, D. 1972. Theoretical and practical aspects of glutaraldehyde fixation. In: *Fixation in histochemistry*, P.J. Stoward (ed). Chapman and Hall, London.

Karnovsky, M.J. 1965. A formaldehyde-glutaraldehyde fixative of high osmolality for use in electron microscopy. *J. Cell Biol.* 27:137A.

Kellenberger, E., Schwab, W., and Ryter, A. 1956. L'utilisation d'un copolymère des polyesters comme materiel d'inclusion en ultramicrotomie. *Experientia* 12:421.

Leong, A.S.-Y., Daymon, M.E., and Milios, J. 1985. Microwave irradiation as a form of fixation for light and electron microscopy. *J. Pathol.* 146:313.

Leong, A.S.-Y., and Gove, D.W. 1990. Microwave techniques for tissue fixation, processing and staining. *Electron Microsc. Soc. Amer. Bull.* 20:61.

Login, G.R., Dwyer, B.K., and Dvorak, A.M. 1990. Rapid primary microwave-osmium fixation. I. Preservation of structure for electron microscopy in seconds. *J. Histochem. Cytochem.* 38:755.

Luft, J.H. 1956. Permanganate—a new fixative for electron microscopy. *J. Biophys. Biochem. Cytol.* 2:799.

Luft, J.H. 1959. The use of acrolein as a fixative for light and electron microscopy. *Anat. Record* 133:305.

Luft, J.H. 1961. Improvements in epoxy resin embedding methods. *J. Biophys. Biochem. Cytol.* 9:409.

Luft, J.H., and Wood, R.L. 1963. The extraction of tissue protein during and after fixation with osmium tetroxide in various buffer systems. *J. Cell Biol.* 19:46A.

McDowell, E.M., and Trump, B.F. 1976. Histologic fixatives suitable for diagnostic light and electron microscopy. *Arch. Pathol. Lab. Med.* 100:405.

Millonig, G., and Marinozzi, V. 1968. Fixation and embedding in electron microscopy. In: *Advances in optical and electron microscopy,* Vol. 2., R. Baser and V.E. Cosslett (eds.), Academic Press, New York.

Mollenhauer, H.H. 1964. Plastic embedding mixtures for use in electron microscopy. *Stain Tech.* 39:111.

Mollenhauer, H.H. 1986. Surfactants as resin modifiers and their effect on sections. *J. Electron Microsc. Tech.* 3:217.

Newman, S.B., Borysko, E., and Swerdlo, M. 1949. New sectioning techniques for light and electron microscopy. *Science* 110:66.

Page, S.G., and Huxley, H.E. 1963. Filament lengths in striated muscle. *J. Cell Biol.* 19:369.

Palade, G.E. 1952. A study of fixation for electron microscopy. *J. Experim. Med.* 95:285.

Pease, D.C., and Peterson, R.G. 1972. Polymerizable glutaraldehyde-urea mixtures as polar, water-containing embedding media. *J. Ultrastruct. Res.* 41:133.

Sabatini, D.D., Miller, F., and Barrnett, R.J. 1963. Cytochemistry and electron microscopy, the preservation of cellular ulstrastructure and enzymatic activity by aldehyde fixation. *J. Cell Biol.* 17:19.

Salema, R., and Brandao, I. 1973. The use of PIPES buffer in the fixation of plant cells for electron microscopy. *J. Submicrosc. Cytol.* 5:79.

Smith, R.E., and Farquhar, M.G. 1966. Lysosome function in the regulation of the secretory process in the cells of the anterior pituitary gland. *J. Cell Biol.* 31:319.

Spurr, A.R. 1969. A low-viscosity epoxy resin embedding medium for electron microscopy. *J. Ultrastruct. Res.* 26:31.

Thorpe, J.R., and Harvey, D.M.R. 1979. Optimization and investigation of the use of 2,2-dimethoxypropane as a dehydration agent for plant tissues in transmission electron microscopy. *J. Ultrastruct. Res.* 68:186.

Tokuyasu, K.T. 1983. Present state of immunocryoultramicrotomy. *J. Histochem. Cytochem.* 31:164.

Wolosewick, J.J. 1980. The application of polyethylene glycol (PEG) to electron microscopy. *J. Cell Biol.* 86:675.

CHAPTER 2

Ultramicrotomy

I. ULTRAMICROTOMES

A. Purpose

An ultramicrotome is designed to cut ultrathin sections (10–100 nm), semithin sections (0.25–0.5 μm), and ultrathin frozen sections (if suitably equipped). These requirements have resulted in the development of two basic instruments over the years: thermal- and mechanical-advance ultramicrotomes. An excellent and concise source of information on ultramicrotomy is the booklet "Ultramicrotomy: Frequent Faults and Problems," by Sitte (1981). A more lengthy discussion of ultramicrotomy can be found in Reid's (1975) book.

B. Design

All modern ultramicrotomes have the following important features:

1. A knife-holding stage with adjustments for lateral swings of the knife, as well as the ability to tilt the knife at various angles. The former feature allows lateral swings of the knife edge to accommodate a block face trimmed to some angle other than perpendicular to its long axis, while the latter feature allows adjustment of the holder for glass knives and diamond knives that manufacturers specify for use at angles between 0° and 10°.
2. A moveable specimen-holding arm, which either retracts or follows a circular arc to permit single-pass cutting. This feature, the first major modification made to the original design of histological-type microtomes, was added because the first microtomes did not have single-pass cutting, which meant that the block face touched the knife edge on the return stroke (after a section had

been cut). At the ultrastructural level, this contact causes damage to the block face and thus the ensuing sections.

3. A knife stage with coarse and fine advance capabilities is necessary in order to approach the block and to cut semithin sections.

4. A specimen-holding arm with ultrafine (nanometer range) advance capabilities, provided by thermal or mechanical advance mechanisms, with or without stepping motors, is necessary to cut reproducible ultrathin sections. The thermal units accomplish this by heating a thermally expansive block attached to the specimen arm with an internal light bulb controlled by a rheostat (Reichert OM-U2 and OM-U3) or by having a set of coils wrapped around the arm connected to a rheostat that allows adjustment of the heat delivered (LKB). Some mechanical advance units have a simple threaded shaft with fine increments that allow direct nanometer-level advance adjustments (Porter-Blum MT-1, MT-2, MT-2B), while others utilize stepping motors to move the arm forward specified distances on a finely threaded shaft (Sorvall MT-5000, MT-6000; RMC MT-7000; Reichert Ultracut, Ultracut-E).

5. Diffuse overhead lighting (fluorescent tubes) is needed to judge section thickness by interference colors produced when sections are floated off on water. The diffuse light source reflects off the surface of the sections and also passes through the section to reflect off the surface of the water beneath. The latter reflected light has a slightly longer path back to the viewer than the former, so an interference color is developed that can be used to determine the section thickness. All ultramicrotomes designed after the 1970s also have some form of backlighting with an incandescent source to aid in approaching the block with the knife.

6. All models produced since the 1980s can be adapted for cryoultramicrotomy. The most usable cryostages available from the three major manufacturers (LKB, Reichert, RMC) are designed to fit only their instruments.

7. All ultramicrotomes have variable cutting speeds for flexibility when dealing with different types of blocks. Thus, the user can adjust the microtome for use with soft biological tissues in epoxide resins, frozen blocks of biological tissues, or resin-embedded materials, such as kidney stones or battery insulator plates.

8. Most ultramicrotomes have a slow cutting stroke and a faster return speed. Some also have variable return speeds. The variable return speed is claimed to improve the quality of ultrathin frozen sections (Sitte, personal communication).

9. Most ultramicrotomes have an adjustable cutting window that allows the user to position the slow cutting stroke portion of the arm movement such that it starts shortly before the block passes by the knife edge and ends immediately after the block has reached a position below the knife edge. This allows the user to work with different-sized block faces as well as blocks oriented in various ways in the chucks used to hold them on the ultramicrotome arm.

10. Cutting force is either supplied by gravity (LKB, Reichert) or the arm is driven (Sorvall, RMC). Both systems seem to work equally well.

C. History

Early attempts toward ultramicrotome design go back to 1939, when Von Ardenne modified a histological rotary microtome to cut wedges, the edges of which were thin enough to allow the electron beam to penetrate them. During the early 1950s, single-pass ultramicrotomes capable of cutting sections of uniform thickness were developed. Porter and Blum in the United States initially produced a thermally advanced unit but found that the slightly later MT-1 design that utilized mechanical advance was superior. This instrument has mechanical advance and is manually operated, with only a single fluorescent overhead light. It remains the simplest device on the market and is still available almost 40 years after its inception (from RMC).

During the same time period, Sitte was working in Austria to develop the Reichert OM-U1. This unit was thermally advanced and cut sections automatically, but Sitte was never very satisfied with its operation. In 1957 the British Huxley MK1, which was a manual instrument with a mechanical advance, was marketed.

In the early 1960s another manufacturer entered the picture, the Swedish company LKB. They marketed the Ultrotome I, which had thermal advance in an automatic machine with a magnetically retracted arm that allowed single-pass cutting.

By 1962, Porter and Blum had developed the mechanically advanced, automatic MT-2, which quickly developed a reputation for reliability and stability that remains to this day. This was followed with the release of the Reichert OM-U2 in 1964, which was a thermally advanced automatic unit with an adjustable cutting window, and variable cutting stroke and return speeds. LKB began marketing the Ultrotome III, which had a novel lighting arrangement, incorporating both an incandescent and fluorescent light in an overhead boom that could be swung over a wide arc from behind the knife to in front of the knife, depending on whether the user wanted to see the interference colors of the sections or a good reflection in the block face during the approach of the knife to the specimen. In 1970, Huxley released the MK-2, which was a mechanically advanced, automatic machine. Shortly afterward, Reichert marketed the OM-U3, which had a backlight for specimen advance and various minor modifications in the knife stage controls and specimen adjustment controls.

In the late 1970s a new generation of ultramicrotomes developed, with emphasis on greater mechanical stability to meet the increasing interest in cryoultramicrotomy, which introduced new design problems that had to be addressed. Along with this development, most manufacturers incorporated methods to have more adjustability in various operating parameters. The MT-2B from DuPont/Sorvall had an adjustable return speed for the first time. Reichert switched from their traditional thermal advance to mechanical advance with the Ultracut (OM-U4) introduced in 1978. Sitte's interest in producing an instrument more suited to cryo work was the impetus behind the development of this ultramicrotome. LKB produced the Nova, which abandoned the vacuum-tube technology of the Ultrotome III for more stable solid-state technology and also introduced the Ultrotome IV as their top-of-the-line instrument. By the late 1970s, Dupont/Sorvall produced the MT-5000, which had an adjustable cutting window for the first time, as well as a stepping-motor controlled mechanical advance for the automatic cutting cycle. In addition, Huxley had bowed out of the ultramicrotome business.

During the 1980s, further refinements of ultramicrotome design, devoted primarily toward the growing cryo market, resulted in the present generation of microtomes that can be cryo adapted: the LKB Nova, the Reichert Ultracut-E, and the DuPont-Sorvall MT-6000.

In 1985, DuPont got out of the microtomy business and passed on most of its product line to RMC. Around 1986 and 1987, a series of corporate mergers resulted in Cambridge Instruments buying Reichert and then also buying LKB's microtomy line. In 1990, Leitz bought out Cambridge Instruments. Since these mergers, the only new ultramicrotome introduced has been the MT-7000 from RMC, which offers few novel features when compared to the MT-6000.

II. KNIVES

A. History

Ultramicrotomy knife development began, as did ultramicrotome development, with the utilization of materials available from histological microtomy. Thus, attempts to cut ultrathin sections began with steel knives that were polished to extremely thin edges. Unfortunately, it was immediately determined that the microcrystalline structure of steel prevented the production of knife-mark-free sections. In 1950, Latta and Hartman introduced glass knives made from plate glass scored and broken into triangular pieces. These had the benefit of having extremely sharp edges that could fairly consistently produce sections of plastic-embedded materials without excessive knife marks. These remain in routine use today and materials cost about 10¢ per knife. Glass knives suffer from fragility, however, and only about 10 ultrathin sections are usually cut on one area of a knife before significant knife marks begin appearing in sections. In addition, after one semithin section is cut on a sharp area of a glass knife, that area will probably be too damaged to cut any further ultrathin sections.

The search for more durable knives was greatly aided by Dr. Umberto Fernandez-Moran (see Hawkes, 1985), who realized that diamonds would be much harder than glass and developed a process for sharpening gem-grade (flaw-free) diamonds, which ultimately had edges in the 1- to 1.5-nm thickness range. He reported on this finding in 1953 and ultimately set up a diamond-knife manufacturing group utilizing gem-grade alluvial diamonds available in his native Venezuela. Over the next 25 years, diamond knives became the tools of choice for cutting ultrathin sections because of their durability when handled properly. Fewer than a dozen manufacturers exist to this day, and the product is relatively expensive ($2,000–2,600 for a 3-mm-long blade), but diamond knives allow workers to cut hundreds of serial sections on one part of the knife edge, and they can be resharpened a number of times for about $1,000 per sharpening.

In the late 1980s, sapphire knives became available as a compromise between inexpensive but limited-life glass knives and expensive but long-lived diamond knives. Sapphire is neither as expensive as diamond, nor as long lasting. On the other hand, a sapphire knife will outlast a glass knife considerably.

When shopping for diamond knives, there are quite a variety available from different vendors. There are knives produced primarily for working with soft biological tissues, knives produced primarily for materials scientists interested in sectioning extremely hard substances, knives without boats for use in cryoultramicrotomy, and knives of lesser edge quality marketed for cutting semithin sections.

B. Glass Knife Manufacturing

Glass knife manufacturing began with and can still be conducted using simple tools available from the glazier's supply house. A simple glass cutter, straight edge, and glazier's pliers can be used to make glass knives. To produce two sharp knives regularly from one small square of glass usually requires the services of a knifebreaker specifically designed for the task. These are available from a number of manufacturers, though the most frequently encountered instrument is the LKB 7800 knifebreaker (recently supplanted by the 2178 KnifeMaker II). It allows the user to easily cut glass strips into glass squares and then to cut the squares into two usable knives.

Glass knives for ultramicrotomy are made from plate glass, preferably "float" glass. This glass is made by floating the molten glass onto a surface, rather than running the molten glass through rollers ("roller" glass) to dimension it. The latter develops many stress lines that may affect the ability to fracture the glass in the plane desired when making knives. All the glass sold by electron microscopy suppliers is float glass and will make suitable knives for sectioning.

There are several steps in glass knife manufacture. The first step is to clean the glass strips, which often have surface dirt and oil. Some workers go to elaborate lengths to wash the glass with detergents, rinse it, and then wipe it with alchohol to make sure that it is spotlessly clean. In our laboratory, we merely rub the strips down with Kimwipes and then visually inspect them for excessive dirt. If the strips look clean, we then use them for knives.

The next step is to score and break squares from the strips. It is important that the knifebreaker be adjusted so that the cutting wheel does not actually cut completely to the edge of the glass. This allows the break, which starts along the score line, to run to the edge through the path of least resistance, thus ensuring a cleaner and sharper break. In addition, the cutting wheel should be adjusted to barely cut the surface of the glass. If it cuts so deeply that small glass slivers or large chips of glass are left on the surface of the glass, the cutting pressure should be decreased. When the cutter is drawn across the glass, a slight "snick" should be heard. Glass scoring should be done quickly, with the cutter being drawn smartly across the glass. Otherwise, the glass may not be adequately scored. Finally, the actual breaking of the knife by bending it across a fulcrum can be done in one of two ways. Some workers adhere to the "slow-break" school and suggest that applying breaking pressure slowly will allow a superior free break to the edge of the glass strip. Others, including our lab, are convinced that a fast break works just as well.

After the squares are cut from the long glass strips, the squares are cut diagonally to produce two knives. The normal adjustment of the knifebreaker will result in the cutter scoring from outside and to the left of one corner to outside and to the right of the diagonal corner opposite (Fig. 49). This results in two knives, each with a sharp edge and a dull edge (Fig. 50).

Some workers spend an inordinate amount of time evaluating their knife edges, while others just use the knives and evaluate the results. The latter method sometimes yields disappointing results, but it probably involves less wasted time than painstakingly examining every knife edge before use. The face-on view of a knife edge (Fig. 51) shows the fracture plane curving to the left side of the edge. Where it intersects the knife edge, the knife is the most dull. Anywhere to the right of that area should be good, though the far-right area of the knife can be seen to have "whiskers" (surface irregularities) when examined with backlighting on a dissecting microscope. According to Sitte (1981), these "whiskered" areas represent stair-stepped glass that will yield good sections except for the first one, which will show the stair steps.

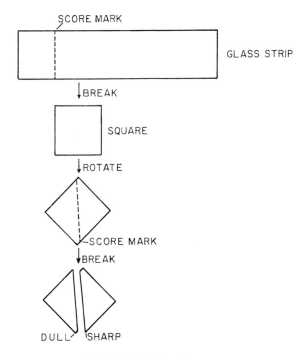

FIGURE 49. Knife-making steps.

FIGURE 50. The final knives.

Students tend to be concerned that glass knives rarely have an edge that is parallel to the bottom of the knife. The edge often visibly rises to a point at one side of the knife (Fig. 52). Other than losing the acute portion of the blade for sectioning purposes, these "spurs" are generally of no concern and are frequently encountered.

Glass knives are typically used with the cutting angle (set on the knife holder) at about 5–6° for biological materials. If undue compression is noted, the knife angle should be increased to eliminate the compression. Needless to say, if the plastic is too soft or the glass knife is too dull, compression will occur.

C. Section Handling

Section collection can be accomplished in a number of different ways. The most common method is to cut sections into a plastic or metal "boat" filled with water that is attached to the knife edge with wax or into a tape boat secured with nail polish (Fig. 53).

Semithin sections are then picked up with a nichrome loop about 3 mm in diameter. The loop is submerged in the water, brought up from beneath the sections, and the drop of water containing the sections is inverted onto a drop of water previously placed onto a glass slide. The sections floating on the drop of water on the slide are then heated on a hot plate until they adhere to the glass, at which time they can be stained. Thin sections can be picked up either from above or below with a freshly cleaned metal grid. Sections picked up by bringing the grid up through the water from below the sections appear to have less wrinkles than if the grid is pushed down on the sections from above the water.

Some workers in production labs that collect a lot of semithin sections merely put

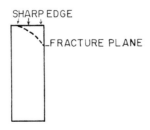

FIGURE 51. Face-on view of knife edge showing sharp edge and fracture plane.

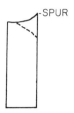

FIGURE 52. Face-on view of uneven knife edge with spur.

a large drop of molten wax directly below the knife edge to hold one large drop of water into which the sections are cut (Fig. 53). Other workers cut semithin sections on a dry knife edge with no boat and roll a dampened applicator stick across the section to pick it up. The stick with the section is subsequently rolled across a drop of water on a slide to transfer the section to the slide.

Excess sections in a water-filled glass knife boat can be quickly removed by taking a piece of standard Scotch tape and touching it to the surface of the water. The sections will adhere to the tape, which can then be discarded.

D. Knife Storage

Glass knives are easily stored in small boxes by affixing in an upright position to double-sided tape attached to the bottom of the boxes. This storage method helps protect them from damage by inadvertent contact of the cutting edge with any hard object, and it also serves to keep dust from settling on the knife edge. Both of these events could result in scratch marks on sections. Many workers contend that since glass is a supercooled liquid, it will "run" fairly rapidly due to gravity, and if a freshly broken

FIGURE 53. Two types of glass knife boats.

knife is not used in a day or so, it will become dull due to glass flow. This has never appeared to be a problem in this laboratory. We have stored knives in boxes for months and still had good luck sectioning with them.

E. Diamond Knives

Diamond knives are clearly superior to glass knives in that they can cut hundreds to thousands of sections in the same place compared with 4–10 on a glass knife. Unfortunately, they are also considerably more expensive because they are made by fracturing diamonds of gem quality in the plane of the crystal and then polishing them and mounting them in a metal boat. Diamond knives, when handled properly, are quite durable, but small mistakes can damage them instantly. The edge of a diamond knife is about 10 nm thick, so it is quite fragile. Any twisting of this edge can fracture the diamond crystal lattice. Exceptionally thick sections can also break off the edge of the knife. Diamond knives can tolerate quite a number of 0.25- to 0.5-μm thick sections, but they will dull the knife more quickly than thin sections. If epoxide sections are allowed to dry on the knife edge, they can become molecularly bonded to the edge, necessitating knife resharpening before sections can be successfully cut on that area of the knife again. The diamond knife washers currently on the market do not seem to be able to correct this problem. Finally, it is always wise to cut semithin sections and to stain them prior to cutting thin sections. Various samples will be found to have some hard substances in them that can be seen to damage knife edges. If possible, it is a good policy to trim away these hard areas prior to thin sectioning to help preserve the edge of the diamond knife.

Manufacturers of diamond knives specify the angle to use for cutting, though this should always be empirically checked by the user. An important warning is always to check the quality of a new or newly sharpened diamond knife immediately upon receipt. Most manufacturers give a limited warranty for a month or so, during which time you can send the knife back if you do not feel it sections adequately. New knives sometimes do not cut well. Make sure you check all your knives before the warranty expiration date.

Some manufacturers say never to clean the knife edge with balsa sticks or styrofoam blocks, while others say it will work. The best procedure is probably to follow the manufacturer's included suggestions for knife care.

III. BLOCK TRIMMING

As pointed out in the section on knives, large block faces cause more stresses on knife edges, resulting in more compression in the sections, more damage to the knife edge, and generally more artifact-laden sections. Thus, trimming blocks to the proper size is an important activity.

A. Trimming Procedures

The first step in trimming blocks is to cut down into the tissue. With the block held in a chuck in a block-trimming holder under a dissecting microscope, it should be fairly easy in most cases to observe the thin slices removed with a razor blade. The slices should contain black material, indicating that you have trimmed down to the specimen.

Semithin sections are usually cut about 0.25 to 0.5 μm thick (green to red in color in the boat under diffuse light). They are used mostly to give us an idea of what is contained in the block face. Thus, it is frequently useful to include as much as is practical of the block face. Most of the loops used to pick up semithin sections are under 3 mm in diameter, which defines the upper limit to section width. The larger the section, the harder to make it lie down flat on a glass slide when heated in the drop of water in which it is placed. If excessive wrinkles are encountered, first try increasing the size of the water droplet into which the sections are placed, then increase the heat of the hot plate slightly. In the end, however, you may have to reduce the overall size of the block face. Whatever the size of the block face, the sides should be cut down fairly obliquely. If the sides are perpendicular to the face, the tip of the block will have a greater potential to flex under the great forces generated during sectioning. If the sides are cut down obliquely, there will be more support at right angles to the cutting stroke (Fig. 54).

Thin sections are usually cut 50 to 80 nm thick. They range from gray to gold in color, respectively, when viewed in the boat with diffuse light. In order to ensure high-quality sections, sections should not be larger than 0.5 mm². It is possible to cut larger sections, but they are more prone to chatter and wrinkles, and will also generally be covered with more stain dirt.

A knowledge of the specimen is necessary in order to trim the block properly and to orient it properly for sectioning. If there is a longitudinal axis (nerve fiber), cut along it; if there is a possible weak point (skin surface), cut at right angles to it. In general, if the block is not square, the face with the smallest width should be presented to the knife edge first (Fig. 55).

If the sample is a homogeneous tissue such as liver, it is probably best to trim away all the plastic surrounding the tissue so that the entire block face consists of tissue (Fig. 56). If, on the other hand, the tissue is nonhomogeneous (skin), and especially if

SIDE VIEW

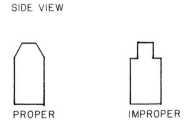

FIGURE 54. Comparison of properly and improperly trimmed blocks.

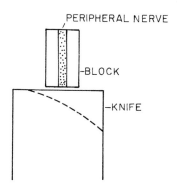

FIGURE 55. Specimen needing specific orientation.

it has a surface of interest, the tissue surface must be surrounded by empty plastic. Enough plastic must be left beyond the surface to be viewed so that it can easily overlap some grid bars. If the surface of a tissue is unsupported by empty plastic spanning an adjacent grid bar, the heat of the electron beam will cause the section to curl up and contract, obscuring detail and making it very difficult to take a photograph (since the specimen will be drifting). An additional problem with some samples, such as skin, is that the plastic embedding medium will tend to become separated from the surface of the sample. This may be due to waters of hydration being tightly bound to the keratinized epidermal surface, blocking proper penetration of resin into the epidermis. In such cases, it is important not only to leave excess plastic beyond the surface of the specimen, but also to section the specimen at right angles to the potential direction of separation (Fig. 57).

It is often useful to have a nonsymmetrical block face, at least when first cutting semithin sections. When semithin sections are picked up from below with a loop that is subsequently inverted onto a drop of water to transfer the sections to a slide, the sections will be upside down. When they are finally viewed with a compound light microscope (that projects an image of the specimen reversed right to left), the image will be properly oriented in relationship to the block face viewed with a non-image-

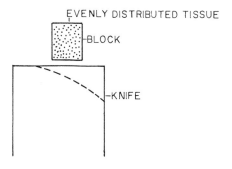

FIGURE 56. Homogeneous sample in block face.

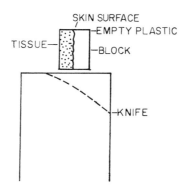

FIGURE 57. Nonhomogeneous sample in block face.

reversing dissecting microscope. Thus, if a corner of the block has been removed or the block face is trimmed into an irregular trapezoid shape, it will be possible to locate specific features noted in the microscopic evaluation of the semithin sections when the block is once again viewed with a dissecting microscope. This will ensure that further trimming of the block face to make it small enough for thin sectioning will not result in the removal of important features in the specimen.

B. Block-Trimming Tools

There are three major methods commonly used to trim block faces:

1. Razor blades, and in particular single-edge razor blades, are most commonly used. It is extremely important always to hold both edges of the razor blade (Fig. 58). If the thumb and forefinger of each hand are used to control the blade, the chance of any serious razor blade wounds is drastically reduced. If only one side is held, there is a good chance the blade will slip, and the back of the other hand will be cut. Various chuck-holding assemblies are provided by ultramicrotome manufacturers to hold chucks containing specimens for trimming.

2. Milling machines are also available. Reichert's TM-60 can be provided with a diamond bit, which can mill the front and sides of a plastic block by passing the diamond bit, rotating at approximately 10,000 rpm, past the block face. The chuck holding the specimen can be moved in various directions for trimming the edges as well as the face of the block. As mentioned in the section on resins, British investigators have suggested that even polymerized epoxide resins are carcinogenic, and that the fine plastic resin dust produced by the milling operation may be dangerous. The main advantage of a milling machine over razor blades is that absolutely parallel top and bottom faces with very smooth edges can be produced. This greatly improves the possibility of producing long ribbons of sections (for serial section work).

3. Microtome trimming can be done with most ultramicrotomes, since most come with a device to hold a chuck in the same track that normally holds the knife-holder

FIGURE 58. Proper method for holding razor blade during trimming.

assembly. The blocks can then be trimmed with razor blades on the microtome. This procedure produces quite a lot of potentially toxic epoxide debris, which also gets into the microtome body itself, and thus is not recommended. A block held in the chuck attached to the ultramicrotome specimen arm can also be trimmed with glass knives, both on the sides and on the face. Again, this produces a lot of mess (potentially toxic), consisting of plastic sections.

IV. SECTIONING PROCEDURES

A. Working Area

Sectioning equipment should be set up in a vibration- and draft-free area (Fig. 59). In addition to the ultramicrotome, the work area should be provided with glass microscope slides, a squeeze bottle with distilled water for filling the knife boat, a tuberculin syringe with a needle to adjust the water level in the boat, a loop on a stick for picking up semithin sections, an eyelash or eyebrow lash on a toothpick or an applicator stick for moving sections around in the boat, fine-tipped forceps for holding specimen grids, grids (200-mesh copper grids for routine work), grid-cleaning solutions, Whatman #1 filter paper in 10-cm disc form, and a small tape dispenser with Scotch tape. If using Spurr resin (which suffers from compression more than most epoxides), you will also need a small bottle of xylene or chloroform, and an applicator stick that can be soaked in the xylene and waved slowly over the thin sections to stretch them out.

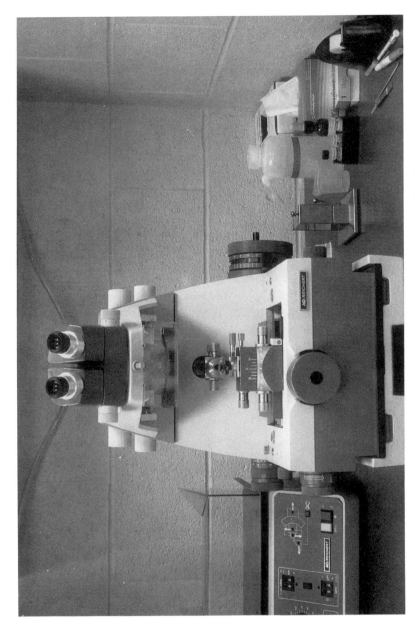

FIGURE 59. View of Reichert Ultracut-E ultramicrotome with tools.

It is most convenient to have a hot plate nearby upon which to dry the glass slides with semithin sections in water droplets. If a sink is directly adjacent to the hot plate, semithin sections can be stained and washed after they have annealed to the glass slides.

B. Sectioning Procedures

Once the block is mounted in a microtome chuck and has been properly trimmed, mount the chuck in the specimen arm and orient it according to the suggestions above. Lock the chuck securely in the specimen arm. Turn on the backlight, if available, or the fluorescent light (both, if the microtome is so equipped). Mount a glass knife into the knife holder. Adjust the knife to an angle of 5–6° and lock it securely into place. Move the knife stage to position the knife as close to the block as possible by looking at the knife and block from the side of the ultramicrotome. Lock the knife stage securely to the bed of the ultramicrotome. Adjust the binocular microscope head so that you can clearly see the edge of the knife. Fill the boat on the knife with distilled water until the surface of the water is convex. While observing the surface of the water and knife edge with the binocular microscope, use the syringe to bring the water level down to the point where it is just barely up to the edge of the knife. At this time, the fluorescent light should reveal a mirror-like area just in back of the edge of the knife. This is the area where sections can be observed to have interference colors. In addition, unless the water level is adjusted this way, it will be difficult to see the reflection/shadow of the knife edge in the face of the block. Depending on the type of ultramicrotome, you may see either a bright reflection or a dark shadow cast by the knife edge on the block face when the two are close together. If this reflection cannot be easily seen in the block face, the delicate approach of the knife to the block face will be extremely difficult.

Next, turn on the backlight and very slowly sweep the specimen arm through its stroke while looking through the binocular microscope to make sure the specimen will not hit the knife edge. Position the specimen arm so that the edge of the block is even with the top of the knife edge in the cutting stroke. Once you have established that there is clearance between the two, begin moving the knife stage forward utilizing the coarse advance controls for the knife stage. Once a reflection is noted, switch to the fine advance control for the knife stage. At this point, rotate the specimen in the specimen arm (as necessary) until the top and bottom of the block are parallel to the knife edge (Fig. 60).

Next, sweep the specimen arm slowly through its stroke while observing the reflection of the knife edge in the block face. The reflection should appear as a band across the block face, which gets narrower and narrower as the knife gets closer to the block. The specimen should be swept through its cutting stroke after each forward adjustment so that you can note if the reflection is present, how wide it is (remember, wider means farther away), and if it is even from right to left. If it is wider on one side of the block than on the other, it means that you should swivel the knife stage toward the block on that side. In addition, if the reflection is wider as the bottom of the block

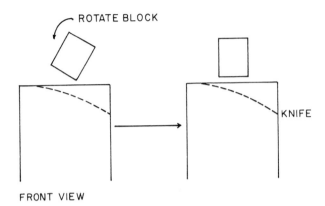

FIGURE 60. Block rotation to get specimen edge parallel to knife edge.

passes the knife edge and then gets narrower as the top of the block passes the knife edge, it means that the top of the block is tilted toward the knife compared to the bottom of the block. This is particularly dangerous if you do not approach slowly, making a pass with each minor adjustment. If the reflection is wider at the bottom of the block than at the top and if the clearance is taken up at the bottom of the block without going through cutting passes to monitor the distance between the knife and the rest of the block, the first cutting cycle would probably break off the top of the block and damage the knife in the process (examine Fig. 61 for examples of misadjustment and corrections needed).

After all the adjustments are made so that the reflection is now even side to side and stays the same width as the block is passed by the knife edge, keep advancing the fine stage control until the reflection is no longer visible. At that point, use the fine stage control to advance the knife about 0.5 μm per cutting pass. When the first semithin section is cut, continue to advance the stage 0.5 μm with each pass until you have the number of sections you want to put on the glass slide (usually three or four).

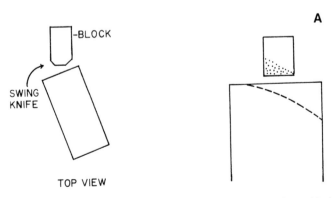

FIGURE 61 A–D. Adjustments to bring knife edge and block together. A: Reflection on block is wide on the left, so knife must be swung to the left.

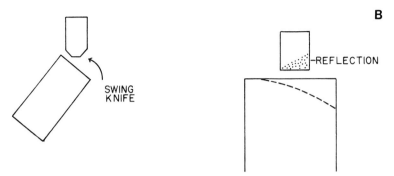

FIGURE 61 (*cont.*). B: Reflection on block is wide on the right, so knife must be swung to the right.

Use the loop to pick them up and place them onto the drop of water on the slide. It is usually best to continue cutting sections until no surface imperfections (left over from razor-blade trimming) remain.

Finally, move the knive stage back somewhat (so that the block is nowhere near the knife edge), move the knife stage laterally, and then reapproach as above. Note that the tilt of the specimen will not have to be readjusted, but the lateral swing adjustments of the knife will have to be redone. After the lateral swing adjustments are made, slowly bring the knife toward the block with the fine advance. Once the reflection is no longer visible, adjust the cutting window so that the cutting stroke ends when the block is just below the edge of the knife, set the cutting speed to 0.5–1.0 mm/sec, and set the

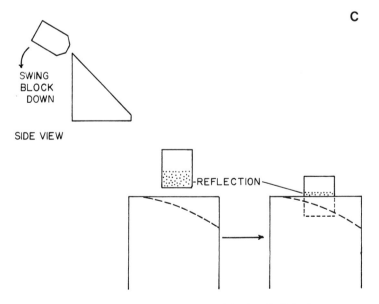

FIGURE 61 (*cont.*). C: Reflection on block is wide when block is above knife but becomes narrow as block passes by knife edge, so the back of the block must be swung down.

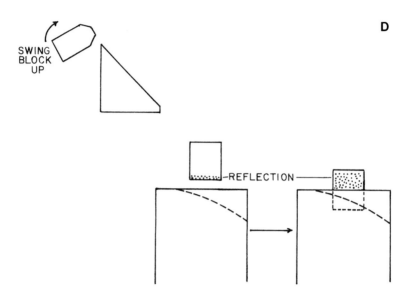

FIGURE 61 (*cont.*). D: Reflection on block is narrow when block is above knife but becomes wide as block passes by knife edge, so the back of the block must be swung up.

cutting thickness to 75–85 nm. Switch the microtome to the automatic mode and continually observe the cutting process. All ultramicrotomes will require minor adjustments to section thickness from time to time.

After sufficient thin sections in the silver-gold range have been cut (usually 5–10/grid), turn off the automatic cycle control and put the arm in its standby position far from the knife edge. Pull the last section from the knife edge with the eyelash tool (this is especially important if using a diamond knife so that the section cannot dry on the knife edge and become permanently annealed to the diamond).

Before picking up sections on copper grids, it is important to clean any oxidation or manufacturing oils from the grids so that the sections will stick properly to them. The easiest method for cleaning the grids makes use of three scintillation-tube-sized containers, the first containing 0.1 N HCl, the second with 95% ethanol, and the third with acetone. The grids are dipped individually into the acid for about 10 sec, followed by three quick dips in the ethanol, and then another three quick dips in acetone. Take a piece of Whatman #1 filter paper and slip it between the jaws of the forceps with which you are holding the grid to absorb any liquid held there. Finally, place the grid down on a clean piece of filter paper while simultaneously pushing it out of the forcep jaws with a clean piece of filter paper. The grid is then ready to pick up sections. The cleaning process should be done just before using the grid, since the copper will oxidize again overnight.

The selection of grids can be somewhat puzzling at first. They come in a variety of styles (square mesh, octagonal mesh, rectangular mesh, holes, slots, bars), mesh sizes (50–1000 mesh), and materials (copper, gold, nickel, stainless steel, beryllium, nylon/carbon, gold), and from various manufacturing methods (punched from screens, photoetched, identical on both sides, or decidedly different on one side from the other). For routine work, copper grids are generally used because of their good conductivity (to allow excess electrical charge from the electron beam to easily pass through to the ground) and low cost. On the other hand, if sections on grids are used in cytochemical or immunocytochemical procedures, it is typical to pick them up on inert grids (stainless steel, nickel, or gold grids). If microanalysis is planned, sections are often picked up on beryllium grids or nylon grids coated with carbon.

Different workers insist that sections should be picked up on either the shiny side or the dull side of copper grids that have sidedness. The arguments used involve the fact that the dull side is not as flat as the shiny side (it is rougher). It makes no practical difference which side you use, but we pick up sections on the shiny side of grids so that we always know on which side of the grids to find the sections for subsequent staining steps. In this way, any worker in the lab can work with a grid prepared by anyone else without having to figure out which side has the sections on it.

V. EIGHT COMMONLY ENCOUNTERED SECTIONING PROBLEMS

1. Sections Sticking to Manipulator Hair

When moving sections around with the eyelash tool, sections get stuck to the hair. Solution: Run your thumb and forefinger down your nose, then pull the hair between your thumb and finger. The transferred nose grease will keep sections from sticking to the hair.

2. Curving Ribbons

While cutting, the ribbon of sections curves to one side or the other of the boat. Solution: Rotate the specimen slightly to get the bottom edge of the block truly parallel to the knife edge. If the problem continues, the top and bottom of the block are not parallel. You will either have to take the block out and retrim it or live with the curving (pick off a few sections at a time so that they do not become balled up at the edge of the knife).

3. Irregularities in Thickness within One Section

Sections exhibit unevenness in the form of bands of different color within each section. Solution: Make sure that all locking knobs are absolutely secure. If you are using a Reichert-type chuck with stepped ledges with a flat block, make absolutely sure that all four corners of the block are in contact with some chuck surface. Make sure that there is not an external source of vibration (a large compressor nearby, workers drilling holes in the wall in the next room, etc.). Any time one section is of uneven thickness, it either can represent a source of vibration or can be a symptom of poor infiltration of the specimen, polymerization of the resin, or merely the difference in consistency between resin-impregnated tissue vs. resin alone.

4. Sections of Differing Thickness

Section thickness varies from section to section, usually alternating from thick to thin to thick to thin, etc. Solution: Make sure all locking controls are secured. This problem usually occurs if something is loose, allowing vibration. Sometimes this problem persists even after tightening everything. At that point, try to increase section thickness for the stroke that cuts the thinner of the two sections (in other words, thicken the thinnest section). This remedy does not work as well when the thicker section is thinned.

5. Sections with Chatter or Compression

Sections have evidence of chatter or compression. These are differing degrees of the same phenomenon. The cutting forces generated by forcing a hard plastic resin past a knife edge result in a compression of the section at right angles to the knife edge. Most of this compression is not noticeable as the section expands in the boat shortly after being cut. If, however, the knife angle is wrong (too steep), the knife is too dull, or the plastic is too soft, compression may become visible. Chatter is compression that is generally not visible except with the electron microscope and consists of very evenly spaced bands of light and dark areas (thick and thin) spaced closely together and oriented at right angles to the plane of sectioning (Fig. 62). Solutions: Make sure that your knife is sharp. To determine if you are using the right angle, cut sections from a known block and examine them. If they show chatter as well, increase the knife angle 1° and cut more sections from the test block. If you still see chatter, increase the angle 1° more. If chatter remains, you can probably assume that your knife is dull. Try another. If the test block cuts wonderfully at the normal knife angle, check to make sure that your new block is properly secured in the chuck. If it is, you can probably assume that your plastic is a bit soft. Put it back in the oven for a few days or try a higher polymerization heat (10–20°C higher than normal).

6. Knife Marks

Knife marks are streaks or tears of uneven spacing and width parallel to the direction of sectioning (Fig. 63). Solution: Change knives or knife area used, since the marks probably represent degraded knife edge that is actually tearing the specimen. Alternatively, a hard material in the specimen can be the source of the knife marks because it is damaging the knife edge. This is the reason that you should always cut semithin sections on glass knives before cutting thin sections with a diamond knife. We have frequently encountered hard materials in cell suspensions embedded in agar for ease in processing. Various sources of agar seem to contain small, hard substances of unknown derivation. When we encounter evidence of these materials in the form of knife marks in our semithin sections, we try to trim that area of the block away before cutting thin sections. If we are unable to do so (because that area is of vital interest), we cut our thins with glass knives and try to work around the knife marks when taking our photographs.

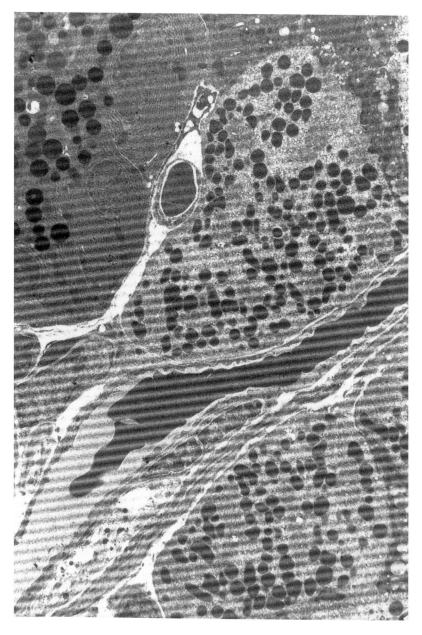

FIGURE 62. Photograph of chatter.

FIGURE 63. Photograph of knife marks.

7. Wet Block Face

The block face becomes wet as the block passes the knife edge. Solution: In most cases, lowering the level of water in the boat slightly will alleviate this problem. Before you begin cutting again, the block face must be blotted dry with a piece of filter paper. The first few passes may pick up water that was dragged down the back of the knife during the original block face wetting. If so, keep blotting the block face just after the cutting stroke until all the water has been removed from the back of the knife. Alternatively, the block face wetting problem may be due to the nature of the block. Acrylic resin blocks are notorious for picking up water from the boat. In that case, it is necessary to blot the block face with filter paper following each cutting stroke. If all else fails, check the knife angle being used. If it was inadvertently set too steeply (say, 0° for a glass knife), it can result in block face wetting (as well as chatter).

8. Section Dragging

Sections are dragged out of the knife boat and down the back side of the knife within the next cutting stroke. Solution: This is really just another form of problem #7. Try to correct it using the methods listed above.

REFERENCES

Hawkes, P.W. (ed.). 1985. *The beginnings of electron microscopy*. Academic Press, New York.
Reid, N. 1975. *Ultramicrotomy*. In: *Practical methods in electron microscopy*, Vol. III, Part II, Glauert, A.M. (ed.), North-Holland, Oxford, pp. 213–353.
Sitte, H. 1981. *Ultramicrotomy: Frequent faults and problems*. AO/Reichert, Vienna.

CHAPTER 3

Support Films

I. PURPOSE

Plastic and/or carbon films are applied to grids to support particulates (viruses or subcellular particles, such as mitochondria or cilia) that would otherwise fall through the openings in the grids. Support films are also used to cover grids with large open areas (e.g., 1 × 2 mm openings in slot grids), which can be used to view an entire 1-mm² section without obstruction from opaque grid bars. Finally, holey films can be prepared that are used for minimizing astigmatism in objective lenses and apertures, and to judge image stability.

Various film types, their preparation, and avoidance of typical difficulties are discussed by Hayat and Miller (1990) and by Faberge (1984).

II. TYPES

A. Nitrocellulose

Nitrocellulose comes as strips of material (under the name Parlodion) that must be dissolved before use in a solvent such as amyl acetate. Alternatively, 1–2% solutions of nitrocellulose in amyl acetate may be purchased under the name *collodion*.

B. Formvar

Formvar (polyvinyl formaldehyde) may be obtained as a 0.25–0.5% solution in either chloroform or ethylene dichloride (2,2-dichloroethane) or may be purchased as a powder. To make up a solution from powder, put 0.25 g in a glass container with a stopper seal that is not soluble in the solvent. Add 50 ml of solvent (e.g., ethylene dichloride), sonicate briefly, and let the solution sit at room temperature. After 24 hr, the powder should be in solution, resulting in a pale straw-colored liquid that will remain stable for months at room temperature. The solutions are not light sensitive, but they will become more concentrated with use as the solvent evaporates. For this reason, it is important not to leave them uncovered in dipping vessels for extended periods of time. Either solvent serves adequately for the preparation of dipping solutions. A Teflon® seal on the storage container (e.g., Teflon tape wrapped around a ground glass stopper) is recommended to prevent the container from becoming permanently sealed by the resin left on the vessel neck during pouring.

Another polymer, Butvar (polyvinyl butraldehyde), is available from some suppliers and is used in a similar fashion to Formvar.

C. Carbon

Carbon may be evaporated onto plastic-coated (collodion, Formvar, Butvar) grids to increase their conductivity and, thus, to increase their stability under the electron beam. This is usually accomplished by placing air-dried, filmed grids into a vacuum evaporator, producing a vacuum of about 5×10^{-5} Torr, and then evaporating about 10–20 nm of carbon onto the surface of the filmed grids at 45–90°. The carbon should make the substrate bearing the grid light to dark gray (see the section on vacuum evaporators for methods of judging carbon film thickness).

III. METHODS

With support films made from any of the substrates mentioned above, it is imperative that the grids be clean so that the materials will adhere readily to the grid surface. The grid-cleaning procedure described in the section on ultramicrotomy is modified in this case to clean as many as 100 grids at a time. Rather than dipping the grids one at a time into the three cleaning solutions (acid, ethanol, and acetone), it is more efficient to put the contents of a vial of grids into a Beem capsule, add 0.1 N HCl, cap the capsule and shake for 10 sec. Next, pipet out the acid and replace it with 95% ethanol, shake for a few seconds, and remove the ethanol. Finally, introduce 100% acetone with a fresh pipet, shake vigorously, remove the acetone with a pipet, and shake the grids onto a piece of clean filter paper in a Petri dish. Once the acetone has evaporated, separate the grids with forceps and line them up face down (shiny side down) on the filter paper in preparation for placing them on a coating film. Do this immediately before using the grids, as they may oxidize again within a few hours.

All film-casting procedures need to be done in a clean glass vessel, such as a rectangular histological staining dish for racks containing 10 slides. This dish is filled with deionized or distilled water and placed over a black surface (such as a bag from a box of photographic paper) in an area with diffuse overhead lighting (fluorescent) so that interference colors can be adequately judged to determine the film thickness after casting. A gold film is very sturdy and can be easily used for most particulates, but a gray-silver film is more appropriate for work with sections. If the thickess of a film supporting sections is too great, it will impede penetration of the electron beam, resulting in more scattering and a reduction in specimen contrast.

A. Droplet

The droplet method is generally used with collodion because nitrocellulose in amyl acetate spreads readily on the surface of water, unlike the other plastic materials in ethylene dichloride or chloroform, which tend to sink to the bottom without spreading completely. A Pasteur pipet is used to drop 1–3 droplets of a 2% collodion solution onto the surface of the water in the film-casting dish. As the solvent begins evaporating, the plastic film will be seen to cover the entire surface of the dish. The film will have areas of differing thickness, as judged by the interference colors. It will also have many evident wrinkles. After the solvent has completely evaporated (1–2 min), the film will cease to change color and can be evaluated. Pick areas on the film of the appropriate color and begin placing freshly cleaned grids face down on the surface. The grids are then picked up by placing a stainless steel screen, pieces of 3 × 5″ card stock, or a previously plastic-coated glass slide (prepared for slide stripping, as described below) on top of the grids. After contacting the grids and the subtending plastic film, push the support matrix (screen, paper, or slide) beneath the surface of the water, turn it over, and bring it out of the water. Thus, the film emerges first, with the grids trapped beneath the film and above the support matrix (Fig. 64). The matrix is then blotted dry from beneath and placed into a Petri dish with the cover propped up slightly with a piece of crumpled filter paper to allow the moisture to evaporate. If the grids are to be carbon coated, wait at least overnight so that the film will be completely dry. Otherwise, the vacuum evaporator will have difficulty developing an adequate vacuum.

B. Slide Stripping

Slide stripping (Fig. 65) is the method generally employed for producing films of Formvar or Butvar (Fabergé, 1984). In this procedure, the greatest difficulty lies in preparing the glass slide to be coated with dissolved resin. If the glass slide is too clean, the plastic film cannot be forced to separate from the glass; if it is too dirty, the film will have numerous imperfections (streaks, holes) visible under the electron beam. Many labs have found that certain types of glass slides are easier to strip than others. Unfortunately, the manufacturers change their methods periodically, so what was once a very suitable slide for stripping may no longer work well. Various methods are suggested by electron microscopists, such as buying only one brand of glass slides, rubbing the slides with nose grease prior to dipping in the resin, washing the slides with various solvents before dipping, or coating them with various detergents in order to ensure good film release. A method that seems to work for slides of any origin is to spray them with 409™ detergent and then wipe them dry with Kimwipes™. Examine the slides closely to make sure that no streaks or imperfections are grossly visible.

After the cleaned slides are prepared, they must be dipped in the plastic resin.

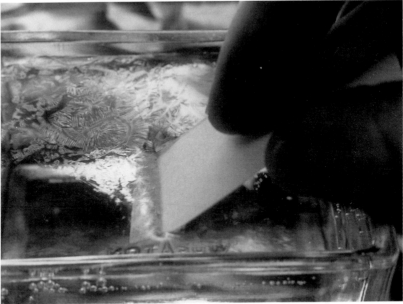

FIGURE 64. Film casting and grid retrieval as used with Parlodion.

FIGURE 64 (*cont.*).

Electron microscopy supply houses market side-arm flasks with elongated thistle tubes attached to the top that cost over $100 and allow the user to pump resin into the covered thistle tube containing a glass slide and then to let it drain back into the flask again. A 30-ml beaker achieves the same end at greatly reduced price. The beaker should be dedicated to slide dipping and kept clean by always rinsing it before and after use with the solvent used with the resin.

Fill the clean beaker with about 20–25 ml of resin mixture and then dip the precleaned glass slides into the beaker for a few seconds. Withdraw the slide slowly for a thinner coating (surface tension will tend to pull off most of the liquid) or remove it quickly to produce a thicker coating. Blot the end of the dipped slide on a clean paper towel and lean up against some vertical surface with the dipped end on a paper towel until dry (1 or 2 min).

Next, rub the edges coated with plastic with a razor blade to remove the plastic coat. Then scrape about 1 mm of the bottom of the plastic film from both sides of the glass slide. Finally, quickly blow on the end of the slide so that the film appears faintly frosted, and quickly touch the coated end of the slide to the surface of the water in the film casting dish. Hold the slide vertically if you desire two films (one from each side of the slide) or hold the slide at about a 30° angle if only one film is desired. As the films begin to release from the slide, slowly immerse the slide in the water until the film is floating free on the surface. Remove the glass slide.

After evaluation of film thickness and quality, place grids face down on the film. The grids and film may be picked up with any of the substrates mentioned for the

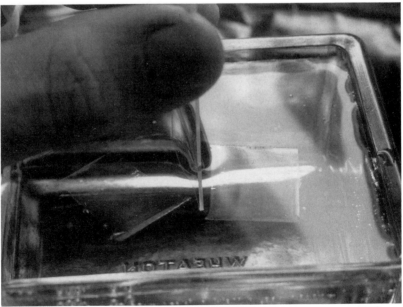

FIGURE 65. Slide stripping as used for Formvar films.

FIGURE 65 (*cont.*).

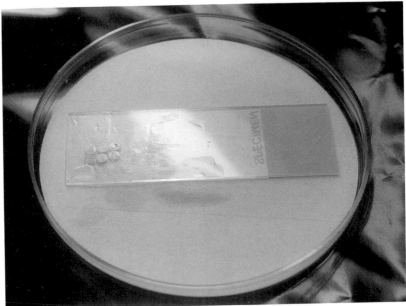

FIGURE 65 (*cont.*).

droplet method, but a glass slide coated with the same plastic resin will tend to retain less water (and thus dry faster) than the other substrates. Do not try to pick up coated grids with an undipped glass slide, because the film will not adhere well, resulting in the grids and film sliding off the glass slide as it emerges from the water.

C. Holey Films

Holey films for microscope alignment can be prepared by adding 1–5% water to a 0.5–1.0% Formvar solution. Shake the solution well, or sonicate it for a few minutes and then immediately dip precleaned slides. The water should produce numerous miniscule holes in the plastic film. Another method is to dip slides in a normal Formvar mixture but to breathe heavily upon them just after dipping to introduce moisture into the film. In either case, strip the slides as usual, put grids upon them, and retrieve them as above. It is also important to carbon-coat them to give them added stability.

FIGURE 66. Apparatus for picking up carbon films.

Freshly filmed grids tend to be fairly hydrophilic and wettable, but grids become hydrophobic after sitting. Carbon coating gives variable results in relation to hydrophilicity. Sometimes the grids become harder to wet, and in other situations they become easier. Hayat and Miller (1990) recommend exposing filmed grids to ultraviolet light for 40 min to increase their wettability. Another method recommended by some workers to make filmed grids more hydrophilic is to glow-discharge filmed grids. Unfortunately, this requires a vacuum evaporator specifically equipped to do the task. Such a device is capable of producing a 5- to 10-kV potential between the coated grids, which are sitting on a grounded platform, and a wire loop (typically) suspended above them at a positive potential. The grids are introduced into the vacuum evaporator, which is then pumped to a minimal vacuum (10^{-3} Torr is adequate), at which time the glow-discharge unit is turned on. The high voltage developed between the grid-bearing platform and the positive pole in the presence of a slight amount of atmosphere results in minor etching of the plastic film, producing a slightly hydrophilic surface. It is recommended that glow-discharged grids be used within hours of manufacture.

D. Carbon Films

Carbon films can be used as the sole support film on grids. Carbon film grids are utilized chiefly for the purpose of supporting materials for microanalytical purposes where the impurities in plastic films would result in spurious readings. These grids are more fragile than plastic-coated grids and are considerably more difficult to produce consistently. This method is more difficult because of the comparative fragility of carbon films. The most common method is to purchase mica sheets, which are split just before use by inserting a razor blade between the crystal layers and prizing them apart. The freshly revealed mica surface is then placed in a vacuum evaporator, and a film of about 20 nm is evaporated onto the mica. The mica is then introduced into a vessel of clean water (as in slide stripping) at about a 30° angle. The carbon film will float off onto the surface of the water. Allow the mica to fall to the bottom of the vessel.

Most workers utilize a device (Fig. 66) with a stainless-steel screen platform on which cleaned grids are held down by an overlying stainless-steel plate with holes. This support with the grids and restraining plate above them is slowly lowered until the grids and plate are submerged in the water. The plate restraining the grids is gently removed, leaving the grids beneath the water surface, supported by the stainless-steel screen support. A carbon film is stripped off of a mica support onto the water surface. The stainless-steel support with grids is then raised slowly until it contacts the carbon film floating on the water. As the support emerges from the water, the grids become coated with the overlying carbon film. The support with the carbon-coated grids is placed into a dust-free chamber until the grids are dry and ready for use.

REFERENCES

Faberge, A.C. 1984. Formvar and butvar support films; some general considerations. *Bull. Elect. Microsc. Soc. Am.* 14:102.

Hayat, M.A., and Miller, S.E. 1990. *Negative staining.* McGraw-Hill, New York.

CHAPTER 4

Transmission Electron Microscopy

I. HISTORICAL REVIEW OF MICROSCOPY (1590–1990s)

When considering the development of electron microscopes, it is useful to recognize the antecedents from the realm of light microscopy. The concepts of electron optics underlying electron microscopes are primarily extrapolations from the physics of light optics.

The first compound light microscope was developed in 1590 by the Janssen brothers and led to the rapid rise in interest in microscopic life forms and microscopic structures found in multicellular organisms, as described in the work of Malpighi, Hooke, and Leeuwenhoek in the 1600s. By the 1800s, countless improvements in optical microscopes had taken place, and important discoveries concerning cell structures had been reported. Brown had identified the nucleus of eukaryotic cells, Schleiden and Schwann had independently posited that all life was made up of cells (the cell theory), and mitosis had been described. In 1886, the maximum resolution available with glass optics and visible light had been achieved by the coupling of the contributions of Carl Zeiss and Ernst Abbé to produce a compound light microscope that was corrected for spherical aberration and chromatic aberration. Careful Köhler illumination and the use of matched substage condenser lenses, oculars, and apochromatic (chromatic aberration corrected for three wavelengths of light) objectives resulted in ultimate point-to-point resolving capabilities of 0.1 μm, as predicted from theoretical calculations of light optics. It was not until 1934 that light microscope resolution capabilities were surpassed with an electron microscope produced by Driest and Muller.

The next step forward in looking for increased resolution came with Knoll and Ruska's (1932) description of the first electron microscope, developed during Ernst Ruska's graduate work. This prototype for our modern instruments had an electron source and two magnifying lenses but no condenser lens, and actually it had less resolving capabilities than light microscopes available at the time. By 1934, Ruska had added a condenser lens to improve illumination. By 1938, von Borries and Ruska reported resolving capabilities of 10 nm. The German electronics firm of Siemens, under the direction of Halske, produced the first commercially available transmission electron microscope (TEM) in 1939 at the same time that Burton, Hillier, and Prebus were producing an experimental electron microscope in Canada. Hillier and Vance built the RCA type B in 1941, which was capable of 2.5-nm resolution. By 1946, Hillier's group had produced an instrument capable of 1.0-nm resolution, which is beyond the practical resolution needs (2.0 nm, or one third of a mitochondrial unit membrane) for almost all biological work with chemically fixed specimens.

During the 1950s and 1960s, power supplies were refined and lens manufacture, vacuum systems, and mechanical controls were improved, while electronic stability was increased. Microscopes were produced with sliding objective lenses (Philips 75); phosphor-coated glass viewing screens pointed almost horizontally at the viewer, reminiscent of television screens (Philips 100); and vertical columns with electronic magnification steps and images projected onto a phosphorescent screen, with images viewed through glass ports in the base of the column. Lens alignments progressed from methods requiring beating on the outside of the electron microscope column with a rubber mallet (RCA 3), to every column element being physically adjusted with set screws (Philips 200), to a mixture of mechanical alignments and electron alignment capabilities (Hitachi HU-11E). Power supplies became smaller and smaller as solid-state circuitry replaced vacuum-tube technology. Column vacuum improved from a level barely in diffusion pump range (10^{-4} Torr) to 10^{-5} or 10^{-6} Torr with the addition of a nitrogen cold trap.

In the 1970s there was further improvement in all of the operational systems of the electron microscope and also the development of high-voltage (1-meV) electron microscopes, microprobe analysis, scanning transmission electron microscopes, and an industry-wide standard of 0.344-nm resolution for TEMs sold in the latter part of the decade.

In the 1980s there were further instrument system improvements (instruments capable of less than 0.2-nm resolution became available), the development of intermediate-voltage electron microscopes (300–400 keV), cleaner vacuum systems (better diffusion pump oils, turbomolecular pumps, ion getter pumps), improved microanalytical capabilities, and new techniques, such as electron energy loss spectroscopy and such distantly related imaging methods as scanning tunneling microscopy (STM) and atomic force microscopy (AFM). Computer-assisted microscopes became a reality, resulting in instruments capable of improved self-diagnostic capabilities, improved ease of alignment in various operational modes, and improved ease of use by minimally trained individuals.

The theoretical resolving capabilities of transmission electron microscopes have been reached, so the future of instrument development appears to be directed toward making instruments more user-friendly through the agency of computer control. In the 1980s, the development of the Philips CM series and the Hitachi H-7000 moved instrumentation from microprocessor control to partial computer control. In the 1990s

JEOL introduced the JEM-1210, which is operated through a software-driven computer to the extent that the few manual controls still remaining are actually working through the computer by converting information from analogue control knobs to digital commands. Improvements in microanalytical capabilities, particularly in the area of quantitative (as opposed to qualitative) analysis, are still ongoing. Intermediate-voltage electron microscopes (IVEM) are being produced at an ever-increasing rate and are shifting much of the work originally performed with a small number of high-voltage electron microscopes to IVEMs.

An excellent source for further information on the history of electron microscopy is *The Beginnings of Electron Microscopy,* edited by Hawkes (1985).

II. THEORY OF ELECTRON OPTICS

A. Light Microscopes versus the TEM

A comparison of schematic diagrams of the image-forming components of a compound light microscope (LM) and a TEM (Fig. 67) shows marked similarities between the two. Typically, they both use a tungsten filament as a source of illumination, though the LM utilizes photons emitted from the heated filament, while the TEM utilizes electrons released at the same time. Both utilize objective, condenser, and projector (ocular) lenses to magnify and focus the illumination source and projected images. The LM uses glass elements to bend the light waves, while the TEM utilizes electromagnetic lenses to bend the electron beam. The TEM requires more lenses than the LM to achieve adequate illumination and high magnification. In addition, the TEM must manipulate the electron beam in a high-vacuum environment within the microscope column, since air would otherwise deflect the electrons, interfering with illumination and the image-forming process. In both microscopes, the condenser lens(es) focus(es) the imaging beam (light, electrons) on the specimen, while the objective lens focuses and magnifies the image of the specimen. Further magnification is provided by the projector (ocular) lens system, which projects the final specimen image onto film, a phosphor-coated screen, a video camera, or directly into the eye.

B. Resolution

Before discussing resolution, it is useful to review the units of measurement that apply to ultrastructural research:

$$1 \text{ Å} = 1 \times 10^{-7} \text{ mm} = 1 \times 10^{-9} \text{ m}$$
$$1 \text{ nm} = 1 \times 10^{-6} \text{ mm} = 1 \times 10^{-8} \text{ m}$$
$$1 \text{ μm} = 1 \times 10^{-3} \text{ mm} = 1 \times 10^{-6} \text{ m}$$

An average person with 20/20 vision can resolve (distinguish) two points located 0.1 mm apart. With the best compound light microscope, a theoretical resolution of 0.1

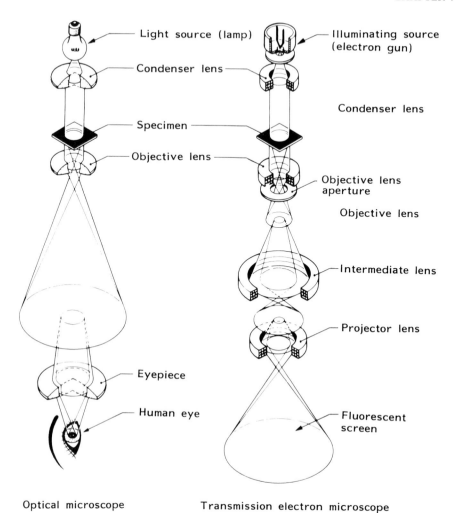

Light source (lamp)

Condenser lens

Specimen

Objective lens

Eyepiece

Human eye

Illuminating source (electron gun)

Condenser lens

Objective lens aperture

Objective lens

Intermediate lens

Projector lens

Fluorescent screen

Optical microscope   Transmission electron microscope

FIGURE 67. A comparison of the optical systems of an LM and a TEM. (Courtesy of JEOL USA, Inc.)

μm can be achieved. Some transmission electron microscopes can discern two points 0.127 nm apart. This means that large macromolecules, such as hemoglobin and certain enzymes, can be viewed directly with the electron microscope. A short film laboriously compiled by Al Crewe and his graduate students shows individual atoms of uranium moving about through a series of time-lapse photographs taken with a transmission electron microscope.

Resolution is generally defined as the ability to identify two points or lines as separate structures. The resolution of films is generally given as lines per millimeter, while point-to-point resolution is used to describe optical systems. The resolution formula can be used to calculate the theoretical resolution (d): in an optical system wherein the refraction of the medium between the illumination source and the lens is defined as n, α is one-half the cone of illumination entering the front of the objective lens after interaction with the specimen, and λ is the wavelength of the image-forming radiation. Objective lenses of light microscopes are usually marked with the numerical aperture (NA), which is equal to n sin α.

$$d = \frac{0.61 \times \lambda}{(n \sin \alpha)}$$

Using the formula to calculate the resolution available utilizing a light microscope equipped with an illumination source of 500-nm wavelength (using a green filter) and a numerical aperture of 1.4 (an oil immersion 100× objective and a 1.4-NA condenser lense with oil between both lenses and the specimen slide), two points 0.2 μm apart could be distinguished.

$$d = \frac{0.61 \times 500 \text{ nm}}{1.4 \text{ NA}} = \frac{217}{85} = 0.2 \text{ μm}$$

In 1924, De Broglie calculated that the wavelength of an electron is approximately 0.005 nm, thus suggesting a theoretical TEM resolution 100,000 times better than that possible with the light microscope. Spherical and chromatic aberration, as well as diffraction artifacts, prevent us from reaching this theoretical resolution limit. Both Meek (1976) and Slayter (1976) have provided excellent in-depth discussions of light and electron optics.

C. Electron Lenses

Electron lenses may be either electromagnetic or electrostatic in design, though the former are the only type encountered in commercially available TEMs. Electromagnetic lenses are similar to the simple solenoids used to operate such items as water shut-off valves in washing machines and starter engagement mechanisms for automobile engines. The basic principle is that a current applied to a coil of wire will produce a magnetic field at right angles to the current flow (Fig. 68). This magnetic field can either move a piece of metal located in the field (opening up the washing machine water source) or deflect small charged particles, such as electrons passing through the field in the electron microscope. The more turns of wire in an electromagnetic lens, the stronger the magnetic field developed (the more flux lines, by which magnetic strength is measured). Wrapping the copper wires conducting the electrical current around a soft iron core will further increase the magnetic field. Thus, there are three ways to increase the strength of an electromagnetic lens: (1) increase the number of turns of copper wire in the lens; (2) increase the amount of iron in the lens, either by enlarging the core or adding a pole piece (see below); and (3) increase the amount of current applied to the windings of the lens.

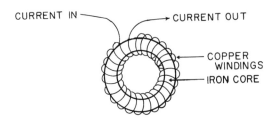

CURRENT IN

CURRENT OUT

COPPER
WINDINGS

IRON CORE

FIGURE 68. A simple solenoid.

D. Properties of Electron Lenses

1. Focal Length

The focal length of a lens is determined by the accelerating voltage of the electron beam (kV) and the magnetic field strength in the lens under consideration. Both of these factors determine where the electron beam comes into focus. At each different accelerating voltage, the lenses of the TEM must be adjusted to maintain the same focus. Modern microscopes have various microprocessors that increase lens current with higher accelerating voltage to keep the focal length constant.

2. Depth of Focus and Depth of Field

Photographers know that using a smaller aperture when focusing on a given object will bring more of the foreground and background into focus than using a larger aperture. This phenomenon is known as depth of field. The mathematical expression for calculating depth of field is as follows:

$$D_{fi} = \lambda/NA^2$$

Meek (1976) calculated that a TEM operating at 50 kV (resulting in a wavelength of 0.004 nm) with an objective lens with a NA of 10^{-3} would provide a depth of field of 4 μm. Since this depth is over four times the thickness of a normal thin section (80–90 nm), every feature in a conventional thin section will be in focus. This great depth of field explains why a slightly bent grid will not be grossly out of focus from one point to another.

A similar phenomenon, known as depth of focus, is also important to consider. The small apertures used in electron microscopes result in the projected specimen image being brought into acceptable focus over a great distance. The formula for determining depth of focus, where M is the magnification, RP is the resolving power, and NA is the numerical aperture, can be used to calculate the depth of focus for a TEM with 0.5-nm resolution at a magnification of 100,000 and a numerical aperture of 10^{-3}, or $D_{fo} = 5,000$ m (Meek, 1976):

$$D_{fo} = \frac{M^2}{NA}$$

This means that TEMs have sufficient depth of focus that an in-focus image can easily be projected onto a 35-mm camera inserted above the viewing screen, onto the viewing screen, onto sheet film in a camera below the viewing screen, and onto a video camera located below the sheet film camera. If illumination is adequate and vacuum could be maintained to keep air from scattering the electrons, the image could be projected below the floor holding the electron microscope. At each viewing level the magnification would be different, but the focus would be the same.

3. Diffraction

A limiting feature of aperture size is the phenomenon of diffraction. When an aperture gets small enough, the wave front of illumination (photons or electrons) bends around the edge of the aperture, causing image degradation. With a light microscope at low magnification (where resolution is not critical), decreasing the aperture of the substage condenser results in greater contrast. However, if the aperture is made too small, the image becomes increasingly grainy and distorted. At high magnifications, at which resolution becomes critical, it is imperative that the substage condenser be used in the wide-open position to eliminate any possibility of image degradation due to diffraction. Diffraction degrades the image in the form of decreased resolution by producing spurious images due to the formation of haloes and fringes. The formation of these fringes in an electron microscope (Fresnel fringes, Fig. 69) around small holes in plastic films is used to correct astigmatism in objective lenses. The effect of different aperture sizes on wave fronts of photons or electrons (Fig. 70) is similar to the effect of openings of different widths at the top of a dam restricting water flow. The objective aperture in an electron microscope is the point at which the effects of diffraction are most easily noted. Electron microscopes are equipped with discrete objective apertures (from 10–100 μm in diameter in most cases), and the smallest of these results in minimal diffraction at the magnifications used with the preponderance of biological specimens.

4. Hysteresis

The strength of electromagnetic lenses is not strictly due to the current flowing through them but also depends, to some extent, on the previous magnetic history of the lens. The soft iron core around which the copper windings are formed retains some of its magnetism, even when current is no longer flowing (remnance). The easiest way to remove the effect of remnance is to reverse the current flow to the lens briefly. Some TEMs have a button labeled *normalizing* that does this. Hysteresis can cause magnification errors of 10% or more in some of the older TEMs, though most modern microscopes are designed to essentially eliminate this problem for practical purposes.

FIGURE 69. A Fresnel fringe.

5. Pole Pieces

A pole piece assembly consists of an upper and a lower short, cylindrical, iron piece separated by a small gap. The lines of magnetic flux are concentrated in the gap between the upper and lower elements and exert a force at right angles to electrons passing through the bore of the lens (Fig. 71). Changing the focal length of a lens is usually accomplished by altering the current flow to the lens, but changing pole pieces is another less convenient method to change lens focal length.

6. Image Rotation

As the image magnification in a TEM is changed, the image rotates slightly at each change in magnification. At certain steps the image may rotate up to 180°. This rotation also occurs during focus shifts, although it is less evident. Newer microscopes are less prone to this behavior because of better corrective circuitry, and some microscopes have the capability to intentionally rotate the image to some extent to better align points of photographic interest with the film orientation.

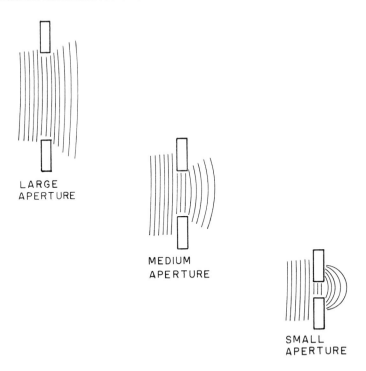

FIGURE 70. The effect of varying aperture size on a wave front.

Image rotation is often utilized during the column alignment procedure to center the optical axis within the column. The phenomenon of image rotation results from the path described by electrons being acted upon by electromagnetic lenses. As the electrons that are initially accelerated in a straight line by the high accelerating voltage encounter an electromagnetic lens, they are directed into a spiral path. Electrons

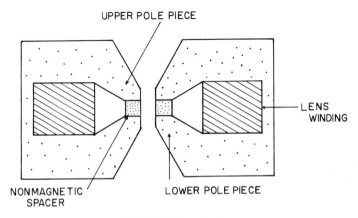

FIGURE 71. A pole piece.

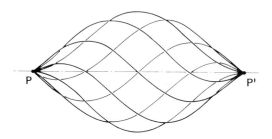

FIGURE 72. The spiral path of electrons induced by an electromagnetic field (lens). (Courtesy of JEOL USA, Inc.)

moving through a magnetic field are exposed to a force at a right angle to the original direction of motion (Fig. 72), resulting in electrons exhibiting a spiral path during the remainder of their journey to the viewing screen or film plane. As the magnification increases or decreases, the diameter of the spiral electron path increases or decreases. The velocity of unscattered electrons remains essentially constant until they are stopped by film or the viewing screen. However, if a lens is strongly energized, it will compress the spiral electron path more than when it is deenergized. An electron traveling in a straight line for 1 m will reach the screen at a certain point. Another electron traveling in a small spiral will arrive at a different point on the screen, while yet a third electron traveling in a comparatively large spiral will hit the screen at still another point. By examining a diagram of these different paths (Fig. 73), it becomes evident why an image rotates on the screen as the current delivered to one of the lenses in the imaging system (objective or projector lenses) is altered. The largest rotations (180°) occur when one of the projector lenses is turned on or off.

7. Lens Strength

As mentioned above, in a vacuum the electron beam does not change velocity while passing through a magnetic field unless it hits something (the specimen). On the other hand, it is refracted by the magnetic field of the lens. The objective lens changes focus by altering the current to the lens, which varies the focal length of the lens (Fig. 74). A more energized lens causes more refraction of the electron beam, resulting in a shorter focal length for that lens.

8. Fresnel Fringes

Fresnel fringes (Fig. 75) occur when the electron beam strikes an opaque or semiopaque edge (e.g., the edge of a hole in a plastic film). When the image is

WEAK LENS

STRONG LENS

FIGURE 73. The effect of electromagnetic lens strength on the position of electrons reaching the viewing screen.

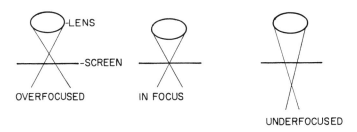

FIGURE 74. Comparison of an overfocused lens, a focused lens, and an underfocused lens.

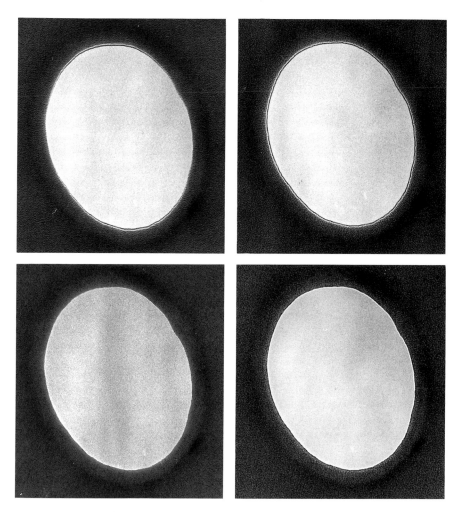

FIGURE 75. A through-focus series illustrating Fresnel fringes. Upper left: astigmatic, overfocus; upper right: overfocus; lower left: in focus; lower right: underfocus.

overfocused, a black fringe is visible around the outside of the hole. When the image is underfocused, a white fringe is visible inside the hole. In addition to Fresnel fringes, an over- or underfocused image has a grainy background. A truly in-focus image has no Fresnel fringe, exhibits little grain, and has minimal contrast. Most users tend to underfocus a bit, because this results in a higher contrast image that is more appealing to the eye. JEOL even provides an OUF (optimal under focus) control to allow the user to take a slightly underfocus photograph after finding true focus. An underfocus image does not provide the highest resolution, however.

9. Spherical Aberration

Spherical aberration is the inability of a lens to focus all the incident beam from a point source to a point. It is important to remember that axial magnification is not equal to peripheral magnification, but all points of the image will be in focus due to the great depth of focus possible in the TEM.

If one examines a simple glass magnifying lens (Fig. 76), it is clear that the light striking the lens edge will be refracted (bent) more than the light striking the center (axial) portion of the lens. Consequently, information from the peripheral part of the light path will focus in a different plane from information from the axial part of the light path. In a high-quality compound light microscope, this problem is compensated for by having convex and concave lens elements (positive and negative lenses) coupled together.

In an electromagnetic lens, the flux lines are strongest near the electromagnet (peripherally) and become weaker near the central axis of the lens. Thus, electrons passing near the edge of the lens are subject to more lateral force from the lens magnetism and are refracted more than those passing through the center of the lens. Once again, due to the extreme depth of focus in a TEM, the whole specimen will be in

LIGHT SOURCE

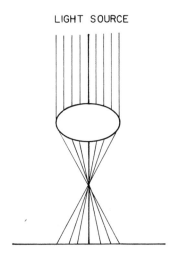

FIGURE 76. Refraction caused by a glass lens.

UNDISTORTED   BARREL DISTORTION   PINCUSHION DISTORTION

FIGURE 77. Undistorted image of a grid compared to barrel and pincushion distortion (spherical aberration).

focus, but the magnification of the image will not be the same for axial vs. peripheral parts of the image. The electronic circuits and lens design of modern TEMs virtually eliminate evidence of spherical aberration, but older instruments still in use can exhibit barrel and pincushion distortion (Fig. 77) at low magnifications. The former results in parallel lines, such as grid bars, converging at the periphery of the viewing field, while the latter is seen as parallel lines diverging at the edge of the field.

10. Chromatic Aberration

This phenomenon can be seen with cheap glass lenses, which give rise to colored fringes in the image due to the different focal lengths for photons of different wavelengths (Fig. 78). The greater the difference in wavelengths, the greater the problem of getting them to focus in the same plane. With light optics, various coatings are employed to minimize the defect. Achromatic objectives are corrected to allow blue and red light to focus in the same plane. Apochromatic lenses allow blue, red, and yellow light to focus in the same plane. One of the reasons for using green filters for black-and-white photomicrography is to provide monochromatic light to the lenses so that chromatic aberration will not be a problem (the second reason is that most panchromatic films are most sensitive in the green region of the spectrum).

Since the electron microscope does not provide color images, it is not intuitive that chromatic aberration can be a problem. However, there are actually three sources of chromatic aberration in an electron microscope (Slayter, 1976). Electrons liberated from an electron source (the heated filament) emerge from its surface with different energy levels (variation in thermal velocity). When the electrons are accelerated by the high-voltage field, they do not all achieve the same velocity (fluctuation in accelerating potential). The practical effect of this phenomenon is that minor fluctuations in the high-voltage supply result in a number of images at slightly different magnifications and focal levels, which are superimposed, with a decrease in resolution. Thus, the sharpest image is produced in the central axis (voltage center) of the lens. To help minimize chromatic aberration due to accelerating voltage fluctuations and to realize the resolution capabilities of contemporary TEMs, it is necessary to have electronic stability in the high-voltage supply, allowing variations of only a few parts per million.

The third source of chromatic aberration in the TEM is the specimen itself (energy loss within the specimen). The process of image formation is dependent on electron

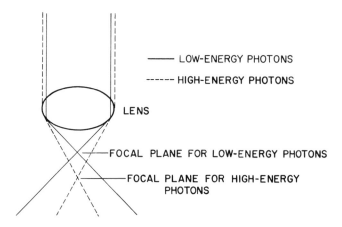

FIGURE 78. Chromatic aberration, resulting in photons of different wavelengths being focused at different planes.

scattering caused by the specimen (particularly the various heavy metals with which it is impregnated during fixation and post-staining). Inelastic electron scattering due to the primary electrons being deflected over wide angles without significant energy loss (usually due to interactions with heavy metal atoms, such as uranium, lead, or osmium) essentially eliminates some of the electrons from the image-forming process (producing dark areas in the projected image, because no electrons reach the phosphorescent screen or film). Elastic scattering of electrons (primary electrons from the electron gun displacing specimen electrons) causes secondary electrons to be liberated from the specimen that have less energy than the primary electrons originally impinging upon the specimen. Even in areas of the sections containing only pure resin with no biological material, some scattering of the electron beam will occur. Thus, some of the primary beam will pass through the specimen undeviated and with undiminished energy, some electrons will never reach the screen (elastically scattered), and still other electrons in the primary beam will produce secondary electrons, which have less energy than the primary beam, but still reach the screen. Those electrons that have lost energy can focus in a different plane from the undeviated electrons. This process, then, results in chromatic aberration. The thicker the specimen, the greater the chance that electrons in the beam will be deviated, thus leading to greater chromatic aberration. Higher accelerating voltages decrease the likelihood of this problem because a higher energy beam is scattered less by the specimen.

11. Astigmatism

One definition for this phenomenon is the inability of the lens to form a point image of an object point. Astigmatism can also be thought of as instrument-induced radial asymmetry. Astigmatism is caused by a defect in the magnetic field symmetry, resulting in different lens strengths being exhibited at different points at right angles to the lens axis (Fig. 79). It can be caused by imperfect machining of pole pieces,

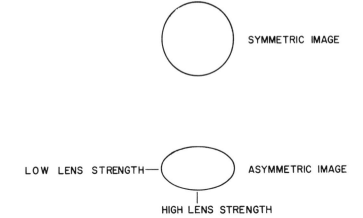

FIGURE 79. Diagram of asymmetric lens strengths.

imperfection of distribution of magnetic material within the lens itself, unevenness in the quality of windings within the lens, or aperture irregularities.

III. FOUR ASPECTS OF IMAGE FORMATION (Meek, 1976)

A. Absorption

Absorption gives rise to amplitude contrast (differences in intensity to which the eye is sensitive). This is the most important image-forming process with light microscopy.

B. Interference

Interference gives rise to phase-shift effect, to which the eye is insensitive. In most cases, interference can give rise to amplitude contrast (phase contrast), thereby contributing to image formation.

C. Diffraction

Diffraction, in general, results in degradation of the image by the formation of haloes and fringes that decrease resolution. With light microscopy at low magnifications where resolution is not critical, the substage condenser diaphragm may be stopped down to increase contrast resulting from induced diffraction. At higher magnifications, this procedure is not advisable because of the concomitant decrease in resolution. In the typical range of magnifications used with biological materials in conjunction with the aperture sizes usually employed, diffraction is not a significant feature in TEM image formation.

D. Scattering

Scattering is, by far, the most important factor in image formation with the TEM (and the least important in light microscopy). There are two types of scattering that take place as an electron beam interacts with the specimen: (1) nuclear interactions (elastic scattering), which result in deflection of the primary beam electrons through a large angle, with little or no energy loss to the primary electrons (the bulk of these electrons do not contribute to information on the viewing screen); (2) electron interactions (inelastic scattering), when the primary electrons encounter electrons within the specimen. The primary electrons dislodge secondary electrons from the specimen. The effect of this interaction is that parts of the primary electron beam are removed from image formation and are replaced by secondary electrons with considerably less energy, which are scattered easily and do not reach the phosphor on the viewing screen. Statistically, the large number of electrons compared to the number of nuclei in a specimen make it most likely that inelastic collisions will occur when the primary beam encounters the specimen. Thus, inelastic scattering is the most important aspect of image formation in a TEM, even though the electrons widely scattered due to elastic scattering also contribute. Electrons scattered by either mechanism do not reach the viewing screen and, thus, a dark spot remains on the screen where the phosphor would have emitted a photon if an undeviated electron had impinged upon it.

IV. GENERAL TEM FEATURES

Now that we have seen how electromagnetic lenses work and have discussed the theoretical aspects of image formation in a TEM, along with the various aberrations that can detract from accuracy and clarity of the image, we can begin looking at the parts and operational parameters of a conventional TEM.

A. Operating Voltage

A conventional TEM operates at up to 100–120 kV, usually with several lower steps (e.g., 40, 60, and 80 kV). Intermediate-voltage microscopes (IVEM) operate in the 200- to 400-kV range, while the high-voltage instruments (HVEM) can generate 1–3 million volts.

B. Resolution

All currently manufactured TEMs guarantee 0.344-nm resolution as judged with a graphitized carbon lattice. Certain instruments are actually capable of resolving less than 0.2 nm, but this is not of much use to most biologists, since conventionally prepared materials rarely yield better than 2-nm resolution (one-third of a mitochondrial membrane).

C. Magnification

Most instruments can provide at least 200,000× magnification, which is necessary to produce the manufacturer's guaranteed 0.344-nm resolution. More expensive units offer magnifications up to 500,000×, though this is of little practical value to most biologists. A picture taken at 35,000× and magnified three times when printed (which is within the enlargement range wherein no image quality reduction will be noted) will resolve a 2-nm-thick one-third of a mitochondrial membrane. Inexpensive microscopes (JEOL JEM-100S) have various steps of magnification up to 50,000×, followed by one step of 100,000× and one of 200,000×. The two highest steps are achieved through the activation of a "mini" lens used only for these magnifications. In operation, this means that the microscope is aligned for use with the lenses for 50,000× and below, and either of the high-magnification steps must be aligned separately. On a day-to-day basis, it usually means that the two highest magnification steps are not used. Despite the fact that the higher magnifications available on more expensive instruments are not of much use to a biologist, the more gradual incremental increases in magnification (more intermediate steps between low and high magnifications) makes them more usable throughout their magnification range.

Another feature of more expensive instruments is a more complex and flexible condenser system. In old microscopes, such as the Zeiss 9A, only one condenser lens was utilized, limiting the amount of illumination available. Microscopes capable of 500,000× magnification need two condenser lenses (some have three) in order to develop adequate illumination, since image intensity is inversely proportional to the square of magnification.

D. High Vacuum, Electronic, Magnetic, and Physical Stability

The higher the vacuum, the less chance for scattering of electrons from gases in the column, resulting in increased illumination and image contrast. As previously mentioned, extremely high stability in power supplies for high voltage and lens currents is necessary to produce an instrument with minimal levels of spherical and chromatic aberration.

V. PARTS OF THE ELECTRON MICROSCOPE: FUNCTIONAL ASPECTS

Periodic reference to Figure 80 will be helpful in the discussion of the TEM column and of the function and adjustment of the various parts.

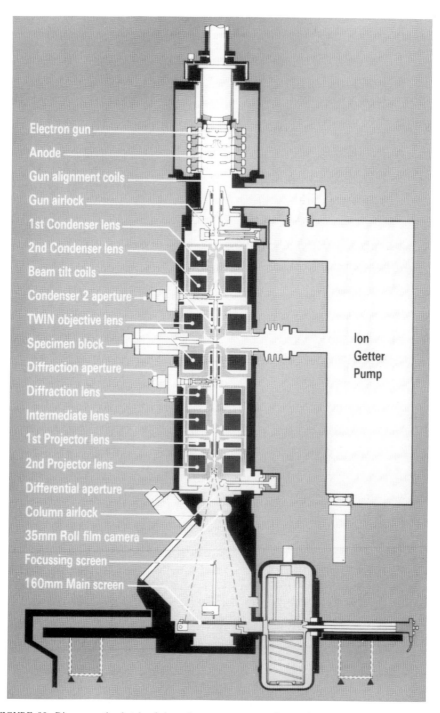

FIGURE 80. Diagrammatic sketch of the column components of a modern TEM. (Courtesy of Philips Electronic Instruments, Inc.)

A. The Electron Gun

There are three common sources for generating electrons that are accelerated by the high-voltage field of the electron gun. The simplest electron source is a tungsten wire (filament) with an acute bend at the tip (Fig. 81), where the highest resistance to current flow is found. The tip of the filament is a small point from which most electrons will be liberated from heat generated when current is passed through the tungsten wire. Some of these filaments may actually be slightly melted during manufacture and pulled out to a thin point to further reduce the cross-sectional profile for an even more discrete source of electrons.

Higher resolution microscopes equipped with extremely high vacuum pumping systems (ion getter pumps) may be fitted with the second type of electron source, known as a lanthanum hexaboride (LaB_6) cathode (Fig. 82). A single, pointed crystal of lanthanum hexaboride is positioned within a cup formed at the end of a tungsten filament so that when current is passed through the filament and electrons are liberated, they transfer their kinetic energy to the lanthanum hexaboride crystal, thereby causing it, in turn, to release electrons. The advantages of this type of electron emitter are that it lasts longer (400–600 hr vs. 100–200 hr for a tungsten filament) and produces a brighter image than that of a conventional tungsten filament. The disadvantages to a LaB_6 emitter are higher costs (about 20 times higher than tungsten filaments) and the need for higher vacuum (10^{-7} Torr or better vs. 10^{-4} or -5 Torr for tungsten).

The final type of electron emitter is a the field emission gun (Fig. 83). It consists of a tungsten filament whose tip has been etched to a radius of 20–30 nm placed near

FIGURE 81. A typical tungsten filament assembly.

FIGURE 82. A lanthanum hexaboride electron source.

an anode at a potential of 3–5 kV. The small radius of the filament tip and the proximity of the positively charged anode cause electrons to be pulled from the filament tip. Thus, they are liberated through electrostatic forces rather than thermionic forces, as with the other two emitter types. The extremely fine tip of the filament allows electrons to be released from a very small area, forming an extremely coherent

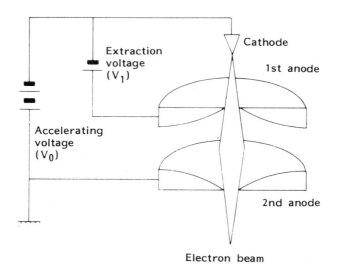

FIGURE 83. A field emission electron gun. (Courtesy of Hitachi Scientific Instruments.)

beam of electrons with a very narrow band width of energies, resulting in higher brightness and less chromatic aberration than the other two types of emitters (Hainfeld, 1977). Once emitted, the electrons are accelerated by the high positive voltage (60–120 kV) of the anode at the bottom of the electron gun. Like the LaB_6 source, very high vacuum must be maintained through the use of ion getter pumps for field emission guns to function properly.

In a conventional tungsten-filament electron gun, brightness is proportional to accelerating voltage, the distance of the filament behind the aperture in the Wehnelt assembly (Fig. 84), and the current passing through the filament. An electron cloud is produced as the filament is heated, with electrons (and photons) being emitted from the heated surface in all directions. Higher filament temperatures (current) dramatically reduce filament life because of evaporation of the tungsten filament. Proper saturation of the filament is essential for optimal illumination coupled with optimal filament life. Saturating a filament may be compared to running water through a garden hose. To be most efficient, the water pressure is turned up as high as possible. If the hose can pass only 1 gal/min, any increase in pressure beyond that flow rate is futile. In fact, further increases in pressure may cause the hose to burst. In a similar fashion, a tungsten filament can pass only so many electrons through it over a given period of time (current). If the current flow is below a critical rate, maximum heating and electron release are not achieved, and if the current is above this critical rate, excessive erosion of the tungsten wire will take place. As a filament ages, it thins due to metal evaporation. As the current-bearing wire thins, its resistance increases. In practical terms, this means that an older filament needs less current to achieve the same illumination it produced when the wire was newer and thicker. It also means that the point of filament

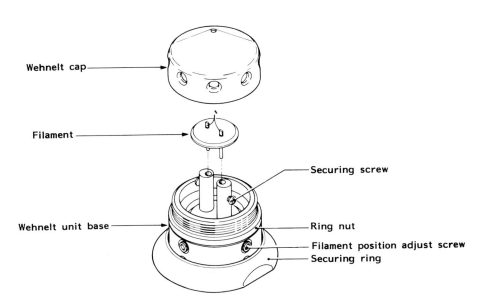

FIGURE 84. A typical Wehnelt assembly with a tungsten filament. (Courtesy of JEOL USA, Inc.)

current saturation (the point at which a further increase in current does not yield an increase in electron emission) becomes oversaturation as the filament ages.

Most contemporary electron microscopes have self-biased electron guns. A resistor is put into the circuit providing current to the filament. It is connected between one leg of the filament and the Wehnelt cap (Fig. 85). As the current is increased to the filament (heating it and causing electrons to be released), a voltage drop occurs at the resistor, producing a slightly negative polarity at the Wehnelt cap. Since the emitted electrons have a negative charge, they are repelled by the negative charge of the Wehnelt cap. Thus, as the current is increased to the filament and more electrons are emitted, the bias voltage (100–500 V) supplied to the Wehnelt cap increases, and the increased negativity forces the electrons toward the center of the Wehnelt assembly. At low beam current (Fig. 86), there is low bias voltage, and electrons from the rear of a filament as well as from the filament tip can be imaged, resulting in multiple spots of illumination on the screen. Further increases in current result in increased bias and a smaller area at the tip of the filament from which electrons can be emitted, producing an image on the screen of a central spot (produced from filament tip emissions) surrounded by a halo (emissions from areas in back of the tip of the filament). At saturation, the spot of illumination is smaller than the original spot and halo, as the two have merged. This is the point at which maximum illumination is achieved and filament structure is no longer visible.

Once the current flow to the filament is properly adjusted, the electron cloud is located primarily at the tip of the filament, which has been precentered over the aperture in the Wehnelt assembly during installation of the filament. Thus, most of the electrons are poised over the aperture and can be propelled through the aperture by the high voltage (40–120 kV) potential generated between the Wehnelt assembly and the anode above the condenser lens assembly.

B. Condenser Lens System

The purpose of the condenser lens system is to focus the electrons produced and accelerated by the electron gun into a coherent beam for illumination of the specimen. In the first microscope designed by Ruska (Knoll and Ruska, 1932), there was no condenser lens system, so illumination was severely limited. Microscopes with only one condenser lens (e.g., the Zeiss 9A) have restricted magnification capabilities

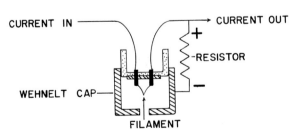

FIGURE 85. A self-biased Wehnelt assembly.

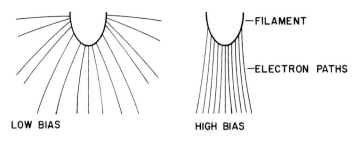

FIGURE 86. The effect of bias voltage on an electron cloud.

resulting from an inability to focus the beam into a small enough spot for high-magnification illumination. As mentioned earlier, illumination is inversely proportional to the square of magnification, so with each doubling of image size, illumination decreases by a factor of two unless the condenser lens system can increase the number of electrons focused on the area of the specimen being viewed. Modern microscopes typically have two condenser lenses. The upper lens (C1) is a strong lens with a fixed aperture and is capable of producing a crossover (finest beam spot) image of the beam of about 9 μm diameter.

The lower lens (C2) projects the crossover image of the first lens onto the specimen plane. The final focused beam can be adjusted to 2–3 μm in diameter. The lower condenser lens has a series of apertures (e.g., 50, 100, 150, and 200 μm) that can be inserted and stigmated.

Under normal conditions, the C1 lens, which has a discontinuous series of settings, is set for the same current for all magnifications; while the C2 lens, which has a continuously variable current range, is adjusted for each magnification setting during use. The two apertures in the system limit the electrons striking the specimen to those from the most coherent part of the electron beam. This protects the specimen from excessive heating and generation of X-rays. The largest aperture gives greater illumination at the expense of image quality if larger than about 250 μm.

C. Deflector Coils

Beam alignment is produced by deflector coils located below the condenser lens assembly but above the objective lens. These are small electromagnets capable of deflecting the focused electron beam and, thus, centering it over the specimen area being viewed. There are typically two sets of deflector coils. If the upper and lower coils are energized the same amount in the same direction, beam shift (pure translation of the beam) occurs (Fig. 87). If the upper and lower coils are not adjusted equally, beam tilt occurs (Fig. 88).

D. Objective Lens

After the beam is produced by the gun, focused by the condenser lens system, and centered with the deflector coils, it interacts with the specimen, which is inserted into

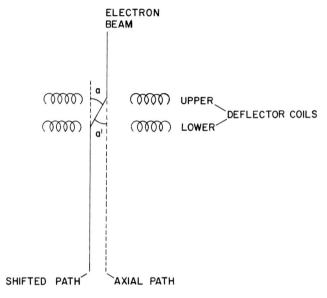

FIGURE 87. Beam translation. The upper deflectors move the beam through angle a, while the lower deflectors move the beam through an exactly equal but opposite angle a′.

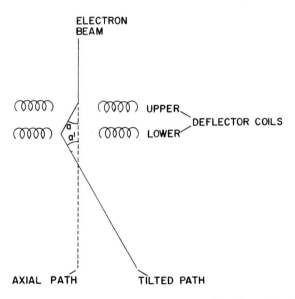

FIGURE 88. Beam tilt. The upper deflectors move the beam through angle a, while the lower deflectors move the beam through an exactly equal angle a′ in the same direction.

the objective lens assembly. The objective lens both focuses and magnifies the specimen image.

Located below the specimen plane, there is an objective aperture assembly that contains a series of user-adjustable apertures of different diameter (e.g., 15, 30, 50, and 70 μm). Large objective apertures give little contrast and are not subject to much contamination. Small objective apertures offer increased specimen contrast but become contaminated more readily. The objective aperture eliminates the most widely scattered elements of the electron beam emerging from the specimen, which would cause something similar to flare in the glass optical system of a 35-mm camera from light being reflected by internal lens surfaces, leading to decreased image contrast. If a plastic section is viewed in a TEM with no objective aperture in place, the contrast will be extremely low, and the section is likely to be severely damaged by the electron beam, often "popping" as it is examined.

An objective lens astigmatism corrector is associated with the objective lens (Fig. 89) and consists of a series (typically eight) of electromagnetic coils that are energized in pairs to adjust for any inherent asymmetry in the lens or the objective aperture being used.

As objective apertures are used, they are subject to minute amounts of contamination from materials that outgas from lubricants and seals within the microscope column, as well as materials vaporized from the specimen itself. If the electron beam is brought to a fine spot on an average plastic section, producing adequate illumination at high magnification, subsequent examination of the exposed area with low magnification will reveal that the area under the intense, focused beam at high magnification appears lighter than the surrounding area. This effect is caused by the electron beam evaporating part of the section, thus thinning the plastic. Some of the evaporated specimen often contaminates the objective aperture, which affects the electron beam, but this contamination can be compensated for by the use of the objective lens stigmator if not too severe.

Two types of stigmators may be encountered. The older, less frequently seen type consists of a set of knobs called the *azimuth* and *amplitude controls*. The former selects the direction of applied astigmatism, while the latter determines the magnitude of the applied effect.

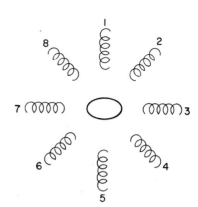

FIGURE 89. Objective lens stigmator. Eight concentrically arranged deflector coils are shown around a beam path that is normal to the page and in the center of the coils. More current is provided to coils 1, 2, 4, 5, 6, and 8, and less to coils 3 and 7 to achieve the asymmetric beam distribution shown.

Most current microscopes have a simple X-Y type of astigmatism correction system. By adjusting two knobs (X and Y), thereby adjusting the amount of applied astigmatism in the X and Y directions, inherent astigmatism is countered with applied astigmatism to produce a symmetric image.

E. Diffraction Lens

The first lens in the projection system (those lenses that magnify but do not focus the image of the specimen) is called a *projector lens,* an *intermediate lens,* or a *diffraction lens.* This lens is capable of producing an extremely low magnification image such that the objective aperture can be viewed. This procedure allows the objective aperture to be centered around the electron beam and is the most common use of the diffraction setting for biologists. In addition, this lens can be used to produce low-angle diffraction patterns of crystalline structures to help identify their chemical composition (see Beeston *et al.,* 1972, for further discussion).

F. Projector System

This system typically is considered to consist of all lenses that magnify the image produced by the objective image. Lenses have names such as diffraction, intermediate, projector, minilens, or supplementary lens. The semantics are less important than the functional characteristics. All of these lenses magnify the specimen image produced by the objective lens. It is important to remember that each lens inverts the image it receives from a previous lens. Thus, as lenses are turned on and off, the image can be rotated up to 180°. Most magnification is achieved by altering the strength of the intermediate lens (usually the first projector lens). High magnification is possible with a TEM because of the multiplying effect on image size of each magnifying lens in the system. With a TEM possessing an objective lens and two projector lenses, the total magnification (MT) can be calculated: $MT = MO \times MP1 \times MP2$, where MO is the magnification of the objective lens, MP1 is the magnification of the first projector lens, and MP2 is the magnification of the second projector lens. If MO is 50, MP1 is 250, and MP2 is 40, then MT is 500,000.

G. Camera Systems

Most TEMs employ a camera system consisting of metal film cassettes designed for $3^{1}/_{4} \times 4$ sheet film with a plastic base or, in some cases, glass plates coated with the same film emulsions. These fine-grain, large-format negatives suffer no loss in resolution when enlarged up to 3–4×, allowing sharp 8 × 10 prints to be produced. The camera is located below the viewing screen and may contain up to 50 sheets of film.

Another camera available for many microscopes is a 35-mm system located above the viewing screen. This system is commonly found on older microscopes with limited sheet-film capacity. The smaller format film cannot be enlarged to 8×10 size without evident loss in sharpness and increased graininess. In addition, defects on the negative surface (scratches, dust) are more evident than with large-format film.

Zeiss developed a fiber-optic plate for their Zeiss 109 introduced in the early 1980s. In conventional camera systems in which the film is put into the vacuum of the column, the gelatin of the photographic emulsion can become somewhat hydrated from the air and cause problems for the high-vacuum system in a TEM. To solve this problem and to maintain an extremely clean column, Zeiss designed their camera system so that the film is outside the vacuum system. Photographs are taken with common 120-size films, allowing quick adjustments for materials of differing contrast (by changing to different types of readily available emulsions). The film is firmly pressed against the fiber-optic plate at the bottom of the viewing chamber and electrons hitting the plate cause photons to be emitted from the bottom of the plate, thus exposing the film.

Finally, various microscopes can now be equipped with video cameras that allow direct acquisition of images that can be stored and manipulated by attached computers. At the present time, it is still impractical to store images by computer because of some decrease in resolution over photographic materials and also because of the amount of space needed in data storage media to record the images. As video resolution improves (more pixels per unit area) and optical disc technology expands, photographic emulsions will probably be supplanted as the major image-storage media.

H. Specimen Holders

Specimens are inserted into the TEM column with either top-entry or side-entry holders (Fig. 90). The former allows easier specimen manipulation within the column, while the latter offers greater stability and resolution. Some microscopes have goniometer-equipped stages that allow tilting of the specimen in one plane up to 60° to produce stereo pairs and to move an oblique section until it is normal to the beam. If the tissue section has a smudged tangential section of a membrane, it can often be tilted enough to give a cross-sectional view that has a clear dark-light-dark "railroad track" membrane image. Microscopes with simple biological stages also usually have provisions to tilt the specimen up to 10°. Goniometer stages can also be provided with specimen holders with the capability to rotate a grid through 360° so that an item of interest can be effectively located at the proper position to take advantage of the tilting plane of the goniometer.

I. Viewing System

Different TEMs have a variety of small focusing screens, large viewing screens, and phosphor screen coatings. Some of the phosphors exhibit finer grain than others; some are yellowish and some are greenish. Screens have scribe marks to show the approximate outline of the boundaries of the film in the camera beneath and various marks to designate the center of the screen, which can also be used as size references

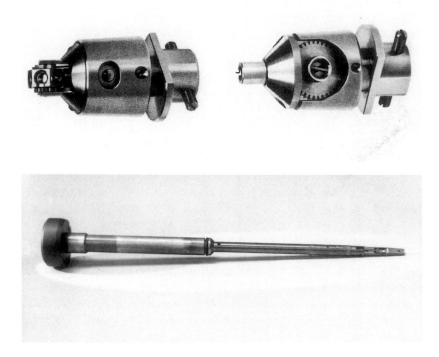

FIGURE 90. Top- vs. side-entry specimen holders. Top: Two top-entry holders available for a Zeiss EM-900 TEM. (Courtesy of Carl Zeiss, Inc.) Bottom: A side-entry specimen holder from a Philips 410LS TEM.

when calibrated properly. Unfortunately, the screens become less bright with time, and the various markings on them can actually become indecipherable. It is recommended that the viewing ports be covered between use to prevent fading of the phosphor screen coating.

The screens are used as meters for the camera system to provide reproducible exposures. In some systems (Philips) the large viewing screen can be used to meter the entire field of view, or the small focusing screen can be used as a "spot" meter to read a specific portion of the field being viewed. The Philips system determines exposure based on electrons hitting the screen and the current they generate as they flow to ground potential. Other systems (JEOL JSM-100S) have photoelectric sensors above the viewing screen that read light emitted from the phosphor on the screen after electrons impinge upon it. Both types of metering systems work quite well.

The binocular microscopes provided for focusing on the viewing screen generally magnify the image projected onto the screen an additional $10\times$. Thus, when the microscope is set for a $31,000\times$ enlargement on the film plane, the magnified image viewed with the binoculars will be almost 310,000 (the image on the viewing screen will not be quite 31,000 because it is above the film plane, which is the basis for the magnification calibration). Unfortunately, the grain size of the phosphor decreases the actual resolution possible with the viewing system. Agar (1957) reported that the actual resolution of the viewing system is limited to approximately 35 μm because of the size of the phosphors on screens. Thus photographs reveal more detail than can be perceived while actually viewing specimens with the microscope.

J. Detectors

A variety of add-on units to detect secondary and backscattered electrons (found on scanning transmission electron microscopes [STEMS]), characteristic X-rays (energy-dispersive spectroscopy), or losses in electron energy after interaction with the specimen (electron energy loss spectroscopy) are placed into various ports on the microscope column and will be discussed in the chapter on microanalysis.

VI. OPERATION OF THE TEM: DECISION MAKING

Now that all the basic components in the illumination, imaging, and viewing system have been described, it is possible to examine the various adjustable components and to decide which settings are useful for general TEM operation when working with typical biological specimens.

A. Accelerating Voltage

With a typical microscope equipped to provide high voltage from 40 to 100 (120) kV, it is useful to consider what will happen when the accelerating voltage is raised or lowered. The highest voltage allows the best specimen penetration, least specimen heating and damage (because of decreased inelastic electron scattering), and brightest screen image, but the lowest contrast with an average specimen. At the same time, there is more possiblity of decreased stability at high voltage because of contaminants within the electron gun, leading to high voltage discharge.

Both the screen phosphors and the photographic emulsions used for electron microscopy are optimally efficient at about 80 kV, so most microscopes are operated at this potential. Thicker specimens often require higher voltage, while low-magnification work with specimens of low contrast (where ultimate resolution is not critical) can be improved by utilizing lower voltage.

B. Choice of Beam Current and Bias

For greatest filament life, the current should be adjusted to a point just slightly below saturation. Most microscopes have self-biased filaments, so the bias will automatically be adjusted properly as the current is increased. If bias is selectable, keep the bias as low as possible while maintaining good filament saturation and adequate illumination. Increased beam current will significantly lower filament life, so minimal current is recommended (10–20 μA). At higher magnifications, where illumination becomes critical, it is often necessary to increase the beam current.

C. Condenser Settings

As previously mentioned, the upper condenser lens is generally set for maximum illumination and the aperture is nonadjustable. The lower condenser (often labeled *intensity* or *brightness*) is continuously variable and is frequently adjusted during microscope operation. As magnification is increased, the lower condenser is adjusted for a smaller spot (of illumination) size. At lower magnification, the spot is spread to cover the screen to prevent undue erosion of the specimen or damage to the screen from excessive beam intensity. The lower condenser lens is equipped with apertures of several sizes. An aperture of about 150 μm provides good image quality and good illumination.

D. Objective Settings

Like the second condenser lens, the objective lens has a series of apertures. They may be small holes bored in molybdenum or platinum strips or discs. Alternatively, they may be plastic films coated with a noble metal (e.g., gold), known as thin-film apertures. The latter type are self-heating and are reputed to stay cleaner longer than metal apertures.

The second to the largest aperture (usually 50–70 μm) is used for most samples, providing sufficient contrast and adequate illumination. If the specimen suffers from too little contrast, a smaller aperture may be selected to increase contrast. If small apertures are used routinely, however, they tend to become contaminated and eventually cannot be stigmated. They will then have to be removed and replaced or cleaned (if metal apertures) by heating in a vacuum evaporator to evaporate contaminants.

Every time the microscope is used it is important to center the objective aperture, because an off-center aperture can cause astigmatism in the image.

The focus controls on the TEM regulate the current to the objective lens. There may be a series of focusing knobs (typically six) with pairs nested together. On one end of the series will be the coarsest focusing knob and on the other will be the finest. A novel system introduced by Philips has only two knobs, one selecting the size of the focus step and the other selecting how many of those steps to make. At low magnifications, large focus steps are selected, while at high magnifications small focus steps are used.

E. Alignment

In order to produce the illumination and image clarity expected from modern high-resolution TEMs, the microscope column must be carefully aligned, particularly after a filament change. Each manufacturer has different instructions for this procedure that are fairly instrument specific, but a general outline of the procedure can be applied to all microscopes. Older microscopes have more mechanical adjustments than current instruments, which have some mechanically prealigned components that are left untouched, except in the unlikely event that the column is disassembled. Modern microscopes are electronically user-aligned on a regular basis, except for mechanical filament and aperture centering. The latest generation of microscopes has computers that are capable of evaluating their alignment parameters and adjusting them appropriately after self-diagnosis. At the least, an attached computer can record all the user-selected operational parameters for the microscope, allowing the users to switch rapidly between apertures, accelerating voltages, and operational parameters while maintaining optimal optical characteristics. This is a vast improvement over older systems in which a user interested in changing accelerating voltage to alter contrast would have to totally realign all the adjustable settings on the microscope visually, which could take an hour or two. A simple aperture change would also require some realignment.

1. Filament Installation

Each manufacturer gives instructions on mounting a new filament within the Wehnelt assembly. The filament height is specified and should be adjusted carefully, as this setting will have an effect on the total illumination possible by determining how many of the liberated electrons will be able to enter the high-voltage field. Centering the filament over the aperture in the Wehnelt cap will also ensure that the maximum number of electrons will be accelerated down the column. Some manufacturers provide precentered filament assemblies, eliminating one of the preliminary alignment procedures for the user.

2. Finding the Beam

After installing a new filament, the high voltage is generally brought up stepwise through the various accelerating voltage settings, waiting until any high-voltage discharges are completed before moving up to the next higher setting. The discharges, which are caused by contamination such as dust in the gun area, are frequently audible but can also be monitored by observing the beam current meter. As the voltage is increased, the "dark" current on the meter will increase, which represents a small current induced in the filament by the high-voltage field. If the meter shows a large increase in current (typically the needle deflects to the stop at the high end of the scale), this large discharge should not continue for more than a few seconds. If it does, then the voltage should be reduced to the next lowest setting until the current stabilizes. A momentary deflection several times over a minute or so is satisfactory, but the voltage should not be raised until the activity stops. Once the voltage has been brought to the maximum value (100 kV) and discharge has ceased, the voltage should be reduced to the operating value (80 kV). Then the beam current is slowly brought up until illumination is visible on the screen. It is sometimes difficult to see any screen illumination because the filament is grossly misaligned or the beam is overly tilted. If the high voltage and current knob are turned up, but the current meter shows no current flow, either the filament is blown or the filament contacts are not properly positioned. If the high voltage is turned on and the beam current meter is showing increasing current flow as the beam current (filament) knob is turned up but the screen still does not show illumination, turn off all the lenses, which should yield a small, extremely intense spot of light projected onto the viewing screen. Do not leave this on the screen any longer than necessary, as it will burn the phosphor coating. Once this spot is located, the lenses can be turned on one at a time from the top of the column down until the spot vanishes again, which indicates that the last lens needs to be realigned until the illumination reappears. After illumination has been established with all the lenses turned on, return to the top of the column to continue the alignment procedures.

3. Beam Alignment

Once illumination is visible, bring the second condenser (intensity, C2, spot size) to crossover (the smallest, most intense spot of illumination visible on the screen) and increase magnification to 3,000–5,000× for better visualization. Adjust the gun shift controls until the illumination spot on the screen is as bright as you can make it, then decrease the beam current until the spot of illumination breaks up into more than one image. Slowly bring the beam current up again until the spots begin to merge. If there is not a central spot with a halo around it, use gun tilt controls to center illumination around the central spot. You may have to increase beam current to keep the nonaxial component of the illumination visible as you center the halo around the central spot (Fig. 91). Once the halo is centered, increase the beam current until the halo converges with the central spot. Decrease beam current slightly until a slight decrease in illumination is noted somewhere in the spot of illumination, which indicates slight undersaturation (thereby ensuring longer filament life). At this point the beam is aligned in relationship to the condenser lens assembly. Spread the beam (C2) until it fills the viewing screen.

4. Deflector Controls

Bring the beam near crossover by adjusting C2 and center the spot of illumination on the screen with the deflector controls.

5. Condenser Assembly

Insert an appropriate condenser aperture into C2 (the second to the largest is usually a good choice). Desaturate the filament (so the halo is once again visible at crossover) and make the image as sharp as possible by adjusting the condenser stigmator controls. Check to see if the condenser aperture is centered properly by sweeping C2 above and below the crossover. If the spot of light on the screen moves from side to side, use the two condenser aperture-centering controls to minimize this motion. The spot should get larger and smaller as you sweep through crossover, but should not move to the side if the aperture is centered. Spread the beam to cover the screen (C2).

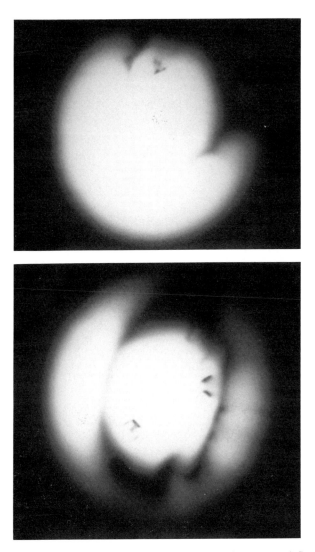

FIGURE 91. Image of an undersaturated filament. Top: Undersaturated, not centered. Bottom: Undersaturated, nearly centered. The dark spots in the center of the filament are defects in the tungsten filament due to aging (it had approximately 130 hr of use).

6. Objective Lens

Insert the aperture of choice (usually the second aperture or one about 50–70 μm in diameter). Next, introduce a specimen of some sort. Bring the beam to crossover. Switch the lenses into the selected area diffraction (sometimes just called *diffraction*) mode. The carbon contained in the specimen will produce a diffraction pattern, which will be a series of concentric rings around a central, bright electron beam. The objective aperture can then easily be centered around the beam, utilizing the centering controls on the objective aperture assembly on the column. Switch back to the non-diffraction (magnification) configuration for normal viewing.

After the aperture is centered, the lens and aperture can be stigmated as a unit. This is typically done at a magnification above which photographs are expected to be taken, so that any minor astigmatism that remains will be imperceptible at the lower magnifications routinely used.

There are two major methods to stigmate the objective lens assembly. The first requires the insertion of a plastic film with minute holes in it (a holey grid, which can be commercially purchased or made as described in Chapter 3). The image is over-focused until a Fresnel fringe is visible around a hole in the plastic, and the stigmators are used to produce an even fringe around the hole.

If the microscope has the azimuth/amplitude type of stigmators, the stigmator is turned off so that the inherent astigmatism in the system is revealed (Fig. 92). Then, the stigmator is turned on and the amplitude is turned up high enough so that the applied astigmatism is evident (Fig. 93). The orientation control is adjusted until the applied astigmatism is at 90° to the original inherent astigmatism. At that point, the amplitude knob is used to reduce the strength of the applied astigmatism until the Fresnel fringe around the hole is the same thickness all around the hole. Bring the image closer to focus (narrower Fresnel fringe) and repeat the process. Ideally, the barest perceptible fringe should vanish uniformly from around the hole with one click of a very fine focus step.

If using an X/Y type of stigmator, the same rules apply, except that the fringe is adjusted with the X/Y controls until it is even around the hole.

Another, more universally applicable, method for stigmation can be done at any time with any kind of specimen. Take the magnification above the normal working level (say, to 100,000×), focus on the grain of the plastic or some graininess in the

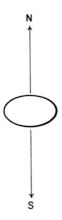

FIGURE 92. Inherent astigmatism.

FIGURE 93. Applied astigmatism.

specimen, and adjust the stigmators until the grain is as sharp as possible (grain focusing). Go closer to focus and repeat. Ideally, the grain should appear as points rather than short streaks.

7. Projector Lenses

In older microscopes, the projector lenses have to be aligned to be in the same axis as the previously aligned gun, condenser, and objective lens assemblies, but most modern microscopes have these elements prealigned at the factory so they do not need adjustment unless the column is disassembled. Any apertures within these lenses are generally fixed and therefore require no separate adjustment. The only exception to this is the diffraction lens aperture, which must be adjusted for critical low-angle diffraction studies.

8. Voltage Center and Current Center

The generalized alignment procedure described above will result in a microscope capable of producing high-quality images. However, without a few supplementary steps to align the voltage center (high voltage) and current center (lenses), the image will suffer from slight chromatic aberration (voltage center misalignment) and spherical aberration (current center misalignment). Under most conditions with biological specimens, these will not be perceptible.

Another set of alignment procedures specified for particular instruments does not actually improve image quality but makes the instrument more user friendly. Thus, when the image is swept through focus, the image stays in the center of the screen, and when the magnification is changed, the image and illumination stay centered.

F. Taking a Photograph

Taking a photograph requires attention to detail, as any omission in the following steps can seriously compromise the quality of the images produced. The first step is to focus the ocular system on some detail (dust) on the focusing screen. Make sure that focusing is *not* done on the specimen image. Next, select the magnification and focus the image using the electronic focusing controls. Wobblers are sometimes useful to determine focus at low magnifications where the eye tends to compensate, suggesting that images are in focus in more planes than they truly are. High magnification focusing is generally not helped by the use of wobblers. Then frame the image within the scribed marks on the viewing screen describing the dimensions of the film format being used. Bring the beam toward crossover with the intensity (spot size) controls so that the periphery of the illumination can be seen, which is centered with the deflection controls. Next, spread the beam until the metering system indicates a proper illumination level for a photograph to be taken. Take the photograph.

Negative density (darkness) can be determined by the exposure setting on the microscope or processing procedures (developer strength, amount of agitation, type of developer, temperature of developer). To assure consistent results, film should be exposed and developed consistently (see section on photography). The microscope meter should be set such that an average specimen (silver-gold section of typical tissue stained in a typical fashion) produces a negative of average density (meaning that it will print well on a middle-grade paper). Since slight variations in staining procedures and section thickness can produce materials with different levels of contrast, a microscope set to produce an "average" negative will yield a product that can be printed with more or less contrast to accommodate these slight specimen variations.

G. Specimen Radiation Dose

Grubb and Keller (1972) examined the beam damage to certain polymers in the TEM and the consequent damage to image quality. Keller was also quoted by Agar (1974) as having determined that 4,000 Mrad of radiation exposure results from the following conditions:

1. 0.01 sec at high beam current in a TEM
2. 20 min at low beam current in a TEM
3. 5 weeks inside a nuclear pile
4. 5 years near a 1-Ci cobalt 60 source of X-rays
5. Exploding a 10-megaton hydrogen bomb 30 yd away

These numbers should make it clear that a TEM is a source of intense irradiation that can cause severe specimen damage. In addition, the production of X-rays makes it imperative that, whenever the integrity of a TEM column is disturbed, a check for radiation leakage is performed immediately.

H. Microscope Calibration

At various times, particularly with older TEMs, it becomes necessary to check the accuracy of the magnification steps on a particular instrument. It is also sometimes necessary to have a specimen that will clearly show if the TEM is exhibiting any slight spherical aberration. As mentioned previously, when a microscope is installed, the usually guaranteed 0.344-nm resolution must be established by photographing a resolution standard (graphitized carbon).

There are a variety of resolution standards (shadowed metals such as gold-palladium, silicon monoxide gratings, biological crystals, graphitized carbon, etc.) available from most of the vendors for electron microscopy supplies. Each vendor supplies detailed instructions on how to utilize the individual standards. The most commonly used ones are described below.

1. Grating Replicas Mounted on Grids

Silicon monoxide grating replicas mounted on grids may be purchased with 1,134 parallel lines/mm, 2,160 lines/mm, or 2,160 crossed lines/mm. The first two replicas are usually photographed at several magnifications, and then the distance between the lines on the negatives is measured with an ocular micrometer to see if the actual magnification agrees with that shown on the magnification readout on the TEM. If the actual values are within 5–10% of those indicated on the TEM, the microscope can be considered to be within normal tolerance limits. If it is suspected that the microscope is exhibiting some sort of electronic stability problem resulting in the magnifications varying from photograph to photograph, suspensions of fragments of the silicon monoxide replicas can be put directly on specimens being examined, and a photograph can be taken showing the item of interest alongside the replica for absolutely accurate magnification calibration directly on the specimen.

The crossed grating replicas will clearly show spherical aberration (pincushion or barrel aberrations will be clearly evident).

2. Catalase Crystals

Suspensions of these protein crystals may be purchased from some vendors and utilized much like the suspensions of silicon monoxide gratings, except that they exhibit 17.2-nm spacings between the lattice elements for higher resolution work.

3. Latex Spheres

These are available in numerous sizes (e.g., 0.087 μm, 0.091 μm, . . . , 25–55μm), but are usually too large to be of much use in TEM work (they are very useful in scanning electron microscopy applications).

4. Graphitized Carbon

Grids, consisting of a polymer film substrate upon which carbon black has been dispersed, may be purchased from various vendors. These specimens exhibit 0.34-nm lattice spacing and are used most frequently to test the ultimate resolution of newly installed TEMs.

REFERENCES

Agar, A.W. 1957. On the screen brightness required for high resolution operation of the electron micro-scope. *Br. J. Appl. Phys.* 8:410.

Agar, A.W. 1974. Operation of the electron microscope. In: *Principles and practice of electron microscope operation,* A.W. Agar, R.H. Alderson, and D. Chescoe (eds.), North-Holland, Amsterdam, pp.166–190.

Beeston, B.E.P., Horne, R.W., and Markham, R. 1972. Electron diffraction and optical diffraction tech-niques. In: *Practical methods in electron microscopy,* Vol. 1, A.M. Glauert (ed.), North-Holland, Amsterdam, p. 444.

Grubb, D.T., and Keller, A. 1972. Beam induced radiation damage in polymers and its effect on the image formed in the electron microscope. *Proc. 5th Eur. Conf. Electron Microsc.,* Institute of Physics, London.

Hainfeld, J.F. 1977. Understanding and using field emission sources. *Scann. Electr. Microsc.* 1:591–604.

Hawkes, P.W. (ed.). 1985. *The beginnings of electron microscopy.* Academic Press, New York.

Knoll, M., and Ruska, E. 1932. The electron microscope. 2. *Physik.* 78:318–339.

Meek, G.A. 1976. *Practical electron microscopy for biologists,* 2nd ed. John Wiley & Sons, New York.

Slayter, E.M. 1976. *Optical methods in biology.* Krieger, Huntington, NY.

Vacuum Systems

Vacuum is crucial to the proper operation of an electron microscope since electrons, the illumination source, are easily deflected by molecules of gas and thus are lost to the image-forming process. In order to have a working understanding of the vacuum system, it is necessary to examine the different types of vacuum sensors (gauges) and vacuum pumps utilized in electron microscopy. This is especially important because various gauges and pumps can sustain damage if connected to the system being evacuated at an improper vacuum level.

To begin our discussion, we need to consult Table 3, which illustrates the various units for vacuum and indicates their relationships. Most electron microscopists measure vacuum in units of Torr, while most vacuum physicists use Pascal units. Two good books on vacuum technology as it applies to electron microscopy are those by Stuart (1983) and O'Hanlon (1980).

A typical transmission electron microscope operates with an ultimum vacuum in the range of 10^{-5} to 10^{-9} Torr. In addition to indicating a high-vacuum (low-pressure) condition, what does this range mean in terms of imaging electrons and of their journey to the viewing screen and the phosphors that they excite? One cubic centimeter of air at 10^{-6} Torr (as found inside an average TEM column) contains 3×10^{10} molecules of various types (Table 4). At this pressure, an electron has a mean free path of at least 50 m at room temperature. In other words, an electron can be expected to travel a distance of at least 50 m without encountering and interacting with any other electrons or nuclei. Of course, when the specimen is inserted into the electron beam, the probability of a specimen/electron encounter is quite high, and the electrons that penetrate the specimen will ultimately encounter the viewing screen. What is important is that random scattering of the electrons is very unlikely under the vacuum conditions described.

I. TYPES OF GAUGES

A. Direct Reading

Direct-reading gauges are used to monitor the water pressure feeding the cooling system for diffusion pumps and electromagnetic lenses. They also monitor the positive

TABLE 3. A Comparison of Vacuum Units

760 Torr = 1 atmosphere
1 Torr = 1 mmHg
1 μ (of Hg) = 10^{-3} Torr
1 Pa = 133.32 Torr

air pressure used to hold various valves open in some electron microscopes (e.g., Philips).

Diaphragm-type gauges have a needle attached to a flat membrane of metal or polymer that moves as pressure increases or decreases. Bourdon-tube gauges are found in the same applications and consist of a curved, thin-walled metal tube attached to a needle. As pressure increases or decreases in the tube, the needle is deflected. With both gauges, the deflection of the needle is read on a scale on the meter face to monitor pressure. Neither of the gauges are instruments of significant precision, but they provide enough information to fulfill the needs of their applications.

B. Indirect Gauges

1. Thermoconductivity Gauges

These gauges measure vacuum by monitoring a pressure-dependent quality of gas. This type of gauge is used at relatively low vacuum and is based on two principles: (1) Heat will be conducted away from a heated wire by molecules of air, and (2) current flow in a wire is inversely proportional to the heat of the wire (a higher heat results in greater resistance and thus less current flow). Thermoconductivity gauges are used to monitor vacuum systems from a pressure of 1 atmosphere down to about 10^{-3} Torr. They are not damaged by contact with atmospheric pressure.

Thermocouple gauges (Fig. 94) are the most simple in design, consisting of a solid-state thermocouple device attached to a wire through which a constant current is being passed. The passage of current through the wire heats the wire. The more air present, the more heat is conducted away from the wire, and the thermocouple conduc-

TABLE 4. Major Constituent Molecules of Air

Constituent	% (of volume)
Nitrogen (N_2)	78.08
Oxygen (O_2)	20.95
Carbon dioxide (CO_2)	0.03
Argon (Ar)	0.93
Other moieties	0.01
	100%

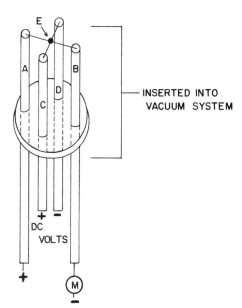

FIGURE 94. Diagrammatic sketch of a thermocouple gauge. The solid-state thermocouple device (E) is electrically connected to terminals A and B. It is also in thermal contact with the wire connected to terminals C and D, which is heated. The microammeter (M) reads the amount of current flow through the thermocouple.

tivity decreases, causing the DC microammeter needle to move to the bottom of the scale. As air is evacuated from the system, the heat of the wire cannot be conducted away as effectively, so the wire temperature increases, the thermocouple passes more current to the DC microammeter, and the meter needle swings toward the high end of the scale.

A slightly more complicated version of the simple thermocouple gauge is the *Pirani* gauge (Fig. 95). This gauge also measures the capability of air to conduct heat away from a heated wire. In this case, the heated wire is one of four resistances (the

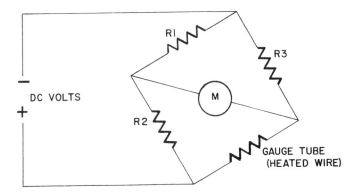

FIGURE 95. Diagrammatic sketch of a Pirani gauge. Resistors at R1, R2, and R3 are balanced by the heated wire in the gauge tube at atmospheric pressure. The microammeter (M) measures current flow as the Wheatstone bridge becomes unbalanced when more gauge tube atmosphere is available to conduct heat away from the gauge wire.

other three are constant-value resistors) in a Wheatstone bridge with an attached DC microammeter. Usually, a constant voltage is supplied to the Wheatstone bridge, and the fixed resistors are selected so that the meter registers very low current flow at high vacuum. When the chamber is at atmospheric pressure, the heat is removed from the wire and transferred to air molecules, which unbalances the Wheatstone bridge. The more air present, the cooler the wire becomes, and the more current it will pass, causing the microammeter gauge to deflect to the high end of the scale.

2. Ionization Gauges (Cold Cathode Ionization Gauges)

Ionization gauges are used to monitor ultimate vacuum in the TEM by measuring another quality of gases: the capability of air molecules to conduct an electrical current. The standard type encountered is the *Penning* gauge (Fig. 96). It is used in high-vacuum range (10^{-3} to 10^{-9} Torr) developed by diffusion pumps, turbomolecular pumps, and ion getter pumps. Exposure to atmospheric pressure can damage an ionization gauge, so it is important to turn an ionization gauge on only after the chamber has been switched over to high vacuum. The gauge consists of a wire anode loop maintained at a potential of about 2–10 kV in relationship to the grounded cathode electrode (typically the outside of the sensing tube). The whole tube is inserted into the vacuum chamber and is surrounded by a strong permanent magnet. The high voltage causes electrons to be emitted from the cathode, which ionize any gas molecules in the tube. The positive ions formed follow a spiral path to the cathode because of the external magnet. This spiral path increases the path length to the cathode, thus causing more collisions with gas molecules and the production of more ions, consequently increasing current flow between the electrodes. As the system is evacuated to ever greater vacuum, ionizable molecules decrease in number, and current flow in the ionization chamber decreases, as shown by the Penning gauge.

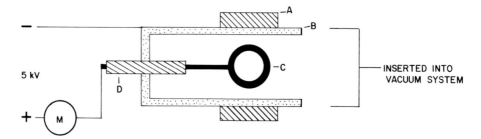

FIGURE 96. Diagram of a Penning (cold cathode discharge) gauge. A permanent magnet (A) surrounds the grounded gauge tube (B) connected to the vacuum chamber. The positively charged wire loop anode (C) is connected to a microammeter and passes into the tube through the vacuum-sealed insulator (D). M = microammeter.

II. VACUUM PUMPS

A. Types of Pumping Systems

Some pumps function by compressing a given volume of gas, reducing its volume, and then expelling the compressed gas through an exhaust port (compressive/mechanical pumps). An example of this type of pump is the rotary (often called *mechanical*) pump, which is used to "rough out" a chamber before connecting the chamber to a high-vacuum pump. Roughing is the process by which a chamber is brought from atmospheric pressure down to about 10^{-3} Torr, at which point a pump capable of achieving a higher vacuum is attached to the chamber immediately after disconnecting the rotary pump.

A second type of compressive pump is the turbomolecular pump, which can reach the ultimate high vacuum level necessary for proper operation of an electron microscope. It is usually "backed" by a rotary pump, which pumps away the compressed gas molecules escaping from the exhaust port of the turbomolecular pump.

Entrapment is a totally different mechanism to remove air molecules from a chamber. Rather than compressing the gas into a small volume by mechanical means and expelling it from the system, air is entrapped within microscopic oil droplets (diffusion pumps) and is ultimately removed by another lower vacuum pump.

Entrainment pumps function by ionizing and accelerating gas molecules with high voltage and causing them to strike and attach to the sides of a chamber at high velocity (ion getter pumps), or by condensing gases onto a cold surface to which they stay attached until the chamber warms up (cryopumps). These three entrapment or entrainment systems require other pumps (usually mechanical pumps) to ultimately remove the entrapped molecules from the entire system. When a pump is used to remove air from another pump, the latter can be said to be "backed" by the former. In some systems, a mechanical pump produces a vacuum on a holding (buffer), tank which is, in turn, used to back a pump producing a higher vacuum on the vacuum chamber.

B. Mechanical/Rotary Vane Pumps

These pumps consist of a rotor in which are housed two spring-loaded scraper blades. The edges of the blades are typically faced with Teflon®. The blade edges run within an eccentric chamber whose walls are lubricated from a reservoir of oil. One side of the eccentric chamber is attached to the chamber to be evacuated, while the side where the gas is exhausted has a small flapper valve, which is opened by the pressure of the gas, thus allowing it to be discharged from the pump.

The pump may have one rotor (single stage), allowing it to achieve a vacuum of about 10^{-3} Torr (Fig. 97). These are typically belt driven from an electric motor.

A more efficient configuration is the two-stage mechanical pump, wherein two rotors are connected in a series, one attached to the chamber being evacuated, which in turn is backed by the second rotor (Fig. 98). Two-stage pumps are typically directly

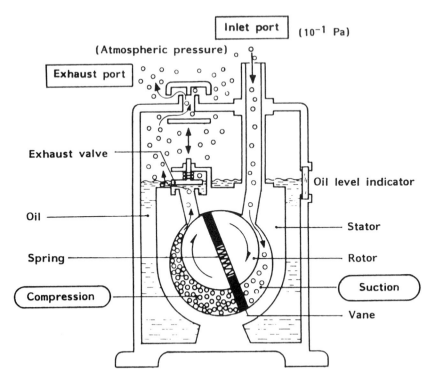

FIGURE 97. Diagram of a single-stage rotary pump. (Courtesy of JEOL USA, Inc.)

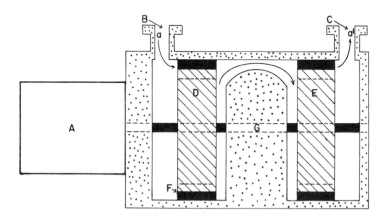

FIGURE 98. Diagram of a two-stage rotary pump. A direct-drive motor (A) drives two rotors (D, E) located in two eccentric chambers containing oil through which spring-loaded scraper blades (F) pass. Gas enters through port B (a), is compressed by the first rotor (D), passes into the second chamber, and is further compressed by the second rotor (E) before being exhausted (a') through port C.

driven by electric motors due to the high load on the motor. They can achieve a vacuum of up to 10^{-4} Torr.

A rotary pump will pump about 500 l/sec from a chamber. Such pumps most frequently contain hydrocarbon oils, but they may have more complex synthetic lubricants containing silicone, chlorofluorocarbons, or fluorocarbons (O'Hanlon, 1980). If the pumps are to be used in situations utilizing oxygen (plasma etching), they must have appropriate oils to avoid the risk of explosion.

Both single- and two-stage mechanical pumps are subject to contamination by solvents being evacuated (acetone, alchohols, water). As the gases are evacuated from the vacuum chamber and compressed within the rotor assembly, they become cooled and can condense on the pumping chamber walls, and finally be mixed with the pump's lubricating oil. The first symptom of a problem is usually less pumping efficiency in the system. If the oil in the pump's sight glass appears turbid or discolored, the oil has become contaminated. If pumping efficiency has dropped but the oil still looks reasonably clean, the pump may be "ballasted" if so equipped. These pumps possess a ballasting valve located just before the pump's exhaust port. The valve is opened slightly for 30 min to an hour to help the system purge the contaminating substances. When this valve is opened a small amount, it admits air just before the exhaust side of the pump, thereby decreasing the ultimate vacuum of the system. However, at the same time, the solvents dissolved in the lubricating oil will be vaporized and exhausted from the system. When it is determined that most of the solvents have been pumped out, the ballasting valve is closed, at which time the vacuum system should return to normal values. If the chamber still cannot be brought to the proper advertised value, the oil may be hopelessly contaminated with solvents, necessitating the draining of all oil from the pump and its replacement with high-quality mechanical pump oil. If a pump has difficulty starting after being stopped even though it has been properly vented, it usually indicates that the oil should be changed even if it looks all right. Always vent mechanical pumps that are not turned on. If they are attached to the vacuum system while at rest, the pump oil will be sucked into the vacuum lines.

A last problem to beware of is "backstreaming." All pumps, when they are close to their pumping limits (have achieved their maximum vacuum on a chamber), are capable of actually releasing their oil into the vacuum chamber in minute quantities. Mechanical pumps with worn scraper blades can backstream badly (indicated by an oil film in the chamber being evacuated). When roughing a chamber, it is a good practice to remove the mechanical pump shortly after reaching its ultimate vacuum to decrease the likelihood of backstreaming.

C. Diffusion Pumps

Diffusion pumps are deceptively simple in construction (Fig. 99). They consist of a metal can containing a stack of inverted funnels held on a central core. Directly below the flange that bolts to the chamber to be evacuated, the pump may have a series of angled baffles surrounded and thermally connected to a water source for cooling. Wrapped around the outside of the pump chamber wall is copper tubing through which

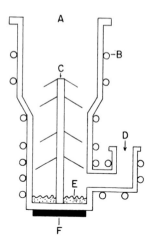

FIGURE 99. Schematic diagram of a diffusion pump showing the top of the pump that would be connected to the vacuum chamber (A), the cooling coils (B) wrapped around the pump and the port (D) to the backing pump, the stack of inverted cones (C) that direct the oil vapors to the wall of the pump, and the heater (F) that boils the oil (E).

runs cooling water or oil. Near the bottom of the chamber is an exhaust port, also usually surrounded by coils containing coolant. This port can be attached to a mechanical pump or buffer tank that backs the diffusion pump. At the bottom of the pumping chamber, typically bolted to the outside, is an electrical heating element.

A diffusion pump contains a small amount of high-grade synthetic lubricant (Table 5). Hydrocarbon and silicone oils are not used much because of potential column contamination.

The oil is boiled by the external heater and rises as a vapor toward the top of the pump chamber. The oil vapor hits the central stack of inverted funnels, and microscopic oil droplets are ejected outward and downward toward the chamber walls at supersonic speeds. Any gas molecules in their path are adsorbed. When the oil droplets with adsorbed gas molecules hit the water- or oil-cooled chamber walls, the droplets condense and run down the walls to the base of the chamber. As the cooled oil with entrapped gas molecules is reheated by the electric heater, the oil vaporizes and releases the entrapped molecules, which are removed from the diffusion pump by the backing system. The cooling coils on the pump wall, the cooled baffle at the top of the diffusion pump, and the coils around the exhaust port usually prevent oil from escaping the diffusion pump. Diffusion pumps have a pumping speed of about 2,160 l/sec, considerably faster than the 8 l/sec of a mechanical pump (Stuart, 1983). They rarely achieve a greater vacuum than 10^{-6} Torr when attached to electron microscopes.

Diffusion pumps should not be used unless the chamber to be evacuated has reached about 10^{-3} Torr. If a heated diffusion pump is brought to atmospheric pressure, the oxygen present can react with the boiling oil, thus oxidizing it. This is known as "cracking" the oil. Most modern oils can withstand brief exposure to atmospheric pressure at operating temperature, but this can eventually result in oxidation of the oil.

TABLE 5. Chart of Some Pump Lubricants and 1991 Prices

Name	Chemical name	Price
Oils for Mechanical Pumps		
Edwards Supergrade A	Mineral oil derivative	$1.20/100 ml
Fomblin Y LVAC 25/5	Perfluorinated polyether*	$286/kg
Silicone Oils for Vacuum		
Evaporators (not for use in		
microscopes)		
Dow Corning 704	Tetramethyltetraphenyltrisiloxane	$27/100 ml
Dow Corning 705	Pentaphenyltrimethyltrisiloxane	$58/100 ml
Oils for Microscope		
Diffusion Pumps		
Fomblin[R] Y HVAC 25/9	Perfluorinated polyether*	$540/kg
Octoil-S[R]	Dioctylsebacate	$24/100 ml
Santovac 5[R]	5-ring polyphenylether	$150/100 ml

*If Fomblin oils are used, they should be used in *both* the mechanical and diffusion pumps. (They do not mix with oils with other bases.)

If the pumping efficiency of the pump drops considerably, check to see if the heater is still heating (the electrical wires to the heaters often corrode and break). If a lack of heat is not the problem, shut the instrument down, vent the pump according to the manufacturer's recommendations, and check that the pump contains a relatively uncolored pump oil (typically 100–200 cc). If negligible amounts of oil are noted, remove the heater from the pump, disconnect cooling lines and the backing line, and then unbolt the upper flange from the vacuum chamber. Then examine the central tower with its inverted cones. If it looks like the bottom of a camping skillet, the oil has been severely cracked. The central core should be removed and cleaned or replaced and then the pump refilled with the proper kind and quantity of oil before being reinstalled.

Most modern diffusion pumps have cooling baffles or cold traps at the upper end of the pump to prevent backstreaming. Even so, if a microscope column is suddenly brought to atmospheric pressure from high vacuum, oil from the diffusion pump may backstream into the column.

D. Turbomolecular Pumps

Turbo pumps have been around since the 1950s but have been incorporated into electron microscope designs for only about one decade. They consist of a series of variably sized and variably pitched blades on a shaft capable of being spun within the pump chamber at 20,000–60,000 rpm, supported by either high-quality metal, magnetic, or air bearings external to the pumping chamber (Fig. 100). The momentum of the rapidly rotating blades is transferred to the gas molecules, and the cascaded series of blades directs the gas toward the exhaust port of the pump while simultaneously compressing the gas into an ever smaller volume. As O'Hanlon (1980) has stated, "The

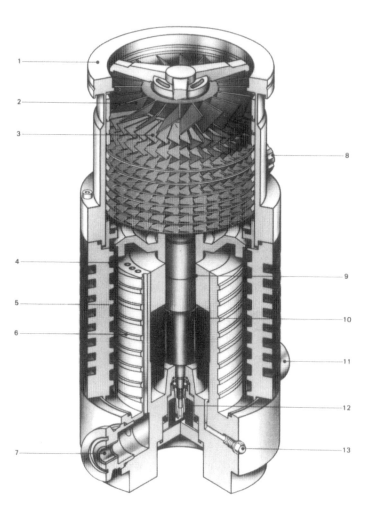

FIGURE 100. Diagrammatic sketch of a turbomolecular pump. (Courtesy of Balzers.) (1) High-vacuum connection; (2) rotor disk; (3) stator disk; (4) molecular stator 1; (5) molecular rotor; (6) molecular stator 2; (7) roughing vacuum connection; (8) venting connection; (9) labyrinth; (10) motor; (11) electrical connection; (12) oil reservoir; (13) sealing gas connection.

relative velocity between the alternate slotted rotating blades and slotted stator blades makes it probable that a gas molecule will be transported from the pump inlet to the pump outlet." These pumps are usually attached to a chamber previously evacuated to 10^{-3} Torr or better, and can achieve a vacuum of about 10^{-6} Torr when attached to an electron microscope. Pumping speed is in the range of 100–500 l/sec.

Turbo pumps are usually used in lieu of diffusion pumps and have the advantage of not being subject to backstreaming or oil cracking. Unfortunately, they are also considerably more expensive than diffusion pumps due to the high precision necessary to sustain the high speeds utilized.

E. Sputter Ion (Ion Getter) Pumps

With the advent of lanthanum hexaboride guns and, more recently, field emission guns, higher vacuum than that achieved by diffusion pumps or turbomolecular pumps alone was needed. Ion getter pumps (Fig. 101) fulfilled this need, being able to achieve 10^{-9} Torr in restricted areas of an electron microscope column (typically in the area from near the specimen up to the top of the column). Ion getters are usually metal chambers lined with titanium containing a high-voltage (5 kV) cathode and surrounded by strong permanent magnets. The theory of operation is virtually identical to that of a Penning gauge, as discussed earlier. Gas molecules within the chamber are ionized by electrons liberated from the cathode. The ionized gas molecles are accelerated toward the cathode and directed into a spiral path by the external magnets. They eventually impinge on the chamber walls, where they become entrained. Some ion getter pumps are designed to be externally heated and baked out periodically to release the molecules trapped by the titanium walls so that they can be pumped out of the ion getter by other pumps. In other cases (Philips electron microscopes), the chambers are merely removed from the microscope when they are no longer working efficiently and are replaced with new/reconditioned pumps. The removed pump is then baked out and reconditioned by Philips.

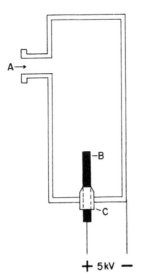

FIGURE 101. Diagrammatic sketch of an ion getter pump showing the inlet port (A) and high-voltage anode (B) insulated (C) from the titanium-lined pump housing cathode.

Ion getters can pump air at about 500 l/sec but work poorly unless the system is prepumped by a diffusion or turbomolecular pump. If these pumps are connected to a system at atmospheric pressure, they will arc audibly. The arcing of electricity from the cathode to the anode surfaces across air molecules will result in excessive heating of and possible damage to the pump.

F. Cryopumps and "Cold Fingers"

Cryopumps are typically filled with liquid nitrogen or the colder but much more expensive liquid helium. They are located below the chamber to be pumped and often used in conjunction with diffusion pumps (being located above the flange connecting the diffusion pump to the chamber being pumped). Liquid nitrogen is contained in a large chamber external to, but in intimate contact with, the top of the diffusion pump. After most of the air is removed from the system by a diffusion or turbomolecular pump, the cryopump is filled and the extremely cold surface produced on the walls of the evacuated chamber (microscope column, etc.) causes any remaining gas molecules to be securely bound until the liquid nitrogen evaporates from the cryopump, at which time all the adsorbed molecules are liberated (and are usually pumped out by the diffusion pump).

Cold fingers are small cages located around specimens within the objective lens in transmission electron microscopes that are in thermal contact with an external source of liquid nitrogen or helium. Any contaminants produced by the interaction of the electron beam with the specimen will tend to condense on the cold surface of the cold finger, where they will remain as long as it is kept cold. When the cold finger comes to room temperature, the trapped gas molecules are released and are evacuated by the diffusion pump.

Some microscopes are equipped with specimen stages that are cooled with liquid nitrogen or liquid helium (cold stages). These are used to maintain frozen sections for viewing in the frozen, hydrated state and are not designed as pumps *per se*.

Cold fingers, cold stages, and cold traps can cause major problems for systems that are brought to atmosphere without first emptying cryogen containers and bringing them to room temperature. If these devices are not warmed up above ambient temperature, they may condense water vapor from the air, which then can result in water droplets in a microscope column. Water droplets or vapor are extremely difficult to remove from the column.

III. SEQUENTIAL OPERATION OF A COMPLETE VACUUM SYSTEM TO ACHIEVE HIGH VACUUM

Figure 102 shows an electron microscope pumping system illustrating the various pumps, pumping lines, and valves. Before beginning pumping on a chamber, the chamber should be isolated from the diffusion pump by the main valve. The mechanical pump should also be isolated from the chamber by a valve. The diffusion pump may

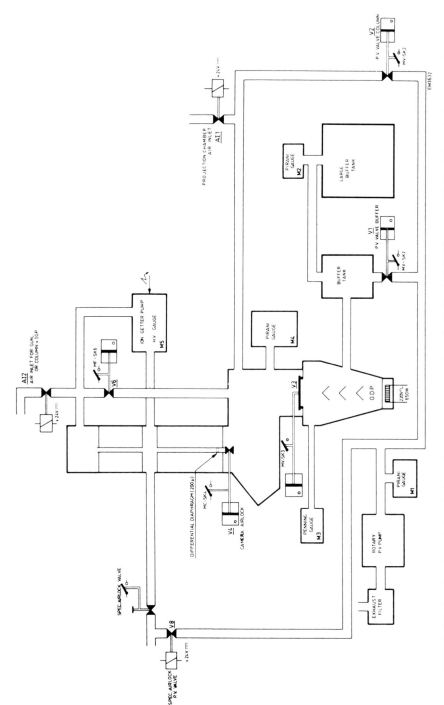

FIGURE 102. Complete transmission electron microscope pumping system of a Philips 410LS. (Courtesy of Philips Electronic Instruments, Inc.)

have some residual vacuum, but the mechanical pump should be at atmospheric pressure.

The typical sequence of events to reach high vacuum from atmospheric pressure begins with the mechanical pump being turned on and attached to the chamber to be evacuated by opening a valve. The mechanical pump is left attached until the chamber has reached approximately 10^{-3} Torr.

In the mean time, the diffusion pump is being backed by either a buffer tank previously evacuated to 10^{-3} Torr or another mechanical pump. Once the diffusion pump has reached 10^{-3} Torr, the heater is activated. The buffer tank or mechanical pump remains actively backing the diffusion pump while it is on. After about 20 min, the oil in the diffusion pump will be boiling. The mechanical pumping on the chamber should then be withdrawn by closing a valve, and the diffusion pump should be connected to the chamber by opening up the main valve. It is imperative that the mechanical pump never be hooked to the chamber at the same time as the diffusion pump, because the higher vacuum developed by the diffusion pump would suck the mechanical pump oil into the chamber.

After the diffusion pump evacuates the system to 10^{-5} or 10^{-6} Torr, the microscope or vacuum evaporator is ready for use, unless it is an electron microscope equipped with a LaB_6 gun or a field emission gun. In that case, once the highest vacuum is reached with the diffusion pump (see above), then the ion getter pump is attached to the gun area through another valve. A differential aperture allows the upper part of the column from above the specimen to the gun area to reach very high vacuum, while the lower part (with its windows with O-rings and its camera with moving parts with O-rings, which are all sources of small, unavoidable leaks) reaches only the vacuum range provided by the diffusion pump directly below the camera and viewing chamber.

Some microscopes are equipped with two diffusion pumps, each pumping on a different part of the column. This increases the overall pumping speed of the system but probably does not affect the ultimate vacuum in most cases.

If a microscope or vacuum evaporator is equipped with a cold trap or cryopump, it should be filled *after* the chamber has been brought below 10^{-3} Torr.

IV. LUBRICATION OF VACUUM SEALS AND LEAK DETECTION

Leaks may be caused by lint, hairs, dust, or small cuts in O-ring seals. Every time a microscope column is brought to atmospheric pressure to change a filament or a camera is vented to change film, there is a possibility that repumping the system will draw dirt from around O-rings under the seal area. The glass viewing ports on a column usually sit on O-rings, and dust is continually falling between the glass and the metal part of the column around the glass. When the system is vented, the vacuum pulling the glass tightly onto the O-ring relaxes and the O-ring rounds up slightly, allowing dust to get under the edge of the ring. When the chamber is pumped down again, the dust may be drawn completely under the O-ring. Repeated cycling as one changes several cameras will probably draw dust under the O-ring and into the camera chamber. You will note that the screen develops lint and dust over time from this process. It is thus important to keep the area around windows and column joints as free of dust as possible.

Moving parts that extend into the vacuum system should be lubricated with silicone-free greases such as Fomblin®. The object is to let the O-ring around the part move freely from lubrication. The grease is not meant to provide a seal, so use it sparingly. Any column component that merely sits on an O-ring and never moves laterally does not need to be greased. If too much grease is used, it will eventually become a contaminant within the column.

Leak detection becomes an issue when the vacuum system takes longer to pump down than usual or does not reach the expected ultimate vacuum. As O'Hanlon (1980) points out in his section on leak detection, keeping a good log for an instrument, including information on normal pumping speed and ultimate vacuum, is a good practice for becoming aware of leaks.

A common confounding factor is wet film. If film is not thoroughly dessicated, vacuum systems can have great difficulties. Microscopes equipped with ion getter pumps are particularly sensitive to this phenomenon. If the film is not held in a separate dessication chamber provided with a Petri dish of phosphorous pentoxide to remove minute amounts of water vapor, a new load of film may make it impossible to use the microscope for hours.

A common method of leak detection is to isolate an area of the microscope through use of the valves and to read the appropriate vacuum sensors to determine if the isolated area is the source of leakage. Alternatively, gaseous freon can be sprayed from a "duster" can at various seals while observing the vacuum gauge to see if it moves. Some workers (Stuart, 1983) recommend using a syringe to apply minute amounts of methanol to suspect seals while observing the vacuum gauge for perturbation, but methanol can actually dissolve paint and plastics in some cases and might be injurious to certain microscope components. Sensitive helium leak detector devices are commercially available but are rarely found in electron microscopy laboratories.

REFERENCES

O'Hanlon, J.F. 1980. *A user's guide to vacuum technology.* John Wiley and Sons, New York.

Stuart, R.V. 1983. *Vacuum technology, thin films, and sputtering. An introduction.* Academic Press, Orlando.

Staining Methods for Semithins and Ultrathins

I. SEMITHIN SECTION STAINING

When we speak of staining semithins, we are generally referring to 0.25- to 0.5-μm-thick sections cut from blocks of epoxide-embedded tissues. There are other resins, such as the acrylic resins (Lowicryls, LR White, and LR Gold), in common use that will have different staining responses from epoxides because of their partial miscibility with water. In addition, semithin frozen sections have still other staining characteristics resulting from their significant hydrophilicity. For purposes of this chapter, however, we will limit our discussion almost exclusively to the epoxide sections, since these are the most commonly used resins for biological work. Any of the stains for these resins will work with much-reduced staining time on more water-miscible sections.

Ideally, staining will produce sections with enough contrast for easy visualization and, better yet, some sort of differential contrast for various cellular components. Histological sections that have been deparaffinized so that only naked tissue remains on the slide accept a variety of polychrome stains, which accomplish both these tasks admirably. Unfortunately, because of the presence of osmium, the heavily cross-linked nature of the epoxide resins, and the hydrophobicity of the resins, it is usually extremely difficult to get reproducible results from the numerous staining regimens used for paraffin sections if applied to epoxy-embedded materials. Lists of various stains with references to the original works may be found in Hayat (1975) and Lewis and Knight (1977). Microwave staining (see Section II.D. below) has been used to achieve shorter staining times but generally does not increase the list of stains that work well on epoxide sections.

Various solvents for epoxides have been used to partially dissolve the resins so that stains could more easily access the tissues, but on a day-to-day basis this procedure generally is more trouble than it is worth. Semithin sections are typically just a step along the path to the finished product, which is photographs of ultrathin sections. Most of these sections are used to determine where to trim block faces so that ultrathin sections will contain materials of interest, so spending a great deal of time producing semithins of publishable quality is usually counterproductive.

A. Toluidine Blue-O

Mixing 0.25 g of toluidine blue-O in a 1% solution of sodium borate will produce a staining solution capable of staining tissues embedded in epoxide resins. This is probably the most commonly used semithin section stain. Various elements of tissues and cells will stain differentially, yielding sections with a variety of shades of blue. Occasionally, lipids may take on a slightly greenish hue, and other areas may exhibit a bit of metachromasia, appearing reddish. These effects can be very pleasing and can help to differentiate between cellular structures, but colors other than basic blue are generally not consistently reproducible.

The standard procedure is to cut semithin sections and to transfer three or four of them with a 2- to 3-mm nichrome wire loop attached to an applicator stick to a 1-cm-diameter drop of distilled water on a glass microscope slide. This slide is placed on a hot plate adjusted to 70–90°C and left until the drop of water has dried completely, allowing the sections to anneal to the glass slide. Next, a drop of stain is applied to the sections, and the stain drop is observed until the edges become gold in color (about 15–30 sec on the hot plate). Then it is important to rinse the stain droplet off into a waste container with a stream of distilled water from a squeeze bottle quickly and completely. It is usually best to direct the stream of water to an area on the slide just above the sections, rather than right at them, to diminish the chance of washing them off of the slide. If the stain dries on the sections, it will produce masses of crystallized stain, rendering the slide unusable (Fig. 103). Once the stain has been washed from the sections, the slide can be reheated for drying, and then it should be removed from the hot plate.

The dried sections may now be observed with the light microscope. Unless a coverslip is used, however, good resolution is not possible, since most light micro-scope lenses are designed to be used with #1.5 coverslips (0.17 mm thick, as marked on most objectives). In addition, any minor section defects (small wrinkles, small knife marks) will be clearly visible, which is not the case after a mounting medium and coverslip are added.

It is important to remove all the water if you intend to use a coverslip on the sections, since the common permanent mounting media are water insoluble (and can be made less viscous with xylene). A satisfactory temporary mounting medium for ob-serving semithin sections is either a drop of water or a drop of oil used for oil-immersion lenses.

If the permanent medium Permount® is used, make sure that the area containing the sections is circled with a marking pen on the bottom of the slide, because the stain usually fades over time and the mounting medium yellows. By marking the slides, the sections can be located again fairly easily with a phase-contrast microscope if the stain

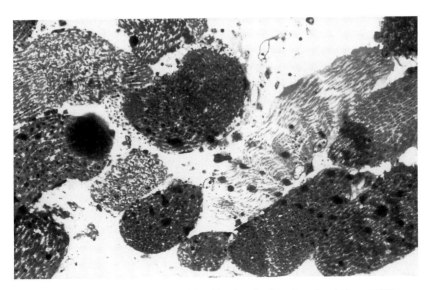

FIGURE 103. Stain dirt on a toluidine blue-O stained section of rat kidney, ×360.

does fade. An alternative is to use Polysciences Polymount®, which handles just like Permount, except that it does not cause the stain to fade and does not yellow with time (Polymount has been used in our laboratory for 7 years, and none of our archived semithin section slides has faded or turned yellow).

If numerous tiny wrinkles (Fig. 104) appear, turning up the hot plate temperature seems to help. If the section dimensions are significantly larger than 1.5 mm^3, large wrinkles are likely to be a problem, and the stain tends to pool underneath the wrinkles, resisting removal by rinsing and making these preparations unattractive. If sections are thicker than recommended or are overly large, they are extremely likely to be washed off the slides during rinsing.

If sections refuse to stick to slides, try washing the slides before use, but this is now an uncommon problem (years ago, the slides had more oils on their surfaces from manufacturing). Another method to improve section adherence to glass slides involves dipping washed, distilled-water-rinsed slides in subbing solution (5 g gelatine dissolved in 250 ml of 100°C water, which is cooled before adding 0.5 ml of a 10% solution of chromium potassium sulfate). The dipping solution does not keep, and the slides must be thoroughly dried before use. The disadvantage of this technique is that the gelatin will become stained along with the tissue.

B. Toluidine Blue-O and Acid Fuchsin

Utilization of these two stains can result in polychromasia (Hayat, 1975; Hoffman et al., 1983). Nuclei stain dark purple-blue, keratinized tissues stain red, collagen

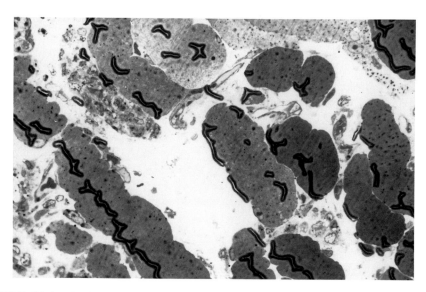

FIGURE 104. Small wrinkles on a toluidine blue-O stained section resulting from insufficient heat when sections were dried onto the slide, ×360.

stains red, and other tissue components stain bluish. Three stocks are necessary: (1) 1% toluidine blue-O in 1% sodium borate (1 g of each in 100 ml of distilled water), (2) 1% sodium borate solution in distilled water, and (3) 0.10 g basic fuchsin dissolved in 100 ml of distilled water. Heat the water to 100°C, mix in the stain, and then filter. This solution keeps for about 1 year at room temperature.

Heat-fixed sections are stained with the toluidine blue as described above and then quickly rinsed. Just prior to use, mix equal parts of basic fuchsin stock and sodium borate stock in a test tube. Put a drop of the mixture on the sections and heat them briefly, then rinse with a jet of distilled water and dry.

As with so many of the polychrome section stains designed for epoxide sections, the results are not consistent (at least in our laboratory), and so this technique is recommended only when a photographable section is needed and time is available to repeat the procedure until the stain produces the desired result. When staining produces poor results, the sections look very much like toluidine blue-stained sections.

C. Basic Fuchsin/Methylene Blue

This technique, as described by Sato and Shamoto (1973), should produce poly-chromatic sections, but unfortunately the results are often erratic. Sections are stained on a heated slide for about 4–5 sec (the staining procedure is essentially the same as for toluidine blue-O). The staining solution contains the following ingredients:

0.5 g monobasic sodium phosphate
0.25 g basic fuchsin
0.2 g methylene blue
15.0 ml 0.5% boric acid
70 ml distilled water
10 ml 0.72% NaOH

The pH should be adjusted to 6.8–7.5. If the pH is above 8.0, sections become excessively blue. Once made up, the solution is stable for several months, though it should be shaken before use. When the stain works properly, it colors mitochondria red; erythrocytes, glomeruli, and kidney tubules pink; collagen brilliant pink; elastic lamina and zymogen granules reddish purple; and nuclei bluish purple.

D. Methylene Blue

Some workers prefer a solution of 0.5–1.0% methylene blue in 1% sodium borate to the toluidine blue-O stain described above. The procedures for use are identical, but the stained sections have a more delicate blue than found with toluidine blue.

E. Periodic Acid/Schiff's Reagent

This staining procedure that has been described by Pool (1973) is good at differentiating polysaccharide-rich areas in epoxide sections. However, this stain cannot be used to definitively locate polysaccharides because it can also stain glutaraldehyde molecules and often stains the same areas whether or not periodic acid is used to open up hydroxyl groups. This stain often produces somewhat inconsistent results, but it can give outstanding results when it works properly.

II. ULTRATHIN SECTION STAINING

A. Purpose

The objective in staining ultrathin sections is to render materials with little inherent contrast visible by causing the materials to scatter the electron beam. It is also useful to be able to selectively stain various cellular components. Unfortunately, the commonly used post-stains (lead and uranyl stains) are only semiselective. Even this partial selectivity depends on a number of factors, such as pH, length of staining time, vehicle, previous chemical exposures during fixation, and the specific formulation used. Typically, the longer the staining time with a given stain, the more elements within the tissue that become stained.

Various cytochemical procedures (see Chapter 13) have been developed to specifically identify various cellular components, but this section will be devoted primarily to the process of post-staining to develop general contrast.

Embedding media and biological materials are chemically similar, containing large amounts of carbon, hydrogen, oxygen, and other fairly electron-transparent atoms, and few atoms of large atomic weight. Since major contrast differences between the specimen and the embedding medium do not exist, other materials must be introduced to help develop contrast. The most effective stains are those containing heavy metals, such as lead, uranium, and osmium, which have sufficient mass to block or scatter electrons that strike the specimen. The larger the mass of the stain molecules, the better they will block the beam, but the size of the heavy metal ions deposited during staining can also affect resolution in certain instances. Some workers also suggest that staining helps to stabilize certain specimen components during bombardment by the electron beam.

The effects of stain interactions within sections have been studied, and it has been noted that dye-stacking effects can actually interfere with resolution, in some cases increasing the size of specific biological structures (see Hayat, 1989, for a discussion of this issue). It has been reported that poliovirus combines with about its own weight of stain before adequate staining is achieved.

B. *En Bloc* versus Post-Staining

An *en bloc* stain is used on wet tissue, whereas post-staining is done on sections of tissue after embedding is completed. Even though osmium tetroxide is usually considered a lipid fixative (post-fixative), it can also be thought of as an *en bloc* stain, since wherever it becomes reduced and fixed into the sample it confers an electron density because of the high atomic weight (190) of osmium atoms (atomic #76).

Another commonly used *en bloc* stain is uranyl acetate. The typical procedure is to rinse samples thoroughly with distilled water after osmication (since the phosphate and cacodylate buffers usually used with osmium react strongly with uranyl acetate) and to place them in 0.5% aqueous uranyl acetate at 4°C overnight. They are then rinsed with distilled water, dehydrated, and embedded. The disadvantages to this technique are that the fixation schedule becomes lengthened, and the effect of the staining is less than found when it is used as a post-stain. Hayat (1989) routinely uses 2% uranyl acetate in the first alchohol or acetone step in his dehydration series for approximately 15 min, which eliminates one of the objections to *en bloc* staining with uranyl acetate.

Post-staining of ultrathin sections has several features to recommend it: (1) It is a simple procedure, (2) staining is more uniform than with *en bloc* procedures because access to the sample is easier, (3) there is less chance of tissue damage than with *en bloc* procedures because structures are stabilized with the resin matrix, (4) stains will not be extracted during subsequent procession steps (*en bloc* stains are subsequently subjected to dehydration agents and the extractive capabilities of the liquid resins themselves), (5) contrast builds faster and generally exceeds that produced by *en bloc* staining, and (6) the effect of different staining procedures can be quickly evaluated on serial sections from the same block, rather than running up numerous samples separately.

C. Commonly Used Post-Stains

1. Uranyl Stains

Uranium atoms with an atomic number of 92 (atomic weight = 238) are the heaviest stain molecules used. Uranyl stains may be used *en bloc* (where they can also be considered to have a fixative effect), as post-stains, or as negative stains (see section on negative staining). Uranyl stains interact with anionic compounds and are known to bind strongly to phosphate groups associated with nucleic acids and phospholipids. They also react with carboxyl groups so that various amino acids of proteins can become stained. Uranyl stains are usually used at pH 3.5–4 and strongly stain proteins. If the pH is adjusted to higher levels, DNA becomes more strongly stained.

One of the most important features of uranyl stains is that they serve as mordants for lead stains. This means that subsequent staining procedures involving lead will interact much more strongly with tissues that have had uranyl exposure. Thus, it is important to perform uranyl and lead staining procedures in the proper sequence to achieve maximum staining.

Uranyl stains used *en bloc* must be handled carefully because they will interact with phosphate and cacodylate buffers, coprecipitating with the buffer compounds and resulting in ineffective staining. When used as post-stains, uranyl solutions are made up in alcohol or water, so this problem does not arise. Uranium is also a radioactive element, even though some vendors sell what they call depleted, nonradioactive uranyl acetate (which appears to be an oxymoron), which should be handled carefully, with the waste viewed as toxic. The staining solutions are photolabile and should be stored in the dark between uses.

The mechanism of staining appears to be ionic, though it is not clearly defined in all cases. Binding of uranyl ions can be reduced as much as 75% in phosphate-depleted cells (Van Stevenick and Booij, 1964).

Uranyl ions form simple salts with phosphate groups of DNA and thus reduce the net charge of DNA molecules, thereby allowing them to clump together to some extent. Bacterial DNA is stabilized, and extraction of eukaryotic DNA from its histone shell is prevented by uranyl acetate treatment (Stoeckenius, 1960). On the other hand, RNA is more reactive with lead stains. Huxley and Zubay (1961) reported that sufficient uranyl ions complex with DNA to almost double its weight.

Uranyl stains are known to reduce the solubility of phospholipids when used *en bloc*. Uranyl ions have also been shown to interact with phosphate groups in lecithin monolayers and to react with phosphate groups in both saturated and unsaturated lecithins, but not with the fatty-acid side chains (Shah, 1969).

Proteins such as histones, ribonucleoproteins, and phosphoproteins stain, with the stain density apparently related to the amount of charge on the protein molecule. The collagen protein stains strongly, and the protein moieties in membranes seem to be largely responsible for cytomembrane staining.

As with all staining regimens, prior treatment of the tissue with other chemical agents can affect staining. In particular, if long osmium treatment is used, uranyl acetate staining is reduced.

There are a variety of uranyl stains utilized for post-staining, such as uranyl acetate, uranyl formate, uranyl magnesium acetate, and uranyl nitrate (see Hayat, 1989 for further discussion). The most commonly used formulations are either aqueous or alchoholic uranyl acetate. Powdered uranyl acetate can be purchased from all the electron microscopy supply houses. All resins, except Spurr resin and a few of the replacement resins for Epon 812, stain readily with a 3–4% aqueous solution produced by adding the powdered uranyl acetate to distilled water in a lightproof container, sonicating it briefly, and letting it sit overnight before use. If Spurr resin is used, either methanolic or ethanolic uranyl acetate must be used to get effective staining. A 4–5% solution is made up with 50–100% alcohol. With all three of the solvents used, a saturated solution is generally the objective, and solubility increases through the series from water to ethanol to methanol. Since the solution is saturated, it is important not to pick up drops of the stain from the bottom of the container, or masses of large, elongate crystals characteristic of uranyl acetate will be deposited on the sections being stained (Fig. 105). Any time elongate crystals are seen on sections, the uranyl acetate staining step should be evaluated.

Since uranyl acetate solutions are photosensitive, some workers stain materials in the dark, but this is not necessary. Typically, a drop of uranyl acetate solution is put onto a fresh square of Parafilm® placed on the bottom of a disposable Petri dish. The

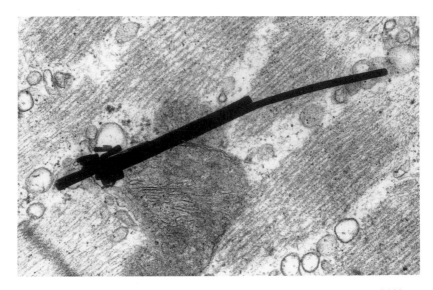

FIGURE 105. Needle-like crystals of uranyl acetate on muscle tissue section, ×47,320.

grid is touched section side down to the drop and then quickly placed at the bottom of the drop, section side up. After 5–30 min (alcoholic stains for 5 min, aqueous stains for longer), the grid is quickly removed from the drop of stain and immersed in a series of three beakers of distilled water (about six dips per vessel), after which the grid is blotted completely dry on Whatman #1 filter paper before proceeding. It is important to remove all the uranyl acetate stain, because the solution is acidic and will coprecipitate with the typically basic lead secondary post-stains.

2. Lead Stains

Lead has an atomic number of 82 (atomic weight = 207) and is used as a post-stain because it is a very effective electron scatterer. It has also occasionally been used *en bloc* (Pisam *et al.*, 1987). Lead salts produced in post-stained sections increase contrast more intensely in the presence of reduced osmium bound to tissue than do uranyl salts. As mentioned in the previous section, uranyl acetate post-staining serves as a mordant for lead stains, resulting in much more intense staining (Fig. 106).

Lead precipitates form quite readily because of the high reactivity of lead with carbon dioxide and oxygen in the air to form lead carbonate ($PbCO_3$), which may appear as finely granular "pepper" precipitates. It may also present as larger granular aggregates (Fig. 107) or, in some cases, as large globose deposits (Fig. 108). Many workers go to great lengths to exclude air from staining dishes or to surround stain drops with carbon dioxide scavengers such as KOH. Lead forms crystals with most anions, and these crystals are highly insoluble because they are hydrophobic. Any lead solution with evidence of precipitate should be discarded because of the likelihood of producing large amounts of stain dirt on sections. Lead staining solutions and waste should be handled carefully because of their well-known toxicity.

Lead staining is influenced by the fixatives used in tissue preparation and the individual cellular constituents being examined. Nucleoli and ribosomes are stained more by lead in formalin-fixed tissue than in tissues fixed with osmium alone, while glycogen stains about the same in both instances.

Lead stains are more effective and more stable at high pH (the commonly used Reynolds' lead citrate has a pH of 12–13). As with most post-stains, prolonged staining results in less specific staining. All of the alkaline lead stains react with similar cellular components, differing only in the speed of staining, ease of formulation, and storage characteristics.

The detailed mechanisms of staining particular cellular components are not clearly understood. Membrane staining is thought to result from lead interacting with the previously bound acidic osmium molecules, which have affinity for positive dye ions such as lead. Glycogen is stained by the attachment of lead to the hydroxyl groups of carbohydrates by chelation, and then additional lead accumulates around the primarily attached lead. Proteins with large numbers of sulfhydryl groups stain readily, as do other proteins containing amino acids with negative charges. Nucleic acids are stained when lead complexes with negatively charged phosphate groups. RNA, in particular, has a high affinity for lead.

A number of lead formulations have been tried over the years. Watson (1958)

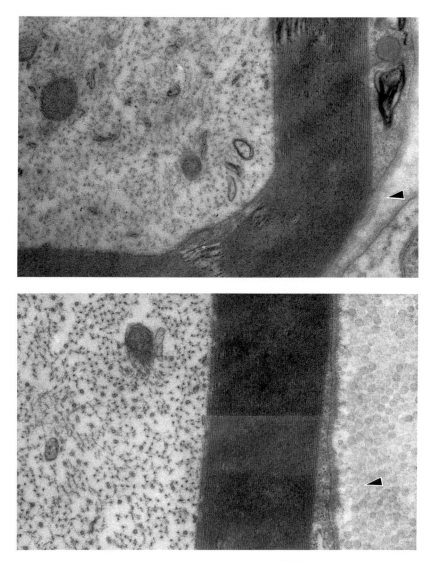

FIGURE 106. Top: Axon of peripheral nerve with no post-stain. Most density due to osmium bound to lipids. Collagen (arrowhead) is unstained, ×34,320. Bottom: Axon of peripheral nerve stained with methanolic uranyl acetate. Microtubules and microfilaments and myelin of axon have more contrast than those seen in above axon. Collagen (arrowhead) has also developed a bit more contrast, ×34,320.

worked with lead hydroxide, which probably is the best stain, but it is tedious to formulate, forms lead carbonate upon reacting with air extremely readily, and seems to contain higher levels of precipitate (probably because of the highly saturated nature of the solution). Lead acetate was introduced by Dalton and Ziegel (1960) with the intention of reducing lead carbonate formation. Lead tartrate (Millonig, 1961) has been

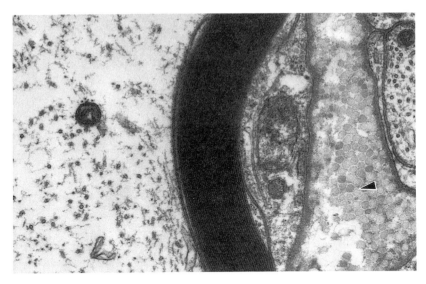

FIGURE 106 (*cont.*). Axon of peripheral nerve post-stained with methanolic uranyl acetate and Reynolds' lead citrate. All cytomembranes, microtubules, and microfilaments have good contrast, and collagen (arrowhead) is clearly visible, ×34,320.

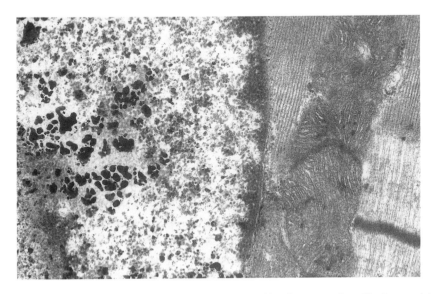

FIGURE 107. Granular lead deposits on sections of muscle resulting from contaminated lead post-staining solution, ×29,120.

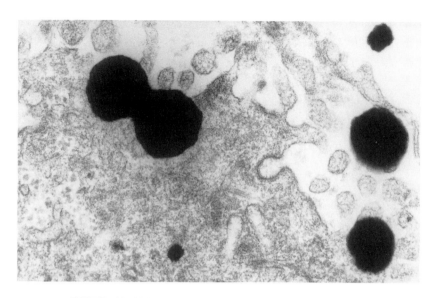

FIGURE 108. Globose lead stain deposits on tissue section, ×39,652.

suggested as a stain less likely to produce lead carbonate deposits because of the chelation of lead by tartrate. Reynolds (1963) introduced the currently most widely used lead stain, lead citrate. In his formulation, lead is strongly chelated with citrate, reducing its tendency to form lead carbonate upon interaction with air while still producing better staining than the lead tartrate formulation introduced earlier. The basic mechanism of action for this formulation is that anionic staining sites have a greater affinity for lead cations than does the citrate portion of the solution, while, at the same time, CO_2 and oxygen have less affinity for the lead.

Formulation of Reynolds' lead citrate is fairly foolproof. Begin by putting 30 ml of distilled water into a scrupulously cleaned volumetric flask, preferably with a snap cap (so that the chemicals added do not interfere with capping the vessel if they stick to the sides of the flask neck). Add 1.33 g lead nitrate and then 1.76 g sodium citrate to the water. Cap the flask and shake for 1 min, followed by another 29 min of intermittent shaking; then add 8 ml of 1.0 N NaOH. Bring the volume to 50 ml with distilled water after the solution has cleared. The final pH will be over 12, and the solution must be totally clear. If it has any turbidity, it cannot be used because it will contaminate sections. The solution can be stored for 2–3 months at 4°C.

Dispose of all lead waste in dedicated containers that can be picked up by toxic-waste disposal personnel. Remember that the solutions are toxic, dried droplets of stain are toxic if inhaled, and lead should not be discarded down drains in any more than trace amounts.

As mentioned previously, uranyl acetate serves as a mordant for lead, so lead post-staining should always follow uranyl acetate treatment (*en bloc* or as a post-stain). The staining procedure for lead used in our lab is quite simple. Place a fresh sheet of Parafilm in the bottom of a disposable Petri dish and quickly place drops of lead citrate onto the film. Place the cover on the dish. When adding grids to the drops, do so by only slightly lifting the dish lid to minimize exposure of the drops to air. Touch the section side of grids to the droplet, quickly turn the grids over and place them section side up at the bottom of the lead citrate drop. Mollenhauer and Morré (1978) suggested that grids floated on a droplet of lead stain will become more contaminated from interactions with CO_2 than those buried beneath the drop. After 8 min, barely lift the plate lid, remove a grid from a lead drop, and quickly immerse it in a beaker of fresh distilled water. Dip it 6–10 times, and repeat in two more beakers of fresh distilled water. Blot the grid completely dry before placing it on a completely dry area of a piece of filter paper in a Petri dish. If a grid is placed in a damp spot, it will usually be contaminated with precipitate.

Double lead staining has been recommended for tissues that do not stain well with conventional procedures (Daddow, 1986). The procedure is to stain sections in Reynolds' lead citrate for 30 sec, rinse in distilled water, stain with uranyl acetate for 1 min, rinse again, and restain with Reynolds' lead citrate.

3. Phosphotungstic Acid

Phosphotungstic acid (PTA) was first used to impart electron density to tissues by Hall *et al.* (1945). It has been used as an *en bloc* stain, but is most commonly employed as a negative stain for bacteria, viruses, and subcellular particle suspensions (see section on negative staining). Negatively charged molecules are the principal sites of PTA binding. Complex polysaccharides with negative charges are stained well. The mechanisms of staining for various cellular components is still uncertain. *En bloc* staining with PTA has been said to result in blocks that are more difficult to section, and much of the bound PTA can be removed by the solvent activity of epoxide resins. Since most conventional fixation methods involve epoxide embedment, the use of PTA as a general *en bloc* stain is not commonly seen. Hayat (1989) provides an extensive discussion of PTA staining of tissues for the interested reader.

4. PTA/Chromic Acid

Roland *et al.* (1972) introduced a novel post-stain composed of PTA and chromic acid that preferentially stained the plasma membranes of plant cells while leaving the other cytomembranes unstained. They did not know the mechanism, nor why it worked with tissues embedded in most epoxide resins, but not Spurr resin. It has been used successfully to analyze populations of isolated cytomembranes to determine the percentage of plasma membranes in their samples.

5. Barium Permanganate

Permanganates have been used as general post-stains in the past, but because of frequent precipitate formation on sections, they are not currently used much. However, Hohl *et al.* (1968) used barium permanganate as a specific stain for cellulose fibrils in the cellular slime mold, *Acytostelium*. A solution of 0.5–1% in distilled water effectively increases the electron density of cellulose if sections are stained for 1–2 min after lead staining (rinse with distilled water between the stains). Stain precipitation can still be a problem, so excessive staining is to be avoided and careful washing is mandatory.

D. Microwave Staining

During the last decade, microwaves have come into use for staining histological sections for light microscopy, particularly with those stains involving silver compounds (Brinn, 1983). Various workers have also utilized microwave staining for ultrathin sections of epoxide-embedded materials, particularly in the setting of human pathology laboratories, where diagnostic materials need to be processed quickly. Estrada *et al.* (1985) described a procedure utilizing a 400-W, 2,450-MHz microwave with a rotating platform to accelerate both uranyl acetate and lead staining. They illustrated their work with photographs of human heart biopsy material fixed with glutaraldehyde and osmium, dehydrated, and embedded in Epon. They cut ultrathin sections, placed them on grids, and then immersed the grids in 3–4 ml of the staining solutions for the microwave step. The two stains used were 4% aqueous uranyl acetate and Reynolds' (1963) lead citrate. If they microwaved the grids for 15 sec in each stain, they achieved excellent results. The normal procedure they had previously used was 30 min in the uranyl acetate solution followed by 10 min in lead citrate. The materials so stained actually appeared slightly overstained in their illustrations, but their main point was that they could cut about 39 min from their procedures for getting sections viewed to make diagnoses. They also mentioned that they used the technique successfully on kidney, liver, nasal brushings, skin biopsies, and tumors.

E. Dark-Field Imaging without Staining

Dark-field imaging is a method that can provide added contrast to low-contrast specimens, though with a decrease in overall image intensity. Rather than utilizing the axial and only slightly scattered part of the beam in imaging, as is done with conventional TEM (bright-field) observation, the widely scattered part of the beam is used for imaging. Since the scattered part of the beam is being used, the electrons have more varied energy levels than with bright-field viewing, leading to greater levels of chromatic aberration and, hence, lower resolution.

There are two ways to produce dark-field images in a TEM. The first and simplest is to manually displace the objective aperture so that its edge blocks part of the electron beam. The second requires a microscope with circuitry designed to perform dark-field imaging. In such a system, the beam is deflected off-axis to produce the dark-field image.

Dark-field imaging in TEM can produce increased contrast levels in unstained materials with very little inherent contrast and is particularly well suited to the examination of biological crystals (paracrystalline arrays of proteins) whose lattices will show up more clearly with dark field. Another advantage of the technique is that the crystals do not need to be stained before viewing, thus avoiding the potential of obscuring structural detail by dye stacking.

In reality, dark-field imaging is of more use to crystallographers, but we occasionally use the technique with selected biological specimens.

REFERENCES

Brinn, N.T. 1983. Rapid metallic histological staining using the microwave oven. *J. Histotechnol.* 6:125.

Daddow, L.Y.M. 1986. An abbreviated method of the double lead stain technique. *J. Submicrosc. Cytol.* 18:221.

Dalton, A.J., and Ziegel, R.F. 1960. A simple method of staining thin sections of biological material with lead hydroxide for electron microscopy. *J. Biophys. Biochem. Cytol.* 1:409.

Estrada, J.C., Brinn, N.T., and Bossen, E.H. 1985. A rapid method of staining ultra-thin sections for surgical pathology TEM with the use of the microwave oven. *Am. J. Clin. Pathol.* 83:639.

Hall, C.E., Jakus, M.A., and Schmitt, F.O. 1945. The structure of certain muscle fibrils as revealed by the use of electron stains. *J. Appl. Phys.* 16:459.

Hayat, M.A. 1975. Positive staining for electron microscopy. Van Nostrand Reinhold Co., New York.

Hayat, M.A. 1989. *Principles and techniques of electron microscopy. Biological applications,* 3rd ed. CRC Press, Boca Raton, FL.

Hoffman, E.O., Flores, T.R., Coover, J., and Garrett, H.B., II. 1983. Polychrome stains for high resolution light microscopy. *Lab. Med.* 14:779.

Hohl, H.R., Hamamoto, S.T., and Hemmes, D.E. 1968. Ultrastructural aspects of cell elongation, cellulose synthesis, and spore differentiation in *Acytostelium leptosomum*, a cellular slime mold. *Am. J. Bot.* 55:783.

Huxley, H.E., and Zubay, G. 1961. Preferential staining of nucleic acid containing structures for electron microscopy. *J. Biophys. Biochem. Cytol.* 11:273.

Lewis, P.R. and Knight, D.P. 1977. Staining methods for sectioned material. North Holland, New York.

Millonig, G. 1961. A modified procedure for lead staining of thin sections. *J. Biophys. Biochem. Cytol.* 11:736.

Mollenhauer, H.H., and Morré, D.J. 1978. Contamination of thin sections, cause and elimination. In: *Proc. 9th Int. Cong. Electron Microsc.,* Vol. II, p. 78, Microscopical Society of Canada, Toronto.

Pisam, M., Caroff, A., and Rambourg, A. 1987. Two types of chloride cells in the gill epithelium of a freshwater-adapted euryhaline fish: *Lebistes reticulatus*; their modifications during adaptation to saltwater. *Am. J. Anat.* 179:40.

Pool, C.R. 1973. Prestaining oxidation by acidified H_2O_2 for revealing Schiff-positive sites in epon-embedded sections. *Stain Tech.* 48:123.

Reynolds, E.S. 1963. The use of lead citrate at high pH as an electron-opaque stain in electron microscopy. *J. Cell Biol.* 17:208.

Roland, J.C., Lembi, C.A., and Morré, D.J. 1972. Phosphotungstic acid-chromic acid as a selective electron-dense stain for plasma membranes of plant cells. *Stain Tech.* 47:195.

Sato, T., and Shamoto, M. 1973. A simple rapid polychrome stain for epoxy-embedded tissue. *Stain Tech.* 48:223.

Shah, D.O. 1969. Interaction of uranyl ions with phopholipid and cholesterol monolayers. *J. Colloid Interf. Sci.* 29:210.

Stoeckenius, W. 1960. Osmium tetroxide fixation of lipids. In: *Proc. Europ. Conf. Electron Microsc.* Vol. 2, Nederlandse Vereniging voor Electronmicroscopie, Delft.

Van Stevenick, J. and Booij, H.L. 1964. The role of polyphosphates in the transport mechanism of glucose in yeast cells. *J. Gen. Physiol.* 48:43.

Watson, M.L. 1958. Staining of tissue sections for electron microscopy with heavy metals. II. Application of solutions containing lead and barium. *J. Biophys. Biochem. Cytol.* 4:727.

CHAPTER 7

Photography

The early history of photography and its antecedents is recounted admirably in the book *The Keepers of Light* (Crawford, 1979). This text also provides details about different types of papers and chemistry used over the last 150 years or so (including recipes and instructions in their use). Other extremely useful texts on technical aspects of photography are Adams (1948), Engel (1968), Lefkowitz (1985), Stroebel *et al.* (1986), and a variety of Kodak publications, some of which are referenced at the end of this chapter. Of particular interest are Kodak Publications L-1, L-9, and L-10, the first of which is a catalog to all of Kodak's photographic information, while the other two list professional and specifically scientific products from Kodak.

Photography is an important partner to any form of microscopy utilized in the sciences. It is rare enough to just observe something in science, but we are compelled by the nature of microscopy to record events and to share them with students and colleagues. The first recording method was to draw freehand the images observed through a light microscope, but the result was frequently inaccurate. In 1807, the physicist and chemist, William Wollaston, invented the *camera lucida*, which projected an image from an ocular lens onto a surface, where the projected image could be traced by the observer. One of the two prime developers of photography, the mathematician William Henry Fox Talbot, was frustrated in his efforts to produce accurate images with a *camera lucida* and began pursuing methods to produce a photographic image. He coated papers with sodium chloride, let them dry, and then coated them with silver nitrate, thus producing a silver chloride emulsion. He placed objects such as leaves and lace on the surface of paper and produced a negative image called a *photogenic drawing*. His writings note the production of a negative photographic image for the first time in 1834 (Crawford, 1979). In the meantime, Louis Daguerre in France was producing positive photographic images, which were first reported to the Academy of Sciences in Paris in 1839, thus starting a long-standing conflict between the British Talbot and the French Daguerre. In 1885, George Eastman introduced film with a cellulose nitrate base that was flexible and could be rolled up inside a camera for multiple photographs (Collins, 1990). At that point, the modern age of black-and-white photography had come into existence.

Electron microscopy utilizes specialized films to record images and various photographic papers upon which the images are printed. Understanding how to use these two products is enough to produce photographs that can be shared, but an important part of scientific communication is to produce photographs of proper quality for slide presentations and poster sessions, as well as for publication in journals. The objective of this chapter will be to introduce the reader to the mechanistic foundation of the photographic process and to introduce a variety of photographic materials and approaches specifically devoted to scientific illustration.

I. EMULSION COMPOSITION

A photographic emulsion consists of silver halide (usually silver bromide) crystals suspended in a matrix of gelatin attached to a substrate consisting of plastic, glass, paper, or resin-coated paper stock. Autoradiographic materials consist of a gel of silver halide dissolved in gelatin, which is then heated by the user and used to coat sections of tissue attached to either glass slides or electron microscope grids. Films often have an antihalation backing to prevent photons of light from bouncing off the camera back and being reflected back through the emulsion, which would produce "fog" (noninformational silver grains in the final image) and reduce resolution.

Quite a number of photographic materials are available for color and black-and-white work. One of the reasons that high-resolution scientific photography is done primarily with monochromatic materials is that images produced on a single layer of emulsion will be slightly sharper than equivalent images recorded on color films (which usually have three layers of emulsion stacked on top of each other). At the present time, color illustrations in publications are fairly rare because they cost more to produce and because of the possible degradation of images if the multiple color printings get out of register. In addition, it is usually not satisfactory to get prints made of photographs of histological preparations stained with various polychromatic stains, because the machines used to print color prints will try to correct any unusual contrast or strong colors (just like the ones we intentionally try to produce to differentiate between various components of tissues and cells). Even a custom hand-printed photograph can have unsatisfactory colors if the printer does not understand what the image is supposed to look like. With this in mind, our discussion will revolve almost exclusively around black-and-white materials.

Fortunately, with black-and-white materials we are only concerned with density and contrast, not actual hue. Even getting these two factors the way you want them can be a problem unless you print your own materials. The rest of this chapter will be directed toward producing your own illustrations with minimal esoteric chemicals or equipment.

II. FILM TYPES

Films for scientific use are available in various formats such as 35 mm, 120, and sheet films of various dimensions. Emulsions can be categorized on the basis of their spectral sensitivities: (1) "ordinary," which is essentially colorblind and sensitive to only blue light; (2) panchromatic, which is sensitive to essentially the same colors as the human eye; (3) orthochromatic, which is sensitive to all colors except red, and (4) color, which is sensitive to about the same spectral bands as panchromatic film and is capable of recording the individual hues themselves. Special-purpose films such as infrared films are also available. It is worth noting that most films have a sensitivity to ultraviolet radiation, which the human eye does not perceive. This explains why those crystal-clear photographs of fall foliage taken on sunny days in the mountains come back from the processor with a somewhat washed-out, lower contrast image than remembered. Ultraviolet radiation decreases the overall contrast of the image on the film compared to what the eye perceives.

Films are also rated as to their speed (ASA or EI). These numbers are a measure of the emulsion's sensitivity to light and are used to set light meters for determining proper exposure. The higher the ASA, the "faster" a film is. This means that less light is required to cause exposure, a feature generally related to the fact that faster films have larger silver halide crystals that are more likely to be hit by a photon than smaller ones. Thus, higher ASA products are associated with larger developed silver grain size and, thus, more graininess in the printed image.

The ASA for each film is determined utilizing very specific exposure, developer, and developing conditions in a laboratory setting. Exposure indexes (EI) are based on different developers and procedures from those dictated for ASA ratings. A film such as Kodak Technical Pan (Tech Pan) can be used for a variety of purposes and has several EIs, even though it can have but one ASA.

Another way to categorize films is to divide them between two major applications: pictorial work (panchromatic films) or graphic arts work (orthochromatic/process films). In reality, some films share characteristics of both groups. Electron microscope films and Tech Pan are two examples that have characteristics of both film classes. They have high contrast and red light insensitivity like graphics films but can produce continuous tone negatives (with numerous levels of gray between black and white), as is characteristic of panchromatic films. When both are developed with high-contrast panchromatic film developer (D-19), they exhibit strong contrast.

Finally, films can be described on the basis of their sensitometric curves (characteristic curves, Hurter & Driffield curves). These curves are published for each film produced (see DeCock, 1985; Kodak #F-5) and provide information that helps determine to which application the film should best be directed. Panchromatic films have gradual curves (Fig. 109), indicating that there is a fairly wide latitude for exposure and development. In practical terms, this means that satisfactory images can be produced even with slightly inaccurate exposure times, developer strengths, and development times, and with slightly uneven illumination. On the other hand, graphic arts films have steep curves (Fig. 110), indicating that precise exposure, even illumination, and

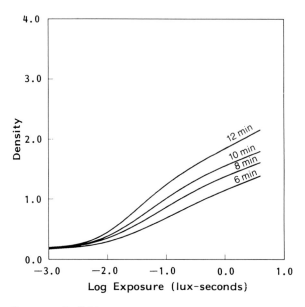

Exposure: Daylight
Process: Small tank, KODAK Developer D-76, 68°F (20°C)
Densitometry: Diffuse visual

FIGURE 109. Sensitometric curve for a panchromatic film. (Kodak T-max 100). Kodak D-76 developer, 20°C. (Reprinted courtesy of Eastman Kodak Company.)

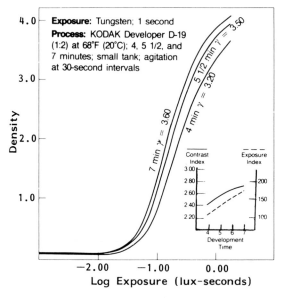

FIGURE 110. Sensitometric curve for a high-contrast film (Kodak Technical Pan). Kodak D-19 developer, 20°C. (Reprinted courtesy of Eastman Kodak Company.)

critical development procedures are necessary. Most panchromatic films must be processed in the dark because of their broad spectral sensitivity, while graphic arts films typically have orthochromatic characteristics and thus may be processed under dim red safelights. Just remember that the term *safelight* is a bit of a misnomer, since all photographic emulsions will eventually become fogged by exposure to safelights. Some films are just more insensitive than others, so they do not usually fog during normal processing times.

III. PRODUCING A LATENT IMAGE

A latent image is formed when a sensitivity speck has been converted to metallic silver. Latent images are produced when film is exposed to light, electrons, radioactive emissions, cosmic rays, certain chemicals, and mechanical pressure. Any form of energy has the potential to reduce ionic silver in the silver halide crystals to metallic silver.

A sensitivity speck is a flaw in the crystal lattice of a silver halide crystal (Fig. 111). An electron is released from bromine (in silver bromide emulsions) when the crystal is struck by a photon or other source of energy. The electron travels through the crystal lattice to the sensitivity speck, where it produces one reduced silver atom. Metallic silver at the sensitivity speck tends to draw other silver ions toward it and reduces them in turn, given enough time.

IV. FILM PROCESSING

Silver bromide will be converted to metallic silver in about 2000 years at an aperture opening of f8 in a camera, even without development (a form of the process known as *printing out*). Chemical development is used to speed up this process about 1 billion times by providing the photographic emulsion with a ready source of electrons (the developer).

Before discussing the chemical components for film processing, a warning must be made concerning the toxicity of the chemicals encountered. Film developers are basic and are thus more reactive with tissues than the acidic stop bath or fixers encountered. Severe contact dermatitis may result from repeated exposure to photographic chemicals. Some workers who have moved prints through trays of chemicals

```
                      Ag+Br—
                  Ag+Br—Ag+Br—
              Ag+Br—Ag+Br—Ag+Br—
          Ag+Br—Ag+Br—Ag+Br—Ag+Br—
      Ag+Br—Ag+Br—Ag+Br—Ag+Br—Ag+Br—
  Ag+Br—Ag+Br—Ag+Br—Ag+Br—Ag+Br—X ←
```

FIGURE 111. Silver halide crystal with a sensitivity speck (arrow).

by hand (without print tongs) for years have no apparent problems, while others may develop severe allergic reactions to the same chemicals after only a few exposures. Some individuals develop skin reactions or respiratory allergies sufficiently debilitating to prevent any further darkroom work.

A. Developer

All developers are complex solutions containing water, developing agents, preservatives, activators, and restrainers. Developer is used to provide electrons to all silver halide crystals containing latent images, catalyzing the conversion of the remainder of the crystal to metallic silver.

Water is used as a solvent for the other constituents and also causes the gelatin of the emulsion to swell, allowing the developer chemistry to reach the silver halide crystals of the emulsion more easily.

The developing agents, as mentioned, are electron donors, almost all of which are benzene derivatives. Developers will ultimately reduce *all* silver ions to metallic silver, but work preferentially on silver halide crystals containing latent images. Film development is performed under controlled time and temperature conditions to selectively develop primarily the silver halide crystals with latent images. Some of the common developing agents are Elon® (metol), hydroquinone, and phenidone. The developing agent(s) used and their concentrations are dependent on the type of emulsion being processed. Panchromatic film developers, such as Kodak D-76 and D-19, are mixtures of metol and hydroquinone. These developers exhibit a short inertial time with gradual grain development. What this means is that very shortly after a film is immersed in the developer, silver grains begin appearing from conversion of all of the exposed silver halide crystals into metallic silver. As time goes on, the crystals enlarge as the process continues. The practical side to this feature is that cutting development time (or lowering development temperature) allows most exposed crystals to begin development, even though most will be slightly underdeveloped (they will be smaller than if fully developed), thus sacrificing the ultimate "speed" of the film (sensitivity) for smaller grain size. Manipulation of development time and temperature are frequently employed to produce images with smaller grain size to improve the enlargement capabilities for a given panchromatic film.

Graphic film developers (such as Kodalith), on the other hand, are composed primarily of hydroquinone. These developers exhibit long inertial time followed by rapid development. Cutting development time will thus yield a loss of apparent film speed with no decrease in grain size or graininess. If the film is underdeveloped significantly, few silver halide crystals will be turned into reduced silver, and so little image will be produced.

Knowing the characteristics of a given film based on its sensitometric curve will give some clues as to the type of developer to use (films with steep curves usually utilize developers such as Kodalith, while panchromatic films identified by shallow curves utilize lower-activity developers) as well as the type of image produced (high contrast or continuous tone, respectively).

Preservatives are included in developers to retard oxidation of the development agent. One typical preservative is sodium sulfite (which is also used as an activator for the developer). Developers are stored in full, tightly stoppered bottles in the dark to help prevent oxidation and photodegradation. As developers become oxidized, they usually become brownish in color. Oxidized developers also develop a characteristic chemical odor somewhat reminiscent of ammonia. Any time a developer has colored noticeably or has a strong smell, it should be discarded. Paper developer is the only possible exception, since it can be tested by printing and developing one photograph, and if it does not develop properly, another can be quickly printed from the negative with new chemicals. Films, on the other hand, offer only one chance for development, and bad chemistry will result in the irretrievable loss of an image.

Restrainers serve to reduce the tendency to develop unexposed silver halide crystals, resulting in fog. Potassium bromide is a common restrainer. The development process itself produces restrainers (bromide ions). As metallic silver atoms are formed, bromide is released from the emulsion into the developer solution. If film is not agitated during processing, two things occur: (1) fresh electron donors (developing agent) do not come into contact with the emulsion, and (2) the bromide ions released from silver halide crystals in the emulsion remain on the surface of the film, thus serving as restrainers for the development process. On the other hand, excessive agitation is to be avoided. The surfaces of plastic film racks, the holes on the edges of individual metal sheet film holders, and the edges of film reels are all places of high-velocity fluid flow during agitation compared to the center of a sheet or roll of film. Excessive agitation, therefore, results in overdeveloped film edges. Such negatives will be difficult to print because the edge of the print will be light (due to excessive negative density on the edge) compared to the central area of the print. Kodak and the other film manufacturers give specific instructions for agitation that should be followed explicitly. Most films should be agitated every 30 sec during development. Reels should be rotated in both directions within the tank of chemistry and inverted once during each agitation cycle, if possible. Sheet film should be raised vertically from the tank and tipped at 45° in one direction before being returned to the chemistry. At the end of the next 30-sec period, the film should be tipped in the opposite direction. Kodak has published a short pamphlet (Kodak, #E-57) describing how to set up a nitrogen-burst system that is often used for electron microscope negative processing to make sure that even development is achieved.

B. Stop Baths

Stop bath solutions usually contain about 4% acetic acid. Kodak Indicator Stop Bath contains a color indicator that turns the fresh solution from yellow-orange to a purple hue as it gets neutralized. The purpose of stop baths is to neutralize any of the basic developer clinging to emulsions prior to exposing the film or paper to acidic fixer. Stop baths are generally needed only for paper development series. Films have plastic or glass backing materials incapable of retaining much developer chemistry, so rinsing with water between the developer and fixer steps is usually adequate (1 min). On the other hand, fiber-based papers soak up significant amounts of developer during the typical 90- to 120-sec development process. If the developer is carried into the fixer, it will quickly render the fixer inactive by raising the fixer pH. The newer resin-coated papers do not retain as much liquid as fiber-based stock as long as minimal chemistry bath times are maintained. Excessive times will allow chemicals to penetrate the cut edges of the resin-coated stock, and it will be difficult to remove the absorbed chemistry during subsequent steps.

C. Fixer

Fixer is made from a solution of sodium thiosulfate (also known as sodium hyposulfate, hence the common name *hypo*). After the silver halide crystals containing latent images are converted into metallic silver, the remaining unexposed silver halide crystals must be removed by the fixer so that future light exposure will not result in their development. Thus, the fixer solubilizes the unexposed crystals and converts them into stable salts that do not decompose during the final washing step and that can be effectively removed from contact with the film or paper. Depleted fixer contains significant quantities of silver and should not be put down the drain to enter the water supply (since silver is toxic to most organisms). There are various businesses that reclaim silver from spent fixer and that can usually be located through local photographic supply houses.

V. DEVELOPMENT CONTROLS

A. Time

With most panchromatic film developers, longer development times will yield greater negative density, greater contrast, greater fog (noninformational density), and greater granularity. If development continued long enough, the film surface would become completely covered with silver grains. Remember, development preferentially develops silver halide crystals with latent images but will ultimately develop *all* silver halide crystals.

B. Temperature

Since film development is a catalytic chemical process, it is obvious that higher temperatures will result in faster development. What is not necessarily intuitive is that more granularity and fog also are generated at higher temperatures. These two features are objectionable in scientific images, where maximum resolution and sharpness are paramount. Most films are processed at 68°C, which allows reasonable development times (4–15 min, depending on film, application, developer, and developer dilution) with production of minimal grain and fog. Films developed above about 75°C can exhibit a phenomenon known as *reticulation*, which is sometimes intentionally produced to make interesting artistic photographs with crazed images somewhat similar to old varnished surfaces. Reticulation is caused by the continuous gelatin layer developing cracks from the elevated temperature. In tropical climates without chilled water supplies, this problem can be dealt with by cooling the developer with ice prior to use. Developer temperature is the only absolutely critical temperature to maintain during processing. Other chemical and washing steps can have temperature fluctuations of several degrees without dire consequences.

C. Agitation

As mentioned in the section on developers, agitation is a critical aspect of film processing. Overagitation will result in uneven development, while underagitation can result in inadequate or uneven development because of the restraining action of bromide ions released from the emulsion.

D. Developer Choice

Section IV.A. on developers describes the general differences between orthochromatic developers for process (graphic arts) films and panchromatic film developers. Within each category, however, several choices can be made. In addition, there are films that share characteristics of both film types, which can be manipulated by developer choice to produce steep-curve, high-contrast, graphic arts types of negatives or relatively shallow-curve, panchromatic-type negatives.

Tech Pan was designed to replace a previous high-contrast, low-grain product. It has the usual panchromatic spectral sensitivity, being a broad-spectrum film with a peak in the green region, similar in response to the human eye. The film has a second sensitivity peak in the red portion of the spectrum, which means it cannot normally be used in portraiture, since Caucasian skin color appears darker than normal.

Shortly after Tech Pan appeared, pictorial photographers discovered its extremely fine grain characteristics and experimented with developers high in phenidone to produce a negative with a broader gray scale (more gradations of gray between strictly black and strictly white). Kodak eventually marketed a phenidone-based developer under the name Technidol specifically for the pictorial use of this film. During the last decade, this film has been used for continuous-tone photomicrography (EI 50, developed in Kodak HC-110 developer diluted 4:246), pictorial work (EI 100, phenidone developer), and high-contrast graphic work (EI 25, Kodak D-19 developer). This film can be manipulated more than most other films, and developer choice is a significant variable that must be considered.

VI. PAPER TYPES

Fiber-based papers are produced by various manufacturers, such as Kodak and Ilford. Each manufacturer has a different set of codes to classify their surfaces. Kodak produces a variety of products, among them F-surface papers (gloss finish), N-surface papers (flat finish), papers with pearl or textured surfaces, single-weight and double-weight papers, papers with blue-black blacks (Kodabromide), and papers with deep sepia tones (Ektalure). Some are even designed to make black-and-white prints from color negatives (Panalure). Fiber-based papers are tray processed and must be handled carefully to make sure that undue amounts of the absorbed chemistry are not carried from the developer into the fixer, resulting in inadequate fixation. Prints must then be washed extensively (about 1 hr with tumbling in running water) or treated with a hypo eliminator that will reduce fixer to sodium sulfate, which is harmless to the silver image. Hypo eliminators (Permawash® or Kodak Hypo Clearing Agent) can reduce washing times considerably and simultaneously improve image permanence. Products such as Edwal Hypochek can be used to check fixer. If the fixer is spent (contains excessive amounts of complexed silver), this agent will produce a precipitate (silver nitrate) that is easily seen.

Resin-coated papers appeared in the 1970s and offered a paper stock that absorbed minimal chemistry, thus allowing relatively shorter processing times in trays. In addition, an activator was incorporated into the emulsion of certain types of papers (e.g., Kodak RC III, Ilford Ilfospeed), allowing them to be quickly processed in machines found in major photofinishing shops and some electron microscopy laboratories (Kodak Royalprint processor; Kodak Dektomatic Processor; Ilford 2050 Processor; Agfa DD5400 processor). These machines should not be confused with processors such as the Kodak Ektamatic, which yields damp, unfixed prints from paper that is dedicated to the machine. The new resin-coated stocks with activator emerge from processors in about 90 sec as dried, fixed prints. The ability to control print density by altering developing time, as can be done with paper-based materials in trays, is no longer available. Of course, prints should actually be developed to completion (unlike the situation for films), so print density should ideally be determined strictly by exposure, not development.

Since the popular market that drives the photographic products industry is based on photofinishing shops dedicated to rapid production techniques, resin-coated papers are most readily available today. The surfaces marketed are not as varied as with fiber-based products, and the long-term archival properties of resin-coated stocks are not as well known (after all, we have fiber-based prints dating from the 1800s).

Both resin-coated and fiber-based papers are available as single-grade papers, each with different contrast characteristics. The Kodak line has five levels of contrast available, from #1 (lowest contrast) to #5 (highest contrast). Each has a slightly different sensitometric curve. When processed in trays, the lowest grade begins showing an image shortly after immersion in the developer and develops continuously until completed at about 2 min. On the other hand, the highest grade (#5) has a long inertial time followed by rapid image development during the last 30 sec or so.

Both major paper types are also available as Polycontrast (Kodak) or Multigrade (Ilford) materials. These papers can produce various degrees of contrast, depending on the colored filter used during printing. Five or six major grades of contrast are available with half-steps between them. The original variable-contrast products did not have quite the contrast range available from the series of single-grade papers, but the current versions of both are virtually identical. The only limitation is that fewer surface types are available in polycontrast papers.

If archival-quality prints with good storage characteristics are desired, prints should be made on fiber-based stock followed by development, stopping, and careful fixation (two baths). The prints then should be washed thoroughly, subjected to a fixer-killing chemical (Kodak Hypo Clearing Agent, Permawash®), and then rinsed again. Finally, increased permanence can be gained by toning the prints in a selenium or gold agent, which will prevent oxidation (tarnishing) of the silver image. In most cases, scientific images do not require this much effort, since they will be used in a publication long before archival quality will be a factor. If, on the other hand, a long-term display of micrographs under continuous institutional lighting is desired, it might be wise to use methods for archival-quality prints.

VII. KEEPING PROPERTIES OF CHEMICALS AND PRECAUTIONS

As mentioned earlier, photographic chemicals, particularly developers, are capable of producing contact dermatitis in some individuals. The liquid solutions should not be allowed to come into extensive contact with the skin. If contact does occur, quickly rinse the exposed surface with water. In addition, the powdered chemicals should be handled carefully when mixing up photographic solutions. Minimize exposure to the chemical dust by pouring the powders slowly into mixing containers. Read all packages for precautionary notes and also for specific mixing instructions, since various chemicals are to be dissolved at different temperatures.

Kodak's darkroom data guide (Kodak #R-18) has guidelines describing the shelf life of chemicals once in solution (as well as information on film types, processing, paper types, etc.). Developer chemicals are the most labile, having a tray life of only a few hours before oxidation degrades them significantly. Most of Kodak's developer solutions have a shelf life of about 2–3 months when stored in full, tightly stoppered bottles to reduce exposure to oxygen in the air. Often, the shelf life is considerably longer than is stated in Kodak literature, and developer that is several months out of date that is uncolored and nonodiferous is most likely usable, but should not be used for negative development. As mentioned, old paper developer is worth trying, since it can be discarded with only the loss of one print if it proves to be bad.

VIII. SHARPNESS

Sharpness, as defined by good contrast and good resolution, is usually the objective in scientific illustration. When processing film, diluted developer used at 68°C will typically produce the smallest grain size for a given emulsion/developer combination. The photographic guides from Kodak (Kodak #J-1, #R-18) and Morgan and Morgan (DeCock, 1985) give details concerning proper developers to be used, dilutions possible, and development times. Smaller grain allows greater resolution and greater enlargement without objectionable graininess. Overexposure in the camera and/or over-development will produce larger grain size and, hence, less resolution. In most cases, the most information can be gained from a negative that is "exposed for the shadows and developed for the highlights" (Adams, 1948). If one uses high-contrast graphic arts materials for producing reproductions of line drawings, slight underdevelopment will result in crisper lines. To minimize background, these thinner negatives (underdevelopment decreases the amount of silver left on the film) should be printed with high-contrast (F5) papers or filters.

IX. FILMS USED IN THE ELECTRON MICROSCOPY LABORATORY

As already stated, our purpose in biological electron microscopy is to record images from light and electron microscopes as we study cells and tissues. We also have to produce 2×2 slides for projection at meetings, as well as for teaching. High-quality prints suitable for viewing at a distance (posters), as well as prints for reproduction in journal articles (publication prints), are also needed. Since a publication or poster may include light micrographs, electron micrographs, line drawings, views of anatomical specimens, radiographs, and photographs of gels containing DNA or proteins, the list of photographic materials utilized in the laboratory can be fairly extensive.

A. Negative-Release Films

These films produce negative images, which must, in turn, be printed to produce a positive image. They encompass electron microscope films, Polaroid products used for SEM and copy work, sheet films for large-format copy work, 35-mm panchromatic films for photographs of experimental setups and anatomical materials, and films for photomicroscopy or for copying high-contrast materials (Tech Pan, Kodalith).

Sheet film for electron microscopy comes in a variety of sizes for the various instruments on the market. Emulsions were originally coated onto glass to produce an extremely rigid, flat surface to prevent image irregularities, which is theoretically possible if the film plane is not precisely flat. Since the projected TEM image will be in focus at a variety of levels above and below the plane of the film, an uneven recording height (slightly buckled film) could produce an image with minute variations in magnification, even though all areas would be in focus, causing inaccuracies in image recording. In practical terms this is of little concern, so most microscopists have discontinued the use of glass plates. The current products are easier to store and less fragile than the old glass plates. Films designed specifically for electron microscopy are available from Kodak and Ilford. Dupont offers a graphic arts film (Cronar COS-7 Ortho S Litho), which was at one time used for electron microscopy, largely because it was less expensive than Kodak products, but the price differential no longer exists. All of the products are high-contrast, fine-grain emulsions coated onto stiff, thick (0.007″) plastic stock. They do not have antihalation backings, as found with films for photography with light, since electrons cannot penetrate the plastic base, much less reflect from the back of the film holder back through the film. Ilford's Electron Microscope film and Kodak's 4489 film have similar characteristics, while Kodak's SO-163 film produces a higher contrast image. Kodak 4489 film is widely used and has a tough Estar base. It should be developed for 4 min at 68°C with agitation every 30 sec in D-19 (diluted 1:2), followed by washing for 1 min, 5 min in full-strength Kodak fixer, a 15-min water wash, 30-sec dip in a tank of Kodak Photo Flo, and drying. This procedure will produce high-quality negatives.

A variety of 35 mm (and 120) format films are available for photomicrography and recording of other subjects examined in the electron microscopy laboratory. T-max films introduced by Kodak over the last few years are designed for minimal graininess for their ASA when compared to the previous products (Pan-X, Plus-X, and Tri-X). Ilford produces a fine-grain Pan F film rated at ASA 50, a slightly grainier film with an ASA of 125 (FP4), and HP5, which is designed for low-light situations and has an ASA of 400. Ilford also markets a chromogenic film, XP-1, which is actually a color negative film requiring development with color negative film developer. Once the negatives are processed, however, they are printed onto black-and-white papers. The advantage to this product is that it can be exposed over a wide range of EI (it is rated at 400 ASA but can be used up to EI 1600) and produces a very fine grain image, so that it can be useful in low-light photomicrography situations as encountered with certain fluorochromes (Texas red). If low-contrast materials are to be utilized in photomicrography, Tech Pan used at EI 50 and developed in HC-110, as previously described, is to be recommended. The negatives can be enlarged up to 28× without objectionable grain (of course, they will become fuzzy because of the empty magnification this represents; see the section below about copying). The only problem with this film is that it is very unforgiving of improper exposure or uneven illumination because of its steep sensitometric curve (Fig. 109). If materials photographed have high inherent contrast, the recorded images may have too much contrast to print well.

Polaroid products, such as Type 55 (4 × 5 single-sheet film) and Type 665 ($3\frac{1}{4}$ × 4) sheet film in packs of eight sheets), are used extensively in SEM work. They offer the advantage of producing an image that can be examined 30 sec after the photograph is made (the positive), as well as a high-resolution negative that can be processed and stored for future use. Other Polaroid products are available that yield only positive images (Type 53, among others), but these do not allow further prints to be made unless the original is copied first.

Various manufacturers, such as Kodak and Ilford, produce both panchromatic and graphic arts (litho) films in 4 × 5 format and larger to produce negatives of continuous tone and line-drawing types of materials, respectively.

B. Positive-Release Films

When these films are developed, a positive image is produced that is suitable for mounting in slide binders and for projection to an audience. Some of these films are quite slow (low ASA) products and so are unsuitable for direct photomicrography and are generally used to copy materials on a well-lit copy stand.

Kodak periodically hosts a biomedical photography seminar series during which they discuss the making of slides for presentations (see also Kodak #M3–106). They describe the ideal format for title slides (Fig. 112) and suggest that all slides should be made to be projected horizontally, since that is usually the way projectionists set up screens at meetings. If slides are projected vertically, they will often go over the top or bottom of the screen. In addition, all the material on a slide should be capable of being read in less than 30 sec. Finally, the type should be large enough to be read with the naked eye without projection. We have all seen slides that do not conform to these rules and recognize that they are difficult to read and understand.

Black-and-white projection slides can be made with Ektachrome (Tungsten 64) with 3200° K lights. The resulting slides will exhibit more contrast than the original prints, but they will seem quite sharp. Unfortunately, any materials copied on a typical light box (radiographs, gels) will often take on the greenish cast characteristic of most fluorescent tubes.

Slides can also be made by photographing materials with a panchromatic film and reversing the image during subsequent processing. Kodak sells kits to do so, but the procedure is quite time consuming (almost 1 hr of work) and yields products with higher contrast than the originals. In addition, line drawings will have quite a bit of background in white areas.

Tech Pan has been used to copy electron microscope negatives on a light box, but because it increases the contrast of the original, a series of negatives made from different sections will be unlikely to have identical contrast. Any slide series made from these negatives may have widely varied contrast and density characteristics.

Polaroid makes a series of products (Polaplan for continuous-tone materials, Polagraph for line drawings, and Polachrome for color materials) that can yield positives for presentation, but the products are relatively expensive and do not seem to offer any superiority to other available products, except in those rare instances when time is absolutely at a premium.

Kodak Rapid Process Copy Film 2064 was designed to copy radiographs. It has a bluish cast (like radiographs), has fairly low resolution, but does yield slides of radiographs that look just like radiographs. It cannot be recommended for any other purposes, since electron micrographs and line drawings appear fuzzy and have excessively blue background.

In the mid-1970s, Brandons (1819 Kings Avenue, Jacksonville, FL 32207–8787) began ordering 1000-foot rolls of Kodak Direct Duplicating Microfilm 2468 (formerly designated 5468 film) after having it perforated for use in 35-mm cameras. Since this film was originally designed as a positive-release microfilm (hence, the lack of perforations), it is capable of very high resolution and is good for line-drawing work with no apparent background. Brandons began giving samples of the film to various laboratories with the suggestion that the ASA was very low and that it would probably be best to develop it in Kodak Dektol (normal paper developer). Upon further investigation in the Botany Department electron microscopy laboratory at the University of Florida, it was demonstrated that the EI for the film if developed in 1:1 Dektol at 68–70°F for 4 min was about 0.6. When originally used, the Dektol was not diluted, but sharper lines on line drawings were noted in slides made with the diluted developer. In addition, it

FIGURE 112. An ideal slide format (2×2) ($5\frac{3}{4}'' \times 8\frac{1}{2}''$) for presentations. If you can't read the slide without magnification, the lettering is too small. Prepare all slides for horizontal projection if possible. Design all slides for easy audience comprehension within 10–30 seconds.

was discovered that this film did not increase contrast over the original being copied and gave true continuous-tone reproductions of electron micrographs, in addition to the excellent copies of typed script and line drawings. The slides have a very slight sepia tone, so scientists devoted to the blue-black blacks typical of Kodabromide papers sometimes have some objections to the film. Unfortunately, we are being forced to accept more and more photographic products with sepia tones, since that is what is demanded by the popular photographic market. With an all-purpose film for making slides that copies contrast and gray scales faithfully, we can prepare a series of well-matched prints from a variety of negatives of different contrasts and then produce a series of perfectly matched 2 × 2 slides that are pleasing to our audiences.

This film must be loaded into cartridges from 100-foot rolls with bulk loaders, but remains inexpensive at under $18.00 per 100-foot roll. The rapid processing time (4 min development, 1 min wash, 4 min fix, 10 min wash, dip in photoflo, hang to dry for about 30 min) allows the user to produce mounted, dry slides in less than an hour if necessary. The cost per exposure for film is under $2\frac{1}{2}$¢ and the chemicals are of little expense, so slide making in the laboratory is not prohibitive in terms of cost or time. The breadth of material that can be reproduced encompasses all the types we normally encounter.

X. COPY WORK

An excellent manual for copy-stand work is available from Kodak (#M-1), which covers copying and duplicating in black and white as well as in color. Copy work is used to reproduce images, often those that are actually composites made from several media (e.g., gels with added lettering). When preparing figures for publication, often you will need three copies of the illustrations (one for the printer and two for the reviewers). If the paper contains a number of illustrations, particularly if they have labeled cellular structures, making multiple copies can be a lot of work. It is usually simpler to make one perfect and complete set of camera-ready illustrations (printed at exactly the size used in the journal) for the printer and then to make a copy negative of them grouped together. The copy negative can be used to make multiple copies quickly and easily for the reviewers (also, it is easier for the reviewers to handle one sheet of photographic paper with several illustrations on it rather than several loose photographs). In addition, if someone requests a preprint of the paper, further copies of the figures can be quickly printed from the copy negatives, rather than having to locate the original negatives and print them all over again, carefully trying to get them to match. Finally, relabeling all the prints is not necessary.

Copy negatives are also useful when the original illustration (a photograph of a gel) is subsequently labeled and then copies of the assembled labels and original material (the gel) are needed.

A. Films

Copy negatives of line drawings can be made with a high-contrast film such as Tech Pan in the 35-mm format, but the camera lens must be stopped down, and the material to be copied must not fill over two thirds of the viewing field. If the lens is used wide open (largest aperture) and/or the material fills the camera frame to the edges, letters in the corners will appear discontinuous, unevenly illuminated, and unsharp when printed because of spherical aberration.

These two cautions (fill only two thirds of the field and stop the lens down to at least the middle of its f-stop range) apply to any film and camera, but are particularly important in the 35-mm format, since the negatives will usually be considerably magnified.

Polaroid types 55 and 665 positive/negative film are ideal for copy-stand work because they are both high-resolution films designed for continuous-tone work. Since they are processed quickly, it is possible to know at once if the copy negative contains the desired features. If the films are used for line drawings, underexposing them a bit and then printing them on a high-contrast paper (Kodak F5 or polycontrast paper with a #5 filter) will result in crisp letters and lines.

B. Improving Copy-Stand Images

In order to produce copy negatives or 2 × 2 slides that contain all the information possible, it is imperative that exposures are accurate and reproducible, and that developing is done accurately and consistently. Several aids are available to help us achieve these ends.

1. Gray Cards

Gray cards, known as 18% gray cards, may be purchased from Kodak to allow accurate metering of light. Light meters, whether hand held or built into single-reflex cameras, are calibrated to average the viewing area and to call for an exposure that would represent an "average" mixture of black and white, with grays in between. This mixture averages out to an 18% gray, and meters are adjusted to see this gray as a midscale reading for a given ASA setting. Metering on copy stands is done by reading reflected light. Clearly, a printed page has much more white (reflectance) under a given set of floodlights than does a continuous-tone electron micrograph that consists mostly of grays and blacks, with very little white (except where holes exist in plastic sections). Since the meter actually is calibrated to see 18% gray, the electron micrograph will be metered properly, while the line drawing will tend to be underexposed to make it 18% gray. To avoid this problem, after the ASA for the film is set on the meter, a gray card is read to determine the f-stop (lens opening) and the exposure time. After this is set on the camera, any automatic metering capability on the camera should be disengaged so that the camera metering system cannot adjust to the levels of reflected light encountered (because the camera would reset its exposure parameters continuously, depending on whether line drawings or continuous-tone drawings were being metered). If this precaution is taken, the camera will be insensitive to the type of material being photographed (since the exposure was based just on the amount of light reflected off the surface of the 18% gray card) and will take consistently exposed photographs of any kind of material (line drawings or continuous-tone photographs) put onto the copy stand.

2. Gray Scale

Gray scales are another aid designed to help determine if all the potential grays available in a photographic print (about eight gradations of gray are usually possible)

FIGURE 113. The effect of aperture size (f-stop) and lens type on sharpness. Top: Portion of dollar bill photographed with 50-mm macro lens at f2.8. Bottom: Portion of dollar bill photographed with 50-mm macro lens at f16.

FIGURE 113 (*cont.*). Top: Portion of dollar bill photographed with 50-mm standard lens with +6 diopter lenses attached, f2.8. Bottom: Portion of dollar bill photographed with 50-mm standard lens with +6 diopter lenses attached, f16.

are being reproduced on a negative. Since various materials have different levels of gray, the scales (available from Kodak) have bars of different density, ranging from black, through a series of gray, to white. If a copy negative showing all the gradations of gray seen on the original scale can then be printed on paper and can reproduce the various grays seen in the test scale, the exposure and development processes for that copy have been successful.

3. Focusing Aids

These devices typically consist of targets of closely spaced lines or other devices containing lines with well-defined edges. Some of these devices are also used to determine lens quality since they can show astigmatism, sharpness, flare (loss of contrast), and spherical aberration. They can be placed alongside the print to be copied and in the same plane so that true focus can be determined by observing the edges of the target lines. Continuous-tone photographs are difficult to focus on at times because they rarely exhibit many sharp, dark lines.

4. Stopping Down for Sharpness

Photographs of high-resolution materials, such as the engraver's lines on a $1 bill (Fig. 113), will exhibit improved sharpness if the lens is closed down to somewhat smaller than midway on the f-stop scale. If the largest aperture (typically around f1.8) is used, spherical aberration may be noted; if the smallest aperture is used (typically f22), diffraction may degrade the image. Midscale is a good compromise.

5. Depth of Field as a Function of f-stop and Magnification

The smaller the f-stop, the greater the depth of field. On the other hand, magnification is inversely proportional to depth of field (the higher the magnification, the less the depth of field). Most copy-stand materials are absolutely flat, since they consist of a print underneath a glass plate. In those cases, the f-stop selected is not critical for depth of field (though it can still affect sharpness due to potential spherical aberration). However, if an illustration is being copied from a journal or book for presentation, any slight curvature may result in part of the image being out of focus compared to another part if the lens opening is large (f1.8) or the magnification factor is large. In addition, materials with three-dimensional aspects may be difficult to get in focus unless the lens is stopped down (Fig. 114).

6. Copy Lenses

Macro lenses are designed to focus on materials close to the lens and are, thus, ideal for copy-stand work. A relatively good aftermarket 50-mm macro lens for a 35-mm camera should cost somewhere between $150 and $175. A 50-mm macro lens, such as the one available from Vivitar, can focus closely enough to produce a 1:1 copy, while manufacturers such as Nikon make a lens that enlarges up to 1:2 but needs an added extension tube to produce 1:1 copies. If you might use a macro lens in a surgical or field setting, a 100-mm lens might be worth considering. They are more expensive than a 50-mm lens, but they allow you to stand back farther from the subject and to gain a bit more depth of field.

Bellows or extension tube sets can also be purchased that are placed behind the primary lens. These decrease the amount of illumination available, so through-the-lens

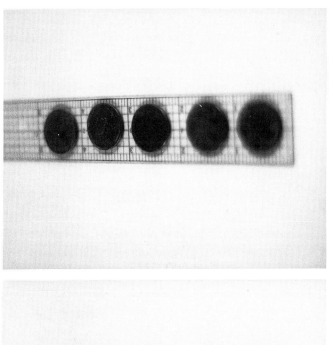

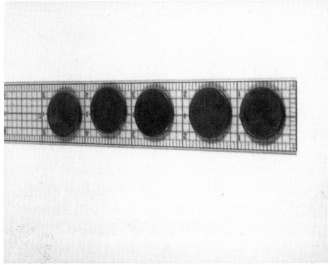

FIGURE 114. Depth of field as related to f-stop. Top: Pennies photographed at an angle with a 50-mm standard lens, f1.8. Bottom: Pennies photographed at an angle with a 50-mm standard lens, f16.

metering is necessary to avoid having to recalculate exposures based on decreased illumination.

Finally, inexpensive diopter lens in sets of +1, +2, and +3 or +6 may be purchased (for under $40). These screw onto the front of the prime lens, may be stacked up in different combinations, do not decrease illumination, but do cause a drop in contrast (caused by flare, since they are not thoroughly coated like the primary lenses) and are not as sharp as a primary lens (normal or macro). Unless photographs

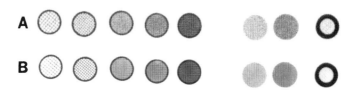

FIGURE 115. An illustration of empty magnification. A: Electron microscope grids enlarged 17.5× from size on negative (taken with 50-mm standard lens, f16). B: Electron microscope grids enlarged 3× from size on negative (taken with 50-mm macro lens, f16).

produced with diopter lenses are placed beside products of a macro lens, it is difficult to detect variations between them in most cases.

It is probably useful to mention the concept of empty magnification at this point. For example, a 100× enlargement can be produced from a negative that contains an image enlarged 50× with microscope optics that is then enlarged 2× further when printed. On the other hand, a 100× enlargement can also be produced by enlarging the same image 10× with a microscope when the original negative was produced, followed by another 10× enlargement when printed. When the two prints are put side by side, the differences in sharpness and resolution will be glaring. Even when viewed alone, it will be evident that the print enlarged 10× cannot be viewed as closely without noting lack of clarity when compared to the 2× print enlargement. The 10× enlargement suffers from empty magnification (Fig. 115). Most black-and-white films can be enlarged 3–5× without any loss in resolution or objectionable increase in grain size at any viewing distance.

XI. TYPES OF ENLARGERS

There are three types of enlargers used in photography: diffusion (cold light), condenser, and point light source. Condenser enlargers are the most commonly encountered, while diffusion enlargers are primarily used for pictorial work (particularly portraiture). Point light source enlargers are most useful in graphic arts and electron microscopy settings.

A. Diffusion Enlargers

This type of enlarger utilizes diffuse light, generally produced by a circular fluorescent tube. There is no lens before the enlarging lens, so the film is illuminated from an unfocused, scattered light characteristic of fluorescent tubes. The image produced is soft and lacks the contrast and sharpness produced by the other types of enlargers. These are obviously unsuitable for the crisp, high-resolution photographs we expect of scientific materials.

B. Condenser Enlargers

These enlargers are equipped with heavily frosted incandescent bulbs (opal bulbs) and condenser lens assemblies designed to optimally focus the beam of light on the negative being printed. Enlargers such as the Bessler 45, which can print negatives of all sizes up to and including 4 × 5, have a moveable condenser lens, enabling the illumination to be properly focused for negatives of various sizes. More condenser enlargers are manufactured and used than any other kind because they offer good illumination (in other words, reasonable exposure times), good contrast (compared to a diffusion enlarger), and moderate expense (since most of them have one condenser lens that is moved around to accommodate various negative formats). Furthermore, relatively inexpensive enlarging lenses may be used with them by stopping them down considerably (to avoid the spherical aberration characteristic of cheap lenses).

C. Point Light Source Enlargers

Of the three enlargers, this type is capable of producing prints with the most contrast and the most apparent sharpness. These enlargers are found almost exclusively in settings where these characteristics are required, so they are rarely found outside of scientific (and, more specifically, electron microscopy) darkrooms. They are typically more expensive than their condenser enlarger counterparts, probably because fewer are produced. They consist of a small, unfrosted incandescent bulb (usually 12 V) whose filament (the "point") is focused by a condenser lens assembly onto the film plane. Because stopping the lens down produces an image of the light source, the lenses must be used at their largest opening. Thus, very expensive enlarging lenses are necessary so that they are optically excellent, even at their largest aperture.

Point light source heads may be purchased for a Bessler 45, allowing easy conversion from a straight condenser enlarger to a point light source enlarger.

Many electron microscopy laboratories still have enormous Durst Laborator enlargers equipped for point light source work (even though the enlarger has been out of production for at least 5 years now). The last one purchased by our laboratory cost more than $10,000 when equipped for all possible formats up to 4 × 5 negatives. The chassis is massively built, and the head contains a front-surface mirror off of which the filament image is projected down through the condenser lens assembly. It is critical that the light source be centered, or uneven illumination will be a problem. Unlike the case for most other condenser enlargers, the Durst has individual upper and lower condenser lenses that are paired, depending on the negative format being used, the enlarging lens being used, and the level of magnification desired for the negative. At the time this enlarger was last produced, the condenser lenses used with 4 × 5 negatives cost over $500 each, but the Durst enlarger is probably the most flexible and precise manual enlarger ever produced. The Durst can also be set up for 110-V operation with an opal bulb as a straight condenser enlarger, which is useful for printing 35-mm negatives. A point light source enlarger will reveal every speck of dirt, every piece of lint, every scratch, and every grain in the emulsion with sufficient enlargement. Used as a straight condenser enlarger, prints show fewer of these common defects. With large-format negatives that typically are not enlarged as much as 35-mm negatives, these defects are generally not noticed.

D. LogEtronics Enlargers

Even though these enlargers are somewhat beyond most laboratory budgets (at around $20,000), the LogE EM-55 Electron Printing System is unique in the enlarger field and should be noted. It consists of a patented MultiDodge exposure control system to produce dodged prints automatically for negatives from 35-mm to 5 × 5. The enlarger can be operated manually or can be set in the automatic mode to read negative density and then to automatically optimize dodging (the process of exposing an unevenly illuminated negative so that an evenly exposed print results) and exposure level for the resulting print. This is accomplished by a cathode ray tube built into the enlarging head, which interacts with a small computer capable of analyzing negative density information and setting up optimal exposures for various points on the negative.

XII. VIEWING A PRINT IN PERSPECTIVE

Kodak's pamphlet #M-15 explains the concept of viewing a print in perspective. What this means is that the use to which a photograph is to be put should be determined before the print is prepared. A print to be published (and viewed from inches away) demands a degree of resolution and clarity that is considerably different from one made for a billboard intended to be viewed at 60 miles an hour from 100 ft away.

The proper viewing distance is determined by the formula:

$$D = F \times N,$$

where D = viewing distance, F = focal length, and N = enlargement. Since the normal reading distance is about 15 in (400 mm), a rearrangement of the formula can determine the optimal enlargement of a photograph taken with a standard lens on a 35-mm camera (50 mm) to be viewed in, say, a journal:

$$N = \frac{D}{F} ; \; N = \frac{400}{50} ; N = 8$$

Thus, a 35-mm negative should not be enlarged more than 8× for normal reading-distance viewing. In normal practice, as previously mentioned, negatives for publica-

TABLE 6. The Relationship between Film Size,
Lens Focal Length, and the Maximum Suggested Print Size

Film	Focal length of lens	Enlargement	Print size
35 mm			
	35 mm	11X	11 × 15″
	50 mm	8X	8 × 10″
	100 mm	4X	4 × 5.5″
4 × 5″			
	100 mm	4X	16 × 20″
	180 mm	2.2X	9 × 11″
	270 mm	1.5X	6 × 8″

Adapted from Kodak Publication #M-15, with permission.

tion are rarely enlarged more than 3–5× to prevent overemphasis of grain structure and to avoid any decrease in sharpness.

In the case of posters, it should be considered that prints will not be viewed from much closer than 6 ft (72″), so a photograph taken with a 50-mm lens (approximately 2″) could be enlarged 36× for use in a poster:

$$N = \frac{72}{2} \; ; N = 36X$$

On the other hand, if the photograph were taken with a 100-mm lens (4″), it could be enlarged only 18× for viewing at that distance:

$$N = \frac{72}{4} \; ; N = 18X$$

Table 6 shows some of the relationships between the focal length of lens using 35-mm and 4 × 5 film and the maximum advisable enlargements, along with print sizes possible.

REFERENCES

Adams, A. 1948. *The negative: Exposure and development*. New York Graphic Society, Boston.

Collins, D. 1990. *The story of Kodak*. Harry N. Abrams, New York.

Crawford, W. 1979. *The keepers of light: A history and working guide to early photographic processes*. Morgan and Morgan, New York.

DeCock, L. (ed.). 1985. *PhotoLab index. Lifetime edition*. Morgan & Morgan, Dobbs Ferry, NY.

Engel, E.E. (ed.). 1968. *Photography for the scientist*. Academic Press, New York.

Kodak Publication E-57. 1979. *Gaseous-burst agitation in processing*. Rochester, NY.

Kodak Publication F-5. 1984. *Kodak professional black-and-white films*. Rochester, NY.

Kodak Publication J-1. 1985. *Black-and-white processing using Kodak chemicals*. Rochester, NY.

Kodak Publication L-1. 1990. *Index to photographic information.* Rochester, NY.

Kodak Publication L-9. 1990. *Kodak professional photographic catalog.* Rochester, NY.

Kodak Publication L-10. 1989. *Kodak scientific imaging products.* Rochester, NY.

Kodak Publication M-1. 1984. *Copying and duplicating in black-and-white and color.* Rochester, NY.

Kodak Publication M-15. 1983. *Viewing a print in true perspective.* Rochester, NY.

Kodak Publication M3–106. 1983. *Making effective slides for lectures and teaching.* Rochester, NY.

Kodak Publication R-18. 1984. *Complete darkroom data guide.* Rochester, NY.

Lefkowitz, L. 1985. *Polaroid 35mm instant slide system. A user's manual.* Focal Press, Boston.

Stroebel, L., Compton, J., Current, I., and Zakia, R. 1986. *Photographic materials and processes.* Focal Press, Boston.

CHAPTER 8

Replicas, Shadowing, and Negative Staining

As has been mentioned previously, TEM images are primarily produced by the electron-scattering properties of a specimen. Contrast occurs when there are specimen areas that stop (scatter) electrons and also areas that let most of the electrons pass through. Thus, the image results from subtractive contrast. We impregnate tissues and sections with a variety of heavy metals to scatter electrons, as discussed in Chapter 6, but we can also surround particulates with heavy metals (negative staining) or cover particulates, cells, and tissues with thin metal films that have areas of differential beam-stopping capability (replicas produced by shadowing). This chapter will discuss these two added techniques for building subtractive contrast, pointing out the commonly used techniques and a few remedies to specific problems that may be encountered.

I. SHADOWING CASTING

Shadowing casting (shadowing) and making replicas both involve coating a specimen surface with metals. A shadowed preparation is one in which metals are evaporated at an angle to the specimen so that the metal is preferentially condensed on the high points of the sample surface toward the evaporated metal source (Fig. 116). If the specimen consists of a thin layer of particulates, such as bacterial flagella, the specimen, with its overlying shadow of metal, may be viewed *in toto*. On the other hand, if the specimen is a relatively thick sample (e.g., a layer of mammalian cells), a conventional electron microscope beam would have difficulty penetrating both the thin film of evaporated metal and the biological material beneath. In this case, after the tissue is shadowed, all biological materials are removed by digestion with acids (concentrated HCl or H_2SO_4) and/or bases (sodium hypochlorite, such as Clorox®). The remaining film is then rinsed thoroughly, picked up on a grid, and dried. The metal film is a

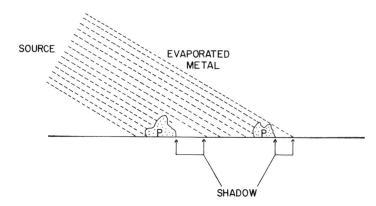

FIGURE 116. Diagram of a sample of particulates (P), showing the deposition of metal on the electrode side of raised portions of the sample, producing a shadow.

replica of the original sample surface that can be penetrated by the electron beam and viewed with the TEM.

Shadowing may be used to develop contrast in specimens of low topography (DNA, RNA, flagella, bacterial pili) or specimens of relatively high topography (freeze-fracture preparations or cellular organelles, such as whole Golgi bodies). Ultimate resolution is typically limited by the grain-structure characteristic of the metal(s) employed in the shadowing process, even though other factors are capable of decreasing resolution, as discussed below.

A. Mechanism

The principle behind shadowing is that if a metal is heated to its boiling point in a chamber held at 10^{-5} Torr or less, the individual evaporated atoms of metal will have a fairly unimpeded path through the chamber until they encounter the specimen (and other surfaces, such as the bell-jar surface and electrode assemblies). When the metal atoms strike the specimen surface, they will condense, since the specimen temperature is lower than that of the recently evaporated atoms. At a vacuum of 10^{-6} Torr, the mean free path of an evaporated metal atom is about 6.8 cm (Stuart, 1983). In basic terms, this means that an atom of metal leaving the heated electrode surface will, on the average, travel 6.8 cm before colliding with any residual gas molecules. Stuart states that "our calculations have included mere elbow brushes and bumps between molecules as collisions," meaning that the mean free path is actually considerably greater than that stated above. Electrodes are usually placed within 8–10 cm of the specimen being shadowed. Hence, the evaporated metal atoms that leave the filament in a trajectory toward the specimen (since metal atoms evaporate from all around the filament and many are not aimed in the direction of the specimen) should arrive and condense onto the specimen surface without any serious deflection from their path from gas molecules within the chamber.

TABLE 7. Metals Commonly Used for Shadowing

Element	Atomic no.	Atomic wt.	Melting point	Boiling point
Gold (Au)	79	197	1065°C	2700°C
Palladium (Pd)	46	106	1555	3167
Platinum (Pt)	78	195	1774	3827
Tantalum (Ta)	73	181	2996	5429
Tungsten (W)	74	184	3410	5900

B. Metals Used

A number of different metals are used to shadow specimens and to make replicas. The metals are chosen on the basis of ease of evaporation, fineness of grains produced on the specimen, and alloying properties with each other as they condense on specimen surfaces. Some of the characteristics of the metals most commonly employed are listed in Table 7.

C. Vacuum Evaporators

Shadowing is performed with vacuum evaporators, which provide the high vacuum (10^{-5} to 10^{-6} Torr) necessary to ensure a largely unimpeded path between the metal source and the specimen surface. The typical apparatus consists of a bell jar, which is ultimately evacuated by a diffusion pump (with or without a cold trap), backed by a rotary pump. Thermocouple or Pirani gauges monitor low vacuum, while a discharge gauge is used once high vacuum is reached. The vacuum chamber contains a number of terminals to which electrodes can be attached, as well as low-voltage terminals to which devices for rotating and tilting specimens can be connected. Some vacuum evaporators are also equipped with a glow-discharge unit, which is used to increase the hydrophilicity of film-coated grids (see Hayat and Miller, 1990, for construction details for a simple bench-top glow-discharge unit).

Evaporators may contain three types of evaporative sources. The simplest consists of two electrically isolated terminal posts between which metal wire or metal foil boats can be connected and heated (Fig. 117). A second type of source has terminals that hold carbon rods. One of the carbon rods is driven by spring pressure toward the other to keep the two carbon rods in intimate contact as they evaporate (Fig. 118). The final source type (electron beam gun) contains a coil of tungsten in close proximity to a carbon rod with a hollowed-out tip within which is fused a metal pellet (Fig. 119).

Some evaporators are equipped with crystal film monitors used to determine precisely the thickness of metal and/or carbon films evaporated onto specimens. These devices consist of a quartz crystal, which oscillates at a specific frequency when supplied with electrical current. As films are deposited on its surface, the frequency of the crystal vibration changes. The frequency of vibration is monitored with an elec-

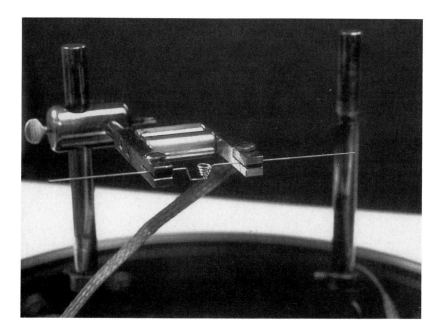

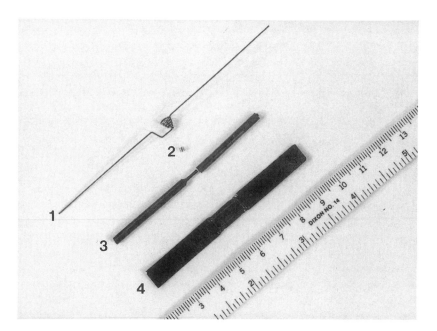

FIGURE 117. Types of electrodes. Top: Wire/boat-holding electrodes. Bottom: Examples of evaporator supplies: (1) tungsten wire basket; (2) coil of platinum wire to be evaporated; (3) carbon rods prepared for evaporation; (4) molybdenum boat for cleaning apertures.

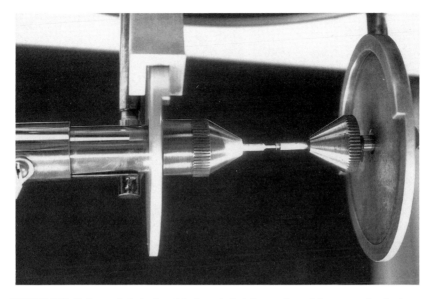

FIGURE 118. Carbon rod electrodes. Attachment of platinum wire coil for evaporation is shown.

tronic sensing circuit so that evaporation can be halted when the film being deposited is thick enough.

If a crystal film monitor is not available, a small piece of predried 3 × 5 card stock can be put into the vacuum chamber along with the specimen. After a typical evaporation run, the card should be lightly brown from the evaporated material. After the actual specimen is examined, future evaporation runs can be matched to the previously successful conditions that produced the card color.

Caution should be exercised with all vacuum equipment to avoid backstreaming. If excessive roughing of a bell jar takes place, there is some danger that oil from the rotary pump will backstream into the chamber, potentially coating the specimen surface and preventing the evaporated metal from adhering to the specimen. The diffusion pump can also produce an oil film on a specimen surface if not backed adequately or cooled properly.

FIGURE 119. Electron beam gun minus shields, aperture, and power supply. The carbon rod (A) has a hole drilled into its end in which is a fused platinum pellet (C). The rod is surrounded by a tungsten coil (B) that is heated with current to produce evaporation of the tip of the carbon rod and platinum pellet.

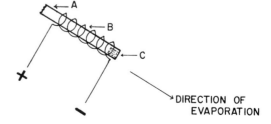

An oil film is used to monitor film deposition in some laboratories. A droplet of diffusion oil is placed on a small cleaned porcelain crucible top near the specimen to be shadowed. As the metal and/or carbon is evaporated, the cap surface becomes darkened, except where the droplet of oil is located. Some vacuum evaporator manufacturers do not recommend this practice because of the chance that the oil may become vaporized by the vacuum developed in the system, thereby contaminating specimen and bell jar surfaces.

D. Electrodes

Electrothermal heating is produced when a current is passed through electrodes composed of a resistant material (carbon rods, tungsten wire, molybdenum boats). If the current is sufficiently high (20–50 A), the resistant material will become hot enough to evaporate.

For carbon evaporation, one of the rods (the nondriven one) is left flat but is polished with 400- to 600-grit emery paper, followed by further polishing on a piece of card stock until it has a mirror-like surface. The driven rod is polished the same way and then either sharpened to a point, or the tip is reduced to a smaller diameter (Fig. 120). The decreased diameter produces increased resistance to current flow, yielding increased heating (and evaporation) in the tip area. The length of the reduced diameter tip can be adjusted to determine the amount of carbon to be evaporated. If metal wires such as platinum are to be evaporated simultaneously with the carbon, they are wound tightly around the tip of the driven rod, heated gently once good vacuum has been achieved so that they fuse with the carbon rod, and then heated more vigorously to evaporate them. If they are heated too rapidly, the wire will suddenly melt and drop off the rod before it evaporates significantly.

Wire electrodes are typically made from tungsten, which is either coiled into baskets or bent in the middle (see Fig. 117). The coiled area or the bend in the wire is the point of highest resistance. Metals with boiling points below that of tungsten are placed in the basket (wires should be balled up first) or wrapped around the bend in the filament. When high vacuum has been reached, the filament current is increased slowly until the noble metal alloys with the filament. Then the filament temperature is gradually increased until evidence of evaporation is noted by darkening of the card stock or activity of the crystal film monitor. When all the noble metal is evaporated, the current is turned off to the electrode. Unless the original source was a carbon/platinum mixture, it is customary to follow the metal evaporation with a coat of evaporated carbon to give greater stability to the metal film.

Metal boats are usually used to clean apertures from the electron microscope. The apertures are placed within the molybdenum boat, high vacuum is achieved, and then current is applied to the electrodes holding the boat until the apertures glow red. After about 1 min, the current to the electrodes is shut off, the apertures are allowed to cool, and then they are removed from the vacuum evaporator. Do not vent the bell jar when the apertures are still glowing, because there will be a risk of oxidation and contamination.

FIGURE 120. Carbon rods prepared for evaporation.

Electron-beam heating employs a source designed somewhat like a LaB_6 gun on an electron microscope. The metal to be evaporated (usually platinum) comes in the form of a small pellet, which is placed in the hollowed-out tip of a carbon rod. The rod is heated under vacuum until the pellet fuses with the carbon. To evaporate both carbon and platinum onto the specimen surface, a heated coil of tungsten (the cathode) emits electrons, which bombard the anode (the carbon rod containing the metal pellet). The kinetic energy of the electrons elevates the anode temperature, causing evaporation of the rod tip (carbon and metal). The rod itself is not subject to current flow, as is found with electrothermal heating of filaments described above. The advantage of electron beam guns is that 10–15 evaporative runs can be made without having to replace the carbon rod/pellet source, while the typical tungsten basket containing metal to be evaporated or the carbon rods driven into each other have to be replaced after every run.

E. Factors Leading to Fine Grains of Shadowed Metal

The evaporative process results in metal deposition (condensation on any relatively cooler surface) followed by nucleation (accumulation of metal atoms around previously deposited metal atoms). There are several factors that must be considered in order to maintain the finest grain size and, thus, the highest resolution.

Low total metal deposition will produce finer grain because the nucleation process will be minimized. Thus, only the minimal amount of metal necessary to produce the image quality desired should be evaporated.

If the current is raised quickly, larger groups of metal atoms can be released from the filament at a time. If the current is increased too quickly, the metal to be evaporated will fall off the electrode filament without actually evaporating. Thus, more gradual deposition is achieved by raising the current slowly to just above the point where evaporation begins and finer grain size will result.

A cooler substrate (specimen) temperature will result in quicker condensation of the evaporated metal. In some instances, the stage surface upon which the specimen is placed may be cooled with water to increase the rate of condensation.

Metals evaporated at a steep angle (normal to the specimen surface) will produce finer grains than those evaporated at a shallow angle. Of course, a steep angle would also produce negligible shadow, the point of the exercise. The shallower the angle, the greater the nucleation that will occur.

Finally, simultaneous evaporation of two metals (or carbon and a metal) will reduce the metal aggregate size by increasing the distance over which an atom must move in order to find a place within the crystal lattice formed.

F. Shadowing Techniques

Shadowing may be directional or rotary, high angle or low angle. If the specimen has relatively high topography (a fractured surface or whole cells), it is customary to use a relatively steep angle (45°) with a stationary specimen. This results in a reasonable amount of metal deposited on surfaces projecting above the specimen support and facing the electrodes. If, on the other hand, the specimen has little height or topography (DNA) and shadowing is used to produce more specimen bulk for visualization, rotary shadowing at a low angle (10°) is employed. With DNA preparations, the relaxed and spread DNA is put on grids, which are then attached to a platform turned rapidly by an electric motor while metal is evaporated onto the surface at a low angle.

G. Sputter Coating

Sputter coating generally produces metal films of greater grain size than those produced by evaporative coaters. Sputterers were developed primarily to coat specimens for relatively low-resolution SEM work. They do not need a high vacuum to operate (a rotary pump producing 10^{-3} Torr is adequate), because they do not need to provide a long mean free path for metal atoms, as is necessary for vacuum evaporation.

A sputter coater (Fig. 121) consists of a vacuum chamber containing a noble-metal-plated cathode and an anode upon which the specimens are placed. A permanent magnet is situated in the center of the cathode to deflect electrons from the specimen

FIGURE 121. A sputter coater (Hummer VI from Anatech) showing the bell jar (A); specimen stage (B), which is the anode; the gold-palladium ring (C) serving as the cathode; and the permanent magnet (D) in the center of the cathode.

surface (to avoid heating the specimen). After a vacuum is achieved, a high-voltage field is produced between the cathode and anode surfaces. Then an inert gas (argon) is admitted to the chamber, allowing current flow between the cathode and anode. The electron flow ionizes the argon gas, and the argon ions are accelerated by the electrical field toward the cathode. As they strike the cathode at high speed, the argon ions dislodge metal atoms (e.g., gold/palladium), which then condense on surfaces inside the vacuum chamber (the specimens). All of this activity takes place below the boiling point of the metals used. Thus, unlike vacuum evaporators, the metal being deposited has little heat to transfer to the specimen surface (since the metal has not been heated to its boiling point), so little chance of thermal damage to the specimen exists. Sputter coating at relatively low vacuum also results in numerous collisions between metal atoms, argon ions, and electrons, so a visible discharge cloud arises. All of these collisions result in a nondirectional metal coating that will evenly coat very irregular surfaces, resulting in no shadows, as produced with vacuum evaporators, even though the specimen is neither rotated nor tilted during the coating process.

The recent advent of high-resolution field emission gun SEMs capable of resolving aggregations of gold palladium deposited by conventional sputter coaters has driven some manufacturers (Denton Vacuum) to produce higher vacuum sputter coaters that coat specimens with chromium. Chromium has a fine grain structure compared to other metals traditionally used, but is subject to oxidation, leading to shorter shelf life for coated specimens.

II. NEGATIVE STAINING

Farrant (1954) introduced negative staining as a method to visualize ferritin particles. Since then it has been utilized to visualize macromolecules (milk proteins, enzymes, hemoglobin), surface components of cells (bacterial pili, flagella), internal cellular organelles (Golgi bodies, nuclear pores, mitochondrial membranes, microtubules, microfilaments), and viruses.

A. Mechanism

Suspensions of the structures to be visualized are supported on a film-coated grid and surrounded by a solution of heavy-metal stain that covers the grid surface and is mostly excluded by the biological material because of surface tension interactions. Thus, the bulk of the grid surface is covered with heavy metal that blocks some of the electron beam, while the specimen, which has excluded most of the stain but has had stain penetration of open irregularities in the particle surface, allows more of the beam to pass through. This procedure produces contrast between the specimen and the background, and also dehydrates the specimen. After a specimen is negatively stained, it should be examined within a day or so, because the preparations do not keep indefinitely. Viral capsomeres that are clearly defined shortly after negative staining with phosphotungstic acid (PTA) will usually be indecipherable 1–2 days later.

The negative staining process is inexact and not reproducible in the normal sense of the word (two grids prepared at exactly the same time in the same fashion often have dissimilar stain distribution). With a typical viral preparation, some of the particles will exclude the stain, which will form a slightly gray background right up to the edge of the viral particle. Other particles will exclude the stain but will have large amounts of stain pooled next to the particle, resulting in a completely electron-opaque zone around the particle. Some of the particles (usually described as "defective") will exhibit leakage, with stain inside the particles making them electron dense. Some areas of the grid may appear to be unstained, with no gray or black stain background and no visible particles, while an occasional grid square will contain such a deep layer of negative stain that the grid square is impenetrable to the beam. All of these situations can be encountered on one grid, which is why workers attempting to do quantitative virology or any study involving counting structures per unit area would be best advised to use another technique for their studies, although the technique does have some supporters (Miller, 1982).

Some staining solutions cause shrinkage, some structures are best fixed first, and some specimens are fragile at certain pH ranges. Hayat and Miller (1990) provide lengthy discussions of artifacts produced during negative staining, along with many other aspects of negative staining.

B. Methods

1. Nebulizers

Horne (1965) introduced a technique using a nebulizer to spray a mist of suspended viral particles onto film-coated grids that were subsequently negatively stained. This procedure has always been discouraged in our laboratory because an aerosol of viruses could have serious repercussions. Many viruses remain viable after drying, exposure to negative stains, and, in some cases, fixation with aldehydes. Even viruses that have no demonstrated human pathogenicity should be handled carefully, since they may cause respiratory problems or replicate in humans without necessarily producing overt disease (Hayat and Miller, 1990).

2. Droplet Method

A droplet of the suspension to be examined is placed on a film-coated grid held in forceps lying on a Petri dish lid under an appropriate hood. After 4–5 min, during which the particulates attach to the film (primarily by electrostatic forces), the grid is wicked almost to dryness with a fresh sheet of Whatman #1 filter paper, which should be disposed of in the autoclavable waste container (as are all waste materials generated by this procedure). Do not let the grid dry completely, or crystals of precipitated media or buffer salts may form on the grid surface. Immediately add one drop of the negative stain solution to the surface of the grid. Any remaining salts from the media or buffer will be sufficiently diluted by the negative stain that they will not precipitate when the grid is subsequently dried. After 30 sec, wick the grid completely dry with a fresh piece of filter paper. Gently blot the surface of the grid with the filter paper and place the grid on a dry piece of filter paper in a Petri dish until it can be examined.

If a brief distilled water rinse is utilized after negative staining, very light staining with excellent resolution can be achieved in some cases. However, this procedure may produce grids with too little contrast to be of any use.

Rather than placing particle suspensions onto grids held in forceps, some workers place the suspensions and the negative stain drops on Parafilm® and then float grids, with coated surfaces down, on the droplets. Some workers leave suspensions on grids for 30 min, but this does not appear to increase particulate attachment to the grid surface appreciably. In some instances, particulate suspensions and negative stain are mixed together, are brought into contact with a film-coated grid for several minutes, and then the grids are blotted dry. Hayat and Miller (1990) suggest that a grid with adsorbed viral particles that could be pathogenic should be placed onto a drop of glutaraldehyde for 10 min to kill the viruses and to reduce the danger of accidental infection. A thorough review of various methods for staining particulates with negative stains may be found in Hayat and Miller (1990).

3. Types of Negative Stains

Solutions of various heavy metals at different pHs have been used effectively to stain biological materials, but the four most commonly used are uranyl acetate, phosphotungstic acid, ammonium molybdate, and methylamine tungstate. Nermut (1982) discussed the advantages of various stains and the appropriate pH at which they should be used to accentuate features of enveloped vs. nonenveloped viruses. He reported that acidic pH is best for membrane structures (such as glycoprotein knobs, or spikes associated with enveloped viruses, such as coronaviruses), and that alkaline pH works well to demonstrate internal proteinaceous components (nucleocapsids). He also suggested that grain size of the final dried stain should be considered, with phosphotungstic acid and ammonium molybdate having fine grain size. Unfortunately, they also have less contrast than the more coarsely grained uranyl acetate.

a. Uranyl Acetate

Uranyl acetate is used as a 0.2–4% aqueous solution with a pH of 4–5. Since a 4% solution is saturated, the stain should not be taken from the bottom of the container to avoid picking up crystals of uranyl acetate. The solution will keep for weeks at room temperature in a dark bottle. The stain is unstable above pH 6.0, and both negative and positive staining can be observed in the same material. Hayat and Miller (1990) suggest that 2% aqueous uranyl acetate is good for viral staining (30 sec to 2 min), because viral structure is stabilized and grids can be observed months after preparation. They also suggest that a 1% aqueous solution is the preferred stain for visualizing macromolecules.

b. Phosphotungstic Acid

Phosphotungstic acid (PTA) is customarily made by dissolving sodium phosphotungstate in distilled water and adjusting the pH to 7.0–7.2 with NaOH. The solution is stable for several years if stored at 4°C. At this pH, viral nucleocapsids are well defined. Hayat and Miller (1990) state that staining time over 30 sec will result in overstaining, but our laboratory has not noted major differences in materials stained from 30 sec to 5 min. This is probably the most commonly used negative stain solution in the literature.

c. Ammonium Molybdate

Ammonium molybdate dissolved in distilled water (3%) will have an unadjusted pH of about 6.5. Ultrathin frozen sections stain well in about 30–60 sec (H. Sitte, personal communication). Large crystals can sometimes form through recrystallization caused by heating from the electron beam.

d. Methylamine Tungstate

Methylamine tungstate was first used by Faberge and Oliver (1974) as a negative stain. Stoops *et al.* (1991) employed this stain to visualize human α_2-macroglobulin. It is complicated to formulate (see Hayat and Miller, 1990) but can be used over a broad pH range and can be mixed with PTA. It is customarily prepared as a 2% aqueous solution at pH 6.5–8.

4. Wetting Agents

As mentioned previously, an unfortunate aspect of negative staining is that the stain does not spread evenly or even predictably in almost all cases. This is caused primarily by the hydrophobicity of the plastic film (Formvar, collodion) that serves as a specimen substrate. Some workers insist that grids should be used immediately after preparation, some say that they will only work well for a few weeks, some say carbon coating improves them, some say carbon coating makes them more hydrophobic, and some say they should be glow-discharged (and used immediately) to make them more hydrophilic or exposed to strong ultraviolet radiation (see Hayat and Miller, 1990, for further discussion). Our laboratory has not been able to demonstrate that carbon coating improves specimen and stain spreading. Fresh films tend to wet better than ones that are over a week old, but the difference between week-old and year-old coated grids seems negligible. The only guaranteed method for spreading the specimen and subsequent stain is to utilize a wetting agent such as the antibiotic bacitracin (500 µg/ml) or bovine serum albumin (0.01–0.02%). Bacitracin is a smaller molecule than BSA, so it usually gives less background.

Bacitracin (Gregory and Pirie, 1973) dissolved in distilled water (1,000 µg/ml) not only allows the stain to spread evenly on the film-coated grid surface, but also can reduce surface tension features associated with the specimen. Dykstra (1976) has studied a protozoan with scales of 4.0 nm × 4.0 µm, which coated the cell surface. When isolated, the scales rolled up tightly into tubes, and normal negative staining procedures did not relax the scales. If the scale suspension was mixed 1:1 with the bacitracin solution and a film-coated grid was then suspended on the resulting droplet for 2 min and subsequently stained with 1% aqueous uranyl acetate, the scales flattened out and were easy to visualize.

REFERENCES

Dykstra, M.J. 1976. Wall and membrane biogenesis in the unusual labyrinthulid-like organism *Sorodiplophrys stercorea*. *Protoplasma* 87:329.

Fabergé, A.C., and Oliver, R.M., 1974. Methylamine tungstate, a new negative stain. *J. Microscopie* 20:241.

Farrant, J.L. 1954. An electron microscopic study of ferritin. *Biochim. Biophys. Acta* 13:569.

Gregory, D.W., and Pirie, J.S. 1973. Wetting agents for biological electron microscopy: I. General considerations and negative staining. *J. Microsc.* 99:261.

Hayat, M.A., and Miller, S.E. 1990. *Negative staining*. McGraw-Hill, New York.

Horne, R.W. 1965. Negative staining methods. In: *Techniques for electron microscopy*, 2nd ed., R.W. Kay (ed.), Blackwell, Oxford.

Miller, M.F. 1982. Virus particle counting by electron microscopy. In: *Electron microscopy in biology*, Vol. 2, J.D. Griffith (ed.), John Wiley and Sons, New York.

Nermut, M.V. 1982. Advanced methods in electron microscopy of viruses. In: *New developments in practical virology*, C.R. Howard (ed.), Alan R. Liss, New York.

Stoops, J.K., Schroeter, J.P., Bretaudiere, J.-P., Olson, N.H., Baker, T.S., and Strickland, D.K. 1991. Structural studies of human α_2-macroblobulin: Concordance between projected views obtained by negative-stain and cryoelectron microscopy. *J. Struct. Biol.* 106:172.

Stuart, R.V. 1983. *Vacuum technology, thin films, and sputtering. An introduction.* Academic Press, New York.

CHAPTER 9

Scanning Electron Microscopy

I. HISTORY

Scanning electron microscopy (SEM) has a history almost as old as TEM, but the development of a commercial product took much longer. Von Ardenne (1938) built the first SEM, and Zworykin *et al.* (1942) produced an SEM with a 50-nm probe. A group in Cambridge, England, headed by Oatley began work in 1948 that led to the first commercial SEM (the Cambridge Stereoscan) in 1965. Pease (1963) achieved a beam diameter of 5 nm, which resulted in 10 nm resolution. With the introduction of the Cambridge Stereoscan in 1965, the biological community immediately began exploiting the tool to examine numerous tissues and cells. During the 1970s, resolution was improved to 5–6 nm, microprobe analysis (energy-dispersive spectroscopy) was applied to SEMs, freeze-fracture methods were explored (Haggis, 1972), and magnifications of over 100,000× became a reality. In the 1980s field emission guns (FEG) were introduced in commercially-produced SEMs offered by Hitachi and then JEOL, which allowed lower accelerating voltages and increased magnification and resolution (under 1 nm), and which were immediately adopted by materials scientists. By the late 1980s, biolologists had discovered FEG instruments and were examining tissues and cells at low voltages with high magnifications previously possible only with sectioned material viewed with a TEM (Ris, 1991). In the 1980s, there were also improvements in probe diameters for microanalysis, along with better instrument sensitivity, allowing the recognition of elements with lower Z numbers than previously possible (see Chapter 12).

In the late 1980s, an instrument called the Environmental Scanning Electron Microscope (ESEM) was introduced that allowed the visualization of hydrated specimens. The advantage of this instrument is that specimens can be dehydrated and hydrated while being observed and some samples can be observed in the native state, without fixation, dehydration, and critical point drying. The pharmaceutical industry has been able to use this instrument for compound evaluation in ways not possible before.

Two excellent reference books on SEM instrumentation and specimen preparation are those by Hayat (1978) and Postek *et al.* (1980).

II. THE USE OF SEM IN BIOLOGICAL RESEARCH AND MEDICINE

The SEM can be thought of as a high-resolution dissecting microscope, since it is used to examine surfaces of specimens. It has been used to study cell-surface receptors for lectins (Kan, 1990), to compare normal tracheal epithelium with tissues from cases of immotile cilia syndrome leading to respiratory abnormalities (Edwards *et al.*, 1983), and the effect of various compounds on cellular motility and cellular polarity. For these types of studies, the SEM has advantages over the TEM because serial reconstructions are not necessary, the sample size is generally much larger, and the three-dimensional view with great depth of field is easy to comprehend, even for viewers not trained in electron microscopy.

Chondrocytes from the growth plate of a bone (Fig. 122) are a good illustration of how SEM can offer understandings not available from other types of microscopy. With light microscopy, chondrocytes appear to be embedded in an amorphous matrix that stains pale blue in hematoxylin and eosin preparations. This matrix appears faintly granular with TEM preparations (Fig. 122, top). When a sample is fixed and dehydrated in the same way as for TEM and is then critically point dried and examined with an SEM (Fig. 122, bottom), it becomes obvious that the "amorphous" material is really quite structured and forms distinct chambers around the chondrocytes. Thus, SEM has given us an understanding of the relationship of the cells to their surrounding matrix not available with other tools.

As Humphreys *et al.*(1975) described, cells can be fixed and dehydrated up to 100% ethanol, quickly frozen in liquid nitrogen, fractured with a cold razor blade, and placed inside a vacuum evaporator bell jar until the ethanol has sublimed. Once sputter-coated, this freeze-fractured material provides excellent intracellular detail, particularly now that FEG SEMs are available.

To make use of another unique capability of the SEM, the vasculature may be perfused with various resins (Suzuki, 1982; Hanstede and Gerrits, 1982). After the tissue is corroded from the sample, casts of the vasculature remain. These can provide three-dimensional information about the smallest capillary beds (Fig. 123) and can also reveal vascular accidents (leakage) and neovascularization associated with certain disease processes. These casts can be easily manipulated and dissected while in liquid, though they become quite brittle when dried.

Evaluation of epithelial layers and their response to injury can be done easily with SEM. Normal chicken tracheal epithelium (Fig. 124, top) possesses ciliated cells interspersed with mucus cells, which have only short microvilli on their surfaces. When challenged by *Mycoplasma gallisepticum*, the epithelial cells become round, lose their intercellular connections, and exfoliate (Fig. 124, bottom). Scanning electron microscopy provides much better evidence of this dynamic process than TEM or light microscopy.

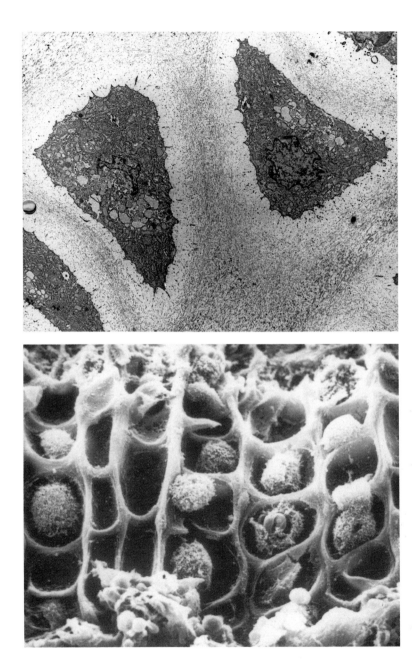

FIGURE 122. Top: Canine chondrocytes from femur growth plate. ×2,500. Bottom: Rabbit chondrocytes from femur growth plate. ×750.

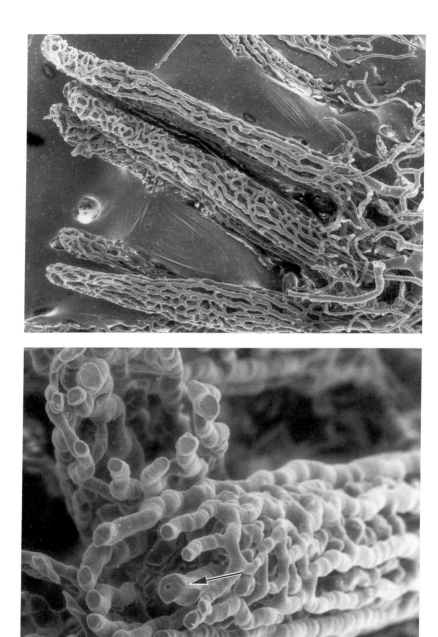

FIGURE 123. Top: Mercox resin cast of microvilli of equine small intestine revealing ultimate capillary bed. ×85. Bottom: Mercox resin cast of microvilli of equine small intestine cut across the tip with a razor blade to reveal the central arteriole (arrow). ×400.

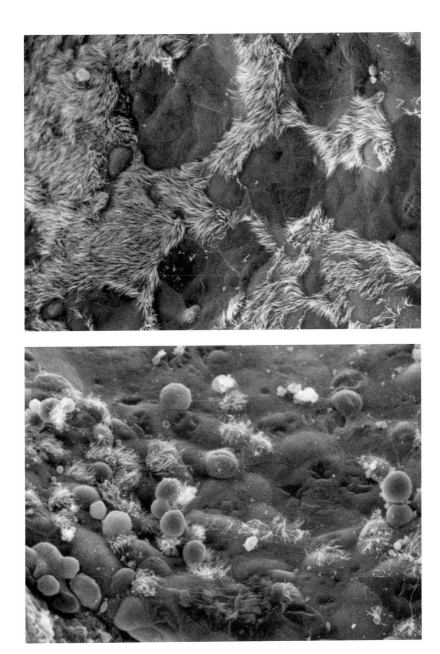

FIGURE 124. Top: Normal chicken tracheal epithelium. ×535. Bottom: Chicken tracheal epithelium exposed to *Mycoplasma gallisepticum*. ×680.

Intricate details of microorganismal surfaces, topographic information about pollen, plant epidermal layers, and bryozoans revealed by SEM may be used for taxonomic, morphologic, and ontogenetic studies (Claugher, 1990).

III. PRINCIPLES OF THE SEM

The SEM employs electromagnetic lenses, vacuum systems, apertures, and electron guns similar to those described in the chapter on TEMs. Figure 125 illustrates the basic parts of an SEM. Unlike the TEM, which passes electrons through a thin specimen, the SEM accelerates electrons and collimates them into a narrow beam that is then impinged upon the specimen surface, producing several different imaging possibilities. Because of the small aperture sizes used and the short wavelength of electrons, tremendous depth of field can be realized. The magnification chosen may be no more than that achievable with a light microscope, but much more of the specimen will be in focus at one time, thereby resulting in one SEM photograph conveying the same information that would require a series of optical sections and computer reconstruction with light microscopy.

A narrow beam of electrons from the electron gun is focused by electromagnetic lenses into a small spot (less than 10 nm) on the surface of the specimen. Deflector coils then scan the beam across the specimen. In the most common imaging mode, secondary electron imaging (SEI), the high-energy primary beam dislodges electrons from atoms near the surface of the specimen (secondary electrons), some of which then strike the SEI collector. Those electrons that contact the collector produce photons

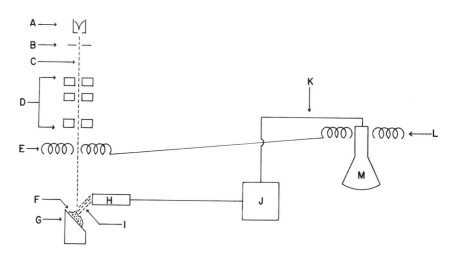

FIGURE 125. Diagrammatic sketch of an SEM showing Wehnelt assembly with filament (A), anode aperture (B), electron beam (C), condenser lenses (D), deflector coils on column (E) connected to deflector coils (L) of CRT (M), specimen (F) on stage (G), producing secondary electrons (I) striking collector (H) connected to amplifier (J) feeding voltage through cable (K) to CRT.

that then are processed by a photoamplifier circuit connected to a cathode ray tube (CRT). An electron that interacts with the collector results in a voltage applied to the gun of the CRT, which produces a point of illumination on the CRT screen. The scan generator that operates the scanning coils controlling the electron beam within the SEM column also is connected to the deflector plates of the CRT. Thus, as the SEM beam is scanned over the specimen, the CRT gun is simultaneously scanned over the CRT screen.

The output of the photoamplifier (voltage) is employed to modulate the brightness of the CRT beam in synchronization with the SEM electron beam. The current of secondary electrons recorded by the collector at a given point in time produces a given voltage after processing by the photomultiplier circuit, determining the brightness of the spot on the CRT. Any variation in elemental composition, texture, or topography can result in variations in the current reaching the collector. Specimen magnification is determined by the relationship between the distance scanned on the specimen surface by the primary beam and the distance scanned on the CRT during the same time period. Unlike the TEM, where the image is recorded over the entire screen simultaneously, the SEM image is collected and displayed point by point.

IV. OPERATION OF THE SEM

An SEM may have any of the three types of sources (tungsten filament, LaB_6, FEG), as described in Chapter 4. The type of gun determines the type of vacuum system needed. Most SEMs have accelerating voltages adjustable from 0 to 35 kV and have three electromagnetic lenses, the first two of which defocus the beam, while the third is used to determine the final diameter of the beam striking the specimen surface. The final lens has apertures of various sizes, the smallest of which gives the greatest depth of field but also decreases the amount of secondary electrons produced (signal). At low magnifications, there is a comparatively high signal-to-noise ratio (SNR), while at high magnification the SNR is lower, resulting in "noisier" CRT images showing "snow" (noninformational, resolution-decreasing emissions).

As stated earlier, magnification is achieved by changing the relationship between the distance scanned over the specimen surface and the distance scanned over the CRT. If 5 μm of specimen surface is scanned at the same time that 5 mm of CRT surface is scanned, the specimen magnification would thus be 1,000.

Working distance can be varied, placing the specimen higher or lower in the column, which also places the specimen nearer or farther from the electron source and collector, respectively. Moving the specimen farther away from the lenses and collector increases the depth of field and can reduce charging to some extent.

Several types of collectors may be located within a given SEM. The most commonly encountered types are designed to be receptive to either secondary electrons emitted from the specimen surface with energy of 50 eV or less (selected by biasing the collector with + 10 kV), backscattered electrons (energies from 50 eV up to the energy of the primary electron beam), or X-rays (EDS collectors).

V. INTERACTION OF THE ELECTRON BEAM AND SPECIMEN

With most biological materials, an accelerating voltage of 10–15 kV is used. This voltage results in the specimen surface being penetrated to about 10 μm and gives rise to five major types of radiation (signal): secondary electrons, backscattered electrons, Auger electrons, X-rays, and cathodoluminescence (Fig. 126).

A. Resolution

The concept of resolution is more complex in the SEM than in the TEM, because time is an integral part of SEM image formation. In addition, the various types of electron interactions with specimen atoms have an effect on SEM images, unlike the relatively simple beam scattering used to produce TEM images.

Resolution is affected in a point-by-point recording system in a temporal sense (*when* a signal is recorded). If the primary beam impinges on a part of the specimen

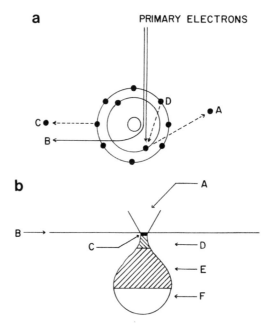

FIGURE 126. a: Diagram of the sources of signal types produced from electron beam interaction with a specimen. The primary electrons may either displace an orbital electron (A), which becomes a secondary electron, or they may pass by the nucleus to become backscattered electrons (B). After a secondary electron is produced, a high-energy outer shell electron (D) will shift to the electron-deficient lower energy inner shell. The excess energy from this shift may be liberated as cathodoluminescence or an X-ray photon, or may cause a further electron rearrangement leading to the release of an Auger electron (C). b: Specimen surface (B) penetrated by an electron beam (A). Auger electrons (C) are produced from near the specimen surface, while secondary electrons (D), backscattered electrons (E), and X-ray photons (F) are produced from deeper in the specimen.

(point A) that results in the release of electrons (or X-rays) that are recorded virtually simultaneously on the CRT, then good resolution has been achieved. On the other hand, if interactions within the specimen result in signals being received by the collector some time after the primary beam has scanned on to a different area of the specimen (point B), the CRT will record an event at point B (Fig. 127), even though the beam interaction that ultimately produced the signal happened at point A. This clearly produces an image of lower resolution than one generated by a signal resulting from the beam interacting with the specimen at point A and being recorded on the CRT at point A.

This problem is of particular significance when working in the SEI mode. When the primary, high-energy beam strikes the specimen surface, it can penetrate up to several micrometers at the 10–15 kV commonly used for SEI work on biological specimens. Secondary electrons dislodged from atomic orbits can be emitted in any direction. Only those electrons that have enough energy to emerge from the specimen surface and reach the collector will be recorded. Thus, only those secondary electrons produced near the specimen surface and directed toward the collector will result in recorded information. The temporal relationship between the production of these secondary electrons and their imaging on the CRT will be almost coincident, leading to high resolution.

Unfortunately, backscattered electrons are also produced at the same time secondary electrons are produced. Backscattered electrons interact with the SEI collector and contribute to noise. Backscattered electrons produced by interactions several micrometers from the sample surface still have high energy and can interact with further atoms until their energy is spent. If they are scattered back toward the sample surface and interact with an atom, causing the release of a secondary electron that escapes from the sample and that can be recorded (Fig. 128), loss of resolution will be a probable outcome. The backscattered electron has traveled some distance through the specimen before causing the ejection of a secondary electron from the specimen surface. Thus, an electron has been released some distance from the original point of primary beam

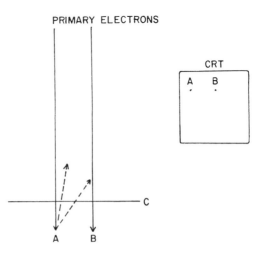

FIGURE 127. The relationship between time and resolution. The primary electron beam interacts with the specimen at point A to produce secondary electrons (dotted arrows), while the beam produces no secondary electrons at point B on the specimen. The CRT has a bright spot at *both* points A and B (corresponding to points A and B on the specimen). The secondary electron emitted almost normal to the specimen surface is recorded at point A on the CRT, while the more oblique secondary electron emerges from the specimen surface at a later time point and is recorded on the CRT at point B, even though no emission took place at point B on the specimen.

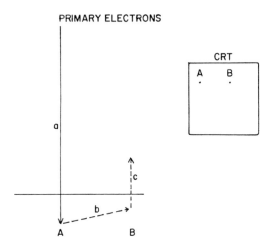

FIGURE 128. Production of secondary electrons by backscattered electrons, resulting in loss in resolution. No secondary electrons are produced from the interaction of the primary electron beam (a) with the specimen at point A. A backscattered electron (b) is produced from the specimen interaction at point A, which travels obliquely through the specimen. At point B, the backscattered electron produces a secondary electron (c) that emerges from the specimen surface and is recorded at point B on the CRT.

interaction with the specimen surface, resulting in a loss in resolution from spatial and temporal inaccuracies. It is important to remember that a number of different signals are being produced simultaneously from an SEM sample, even though only one or two are usually recorded at a time. The fact that events not being recorded can affect those that are being recorded should not be overlooked.

B. Secondary Electrons

Secondary electron imaging is the main method for examining structural characteristics of specimens. Inelastic collisions between the high-energy primary electron beam and the specimen result in orbital electrons being ejected from atoms located no more than 5–10 nm from the specimen surface. The latest generation of FEG SEMs are capable of 0.5-nm resolution under carefully controlled conditions. By convention, secondary electrons are those with less than 50 eV of energy (the upper limit set for the secondary electron collector), though most fall within the range of 2–5 eV, and about 75% are less than 15 eV (Hayat, 1978). The number of secondary electrons emitted depends on the voltage of the primary electron beam, the angle of impingement on the specimen surface, and the electron density of the specimen (based on its elemental composition). The number of secondary electrons emitted is directly proportional to the number of primary electrons contacting the specimen. The probability of inelastic scattering varies inversely with the accelerating voltage, with the highest yield potentially at 1 kV. Higher accelerating voltages penetrate more deeply into the specimen, and any secondary electrons produced at more than 10 nm from the surface have little statistical probability of having sufficient energy to emerge from the specimen for subsequent collection.

The secondary electron collector (scintillator) has a slightly positive bias to attract secondary electrons, but this bias is not strong enough to collect backscattered electrons. The surface of the scintillator is coated with aluminum to reflect photons produced by the secondary electrons toward the photomultiplier tube to which the scintillator is affixed.

Pawley (1988) described the importance of specimen coating, charge accumulation on specimen surfaces, and surface topography on imaging for low-voltage SEM:

> Although there is a danger that this metal coating may decorate or obscure small details on the specimen surface, it does increase the amount of signal produced near the point of beam impact, and it also provides a degree of conductivity which prevents the accumulation of a negative surface charge which might displace or defocus the beam or interfere with the secondary electron collection field . . . Charge accumulation occurs because on most specimens viewed at 10–20 kV, more electrons enter the surface than leave it. Because on flat samples, the total secondary and backscattered electron yield increases to unity at lower voltages, it is widely believed that charging artifacts should always be less important at low voltage . . . it is not true for geometrically complex samples. On these materials, even though the absorbed current averaged over the area viewed may equal zero, marked variations within and adjacent to the scanned area can produce serious charging effects . . . To sum up, coating is needed at low voltage for conductivity and at high voltage for contrast.

1. Factors Increasing Signal (2° Electrons Collected)

The lower (more oblique) the angle of primary beam impingement on the specimen surface, the more secondary electrons will be emitted. If the specimen appears to have little contrast, tilting it in relationship to the primary beam will increase the signal, resulting in improved contrast.

More electrons will be emitted from a specimen surface that is rough or pointed, with edges producing strong emissions (relief contrast). A flat epidermal surface will produce little signal compared to a pointed surface (e.g., villi in small intestine, trichomes on plant leaves).

More secondary electrons will be produced from a negatively charged area than from a positively charged area within a specimen (potential contrast).

The closer the specimen is to the collector, the more secondary electrons will be recorded. Variation of the specimen working distance allows adjustment for this factor, depending on the amount of signal emanating from the specimen.

Lower accelerating voltages yield greater emission of secondary electrons.

In all SEM imaging modes, more signal is produced by concentrating the beam to produce a smaller spot on the specimen. With SEI, when working at low magnifications, higher beam collimation (achieved by increasing the current to the final condenser lens) results in more contrast and more noise. For high magnification, there is inherently less signal and less contrast if the condenser lens is left at the setting used for low magnification. As a rule of thumb, as magnification increases, the current to the final condenser lens should be increased and the SEI collector contrast should be increased. As the magnification is decreased, the condenser lens should be deenergized (spread) and the SEI collector contrast should be decreased.

Field emission gun (FEG) filaments have a tip diameter of only 3–5 nm, producing a beam that is 1,000 times brighter than a conventional tungsten filament. The same number of electrons are generated in both cases, but much higher electron density is possible with the FEG. In addition, the energy spread in an FEG beam is much narrower than that of a conventional tungsten filament, resulting in less chromatic aberration in the former (Hainfeld, 1977). The superior brightness of the FEG produces higher signal from a given specimen, and the lower chromatic aberration allows the utilization of lower accelerating voltages.

C. Backscattered Electrons

Backscattered electron imaging (BEI) records events due to elastic collisions. When electrons of the primary electron beam entering the specimen surface are deflected through large angles as a result of electron/nuclear interactions without energy loss, the electrons may escape from depths of several micrometers. Resolution in BEI mode is about 1,000 nm (Postek *et al.*, 1980), considerably less than that possible with SEI. Backscattered electrons may have energies equal to the accelerating voltage of the primary beam or lower. By convention (and adjustment of the BEI collector circuitry), any electron with more than 50 eV of energy is classed as a backscattered electron. Composition (contrast between elements of significantly different atomic number) and topography (contrast between particles with different surface structures) are the two imaging modes used.

1. Factors Increasing Signal

With BEI, the elemental composition of the specimen is a major factor in developing contrast. Major differences in atomic number and their effect on contrast can be utilized in cytochemical procedures. For example, if it were useful to localize lysosomes in a monolayer of cells attached to a coverslip, a lead-capture procedure could be used on aldehyde-fixed cells to label the hydrolytic enzymes present. After dehydration and critical-point drying, the cells could be coated with carbon so that the only high-atomic-number moiety present would be lead in lysosomes. In the BEI mode, primary electrons would penetrate several micrometers beneath the cell surface and would produce a strong backscatter signal from the lead.

As with SEI imaging, the angle of the primary beam striking the specimen can be manipulated by tilting the specimen stage. A higher beam angle produces more image contrast in BEI.

Changes in topography in adjacent areas also contribute to the development of contrast. Edge effects result in more signal from edges and pointed structures within a specimen.

Some workers impregnate the specimen with a heavy metal, such as osmium or osmium/thiocarbohydrazide combinations, to increase the overall backscatter signal.

Backscatter electron imaging has less resolution than SEI because of the depth from which signal can be produced. High-energy backscattered electrons passing through the specimen at a very oblique angle (Fig. 129) can emerge from the specimen

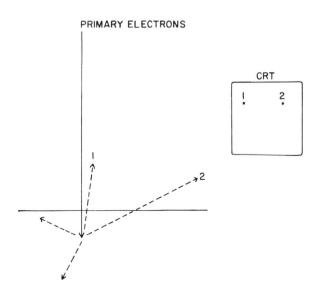

FIGURE 129. Oblique path of backscattered electrons, resulting in loss of resolution in the backscatter mode. All of the dashed lines with arrows represent backscattered electron paths resulting from the interaction of the primary electron beam with the specimen. Backscattered electrons 1 and 2 will be recorded on the CRT, though at different times during the CRT scan.

surface and be recorded a short time after the primary beam interaction with the specimen. However, during this time, the primary beam (and the CRT scan) has moved on to a new part of the specimen. Thus, the obliquely emergent electron will result in a bright spot on the CRT when the scanning beam may actually be over a specimen area producing no signal, thus causing a spurious signal (loss in resolution).

D. Auger Electrons

When a primary electron strikes an atom and causes an electron from an inner orbit to be ejected, an electron from a higher level fills the resulting void. When an electron from a higher-energy outer shell moves to a lower-energy inner shell, an amount of energy equal to the difference in energy levels between the two shells must be released, as dictated by the laws of physics. This can be accomplished by emission of an X-ray or a photon of light (cathodoluminescence) or the release of a low-energy Auger electron.

Auger electron imaging (AEI) measures low-energy electron emissions. The average depth below the specimen surface from which Auger electrons can escape without appreciable energy loss is between 0.5 and 2 nm. Similar to the other signals considered, numerous events take place deep within a specimen that are not recorded because the emissions never escape the specimen surface to strike a collector.

Hayat (1978) defines Auger electrons as secondary electrons with energies from 1 to 2,000 eV released from the specimen surface when the primary beam is set at about 3 kV. The signal is measured in electron volts and allows the determination of elemental composition (anything above lithium in the periodic chart can be detected).

A miniscule quantity of backstreaming diffusion oil can result in a total loss of Auger electron signal, since any signal produced would originate in the thin film of oil deposited on the specimen surface rather than from the sample itself.

E. Energy-Dispersive Spectroscopy (EDS)

In almost all cases, energy-dispersive spectroscopy (EDS), which records characteristic X-rays emitted by specimens exposed to a high-energy electron beam, can provide elemental analysis that is more easily accomplished than with AEI. As described above, displacement of an inner-shell electron results in an electrically unbalanced atom, leading to shifts of outer-orbital electrons to fill inner shells. As these shifts take place, energy is often released in the form of X-rays. If an electron is dislodged from the innermost shell of an atom with three filled electron shells, a variety of rearrangements can take place (Fig. 130). The innermost (K) shell will be filled by an electron shifting inward from the L shell, resulting in energy being released in the form of a characteristic X-ray. The hole left in the L shell will be filled by an electron from the M shell, with an energy release in the form of an X-ray with an energy level characteristic of that orbital shift.

Thus, the bigger the atom, the more possible characteristic X-rays that can be

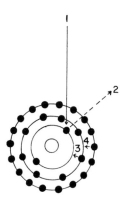

FIGURE 130. Orbital rearrangements following an inelastic collision between the primary electron beam and an electron in the K shell of an atom. The primary electron beam (1) causes the ejection of a secondary electron (2) from the innermost orbital, followed by the shift of an electron (3) from the next orbital to fill the vacancy, resulting in a shift of an electron (4) from the next orbital to fill the vacancy caused by the shift of electron 3.

produced, depending on the orbit from which the secondary electron is dislodged. Chapter 12 will discuss the utilization of this type of imaging in detail.

F. Cathodoluminescence

In order to detect photons released as a result of atomic rearrangements following secondary electron emission, either the scintillator over the SEI photomultiplier tube can be removed or a second photomultiplier tube can be incorporated in the SEM. Cathodoluminescence is weak, requiring a great deal of signal amplification, but Postek *et al.* (1980) state that 80 nm resolution is possible. They suggest that the primary use for this technique is to examine compounds for impurities. Hayat (1978) provides a series of references for the use of cathodoluminescence in biological research while pointing out the technical difficulties in some of these applications.

VI. SPECIMEN PREPARATION

A. Fixation

Sample preparation for SEM requires the same thought and consideration that were described for TEM samples in Chapter 1. Aldehyde fixation is frequently adequate, but some investigators use a post-fixation with osmium to try to improve bulk conductivity of the specimen. Another technique (Kelley *et al.*, 1975) utilizes a rinse in thiocarbohydrazide (TCH) prior to osmication, because the TCH serves as mordant for osmium, allowing increased osmium binding to the tissue.

Samples should be made as small as is practicable to reduce dehydration and drying problems, as well as to reduce the thickness of the tissue attached to the grounded specimen support. Dried tissues and cells act as electrical insulators under most circumstances. When the primary electron beam enters the surface of the specimen, much of the electron-beam energy encounters difficulty reaching ground potential. In extreme cases where specimens are quite thick, the specimen can be seen to bulge under the SEM beam as the subsurface tissues heat up and expand from the electron-beam energy that cannot reach ground and be dissipated. Osmium or osmium/TCH treatment, along with reduction in specimen thickness, can help reduce this problem.

Complex spore-bearing structures found in fungi are typically difficult to fix without losing the spores or developing handling artifacts. In particular, masses of fungal hyphae and sporophores bearing spores can be extremely difficult to wet adequately. Vapor fixation with osmium prior to immersion in aldehydes can help preserve these delicate structures (King and Brown, 1983).

Needless to say, if the specimen to be examined is already naturally dry (e.g., an aged seed, fossil, or egg-shell surface), fixation is generally unnecessary. In some cases, fixation might even be contraindicated because it would change the character of the specimen to rewet it with processing fluids.

B. Dehydration and Transition Fluids

Dehydration may be accomplished with the same agents typically employed for TEM preparations (ethanol, acetone), but the final method for drying the specimen should be considered before choosing the agent. If the specimen is to be critical point dried, ethanol is usually the best choice because some of the seals in most critical point dryers may be damaged by acetone. In some cases, a transition solvent, such as amyl acetate, may be used between the dehydration agent and the critical point dryer fluid (e.g., CO_2 or freon). This was originally done so that the investigator could tell when the dehydration agent in the specimen had been completely purged and replaced by liquid CO_2 in the dryer (the banana smell of amyl acetate would no longer be present, as the dryer chamber was vented and drained during the purging process).

C. Drying

The conventional SEM operates with a vacuum in the diffusion pump range or better (depending on the type of gun used), so it is necessary to dry a specimen thoroughly before introducing it into the high-vacuum system. In the early days of SEM, certain insect samples were examined live without fixation and dehydration, because insects with thick cuticles and air passageways (spiracles) that can be closed have an unusual ability to withstand the harsh environment within an SEM for a limited time. The only instrument that is designed to examine wet specimens is the Environmental Scanning Electron Microscope.

Samples that are air dried may be exposed to surface tensions during drying as high as 2000 lb/in², which can result in up to 45% shrinkage of embryonic and fetal tissues (Hayat, 1978). Hayat also showed two figures of slime mold spores (*Badhamia utricularis*) to support the thesis that air drying can crush biological structures. Unfortunately, his example illustrates a totally different point as well. Slime mold spores in nature become air dried as they mature and are then dispersed by breezes. If field-collected material is examined without mounting in any fluid, spores are collapsed. Hayat shows these as an example of an air-drying artifact. Hayat's second figure shows perfectly inflated, round spores following fixation and critical point drying that he identified as a superior image. The second case is clearly an artifact of specimen preparation. This example is offered to make the reader aware that an examination of a specimen by light microscopy in its native state may be of use prior to SEM sample preparation so that such preparation-induced artifacts can be avoided.

1. Critical Point Drying

Critical point drying (CPD) is based on the principle that under certain temperature and pressure conditions, a fluid and its overlying vapor phase will become indistinguishable (the critical point). At this point, the surface tension on a specimen originally in the fluid phase will be zero. As pointed out by a flyer describing the EMS850 critical point dryer (Electron Microscopy Sciences, Fort Washington, PA), our main concern is to remove water from a biological sample, however " . . . the critical point for water of $+374°C$ and 3212 psi is inconvenient, and would cause heat damage to the specimen."

To avoid these harsh conditions, samples may be dehydrated up to 100% ethanol, put into a critical point dryer, and then purged repeatedly with liquid CO_2 until all the ethanol has been replaced with CO_2. Next, the chamber is sealed and heated until the critical point for CO_2 is reached ($31°C$ and 1072 psi).

As the chamber temperature is raised above ambient, the liquid CO_2 begins to evaporate at a greater rate. At the same time, the pressure inside the critical point dryer increases. The increased chamber pressure causes the vapor to be condensed back into liquid. The process is based on two principles: (1) Higher temperatures turn liquids into vapors, and (2) higher pressures turn vapors into liquids. Above the critical point (critical temperature and pressure), the density of the liquid phase becomes equal to that of the vapor phase, and thus only one phase remains, with no surface tension. If the temperature of the chamber is maintained above the critical temperature (usually $36–38°C$), the chamber can be vented to release CO_2 gas without any danger of recondensation into liquid CO_2 to produce surface tension on the specimen. The chamber should be vented slowly and the temperature gauge must be closely observed. Allowing gas to escape through a small orifice (the venting valve) at high velocity will cause a temperature drop. If the temperature drops below the critical point, specimen surfaces can become rewetted and suffer surface tension damage.

Critical point drying is not a perfect technique. Shrinkage is still observed, resulting in 10–15% size reduction in central nervous system materials, 12–13% shrinkage with embryonic and fetal tissues, and up to 20% shrinkage for lung, kidney, liver, and skin (Hayat, 1978). Much of this shrinkage may be due to the fixation and dehydration process itself. One way to monitor this potential problem is to cut frozen sections of the same material with a cryostat, omitting dehydration and drying. If the dimensions of structures in frozen sections and dried specimens are similar, shrinkage is probably not a serious problem. If they vary considerably, a stepwise comparison of both frozen and dried sections of tissue at each step of the preparative process may yield some clues about methods to circumvent the difficulty.

Once samples are dried, they should be mounted on specimen stubs, coated, and examined as quickly as possible. If the samples must be stored, they should be put in chambers with Drierite® to minimize rehydration from atmospheric humidity. In particular, a coated specimen is subject to decreased image quality if the specimen swells hygroscopically and develops hairline fractures in the thin metal coating sputtered onto it to increase conductivity.

Critical point dryers are usually provided with various safety interlocks to prevent overheating or overpressurization of the chamber. It is possible to overpressurize and explode the chambers (hence, the common nickname *bomb* often applied to CPD devices). Thus, it is important never to leave a critical point dryer that is in operation and to monitor chamber pressure at all times. Sight glasses in the chambers have been known to fail explosively, so it is a good practice not to sit directly in front of the chamber without some sort of plexiglass safety barrier in place. I have been unable to document any actual cases of CPD failure resulting in serious injury, but failure resulting in unpleasant moments for investigators can be documented from my own laboratory.

2. Freeze Drying

Freeze drying was introduced because, in theory, less shrinkage should occur than with CPD. Also, freeze drying avoids the perceived danger of the CPD process.

The schedule is identical through dehydration. Once the specimen is in 100% ethanol, it may be frozen in liquid nitrogen-cooled freon 12 ($-158°C$), or the sample may simply be cryofixed from its native state (see Chapter 10). The sample is then placed on a precooled specimen stage held at -40 to $-80°C$, put under mild vacuum (10^{-3} Torr or less), and left for 6–8 hr with most specimens. Monolayers need only 1–2 hr, while a 1- to 3-mm-thick specimen might need 3 days. When frost is no longer visible on the cooled stage (indicating that all the fluids have sublimed), the specimen is warmed gradually over a period of 6 hr or so. When the stage and specimen temperatures are slightly above ambient, the chamber may be vented. The specimen should be mounted on a specimen support, coated, and examined as soon as possible.

3. Specialized Drying Agents

Over the last 9 years, two drying agents have been introduced that promised to eliminate the need for capital investment in a CPD or freeze dryer, as well as the need for facing the "bomb." The published photographs of materials produced with hexamethlydisilizane (Nation, 1983) and Peldri II® (Kan, 1990) are quite convincing, but results in our laboratory have been somewhat variable with both techniques.

The hexamethyldisilizane technique requires replacement of the final dehydration wash (100% ethanol) with hexamethyldisilizane, which is then allowed to evaporate at ambient temperature before the specimen is mounted (drying under a fume hood is completed overnight).

Peldri II® is a proprietary fluorocarbon available from Pelco that is solid at room temperature. Specimens in 100% ethanol are placed in a 1:1 mixture of alchohol and Peldri II® for a period of an hour or so, followed by several changes in 100% Peldri II®. In order to transfer and to mix the Peldri II® with other fluids, the temperature must be slightly elevated (to 28°C or so). The tissue may then be placed in a small glass Petri dish under a fume hood at ambient temperature until dry (overnight) and then mounted and coated.

D. Mounting Specimens

Specimens must be mounted onto holders for the specific SEM being used. Most specimen mounts are made of aluminum (or brass, although brass is more expensive than aluminum and confers no added advantage). If microanalysis of samples is desired, specimens are usually put on pure carbon stubs to avoid spurious readings that might be produced from metal stubs.

Particulate samples are best attached to poly-l-lysine coated coverslips (Mazia *et al.*, 1975), which are, in turn, attached to stubs. If microanalysis is to be performed, glass is contraindicated as a support because its numerous impurities might give rise to spurious readings with EDS.

Specimens or coverslips with adherent material may be attached to stubs with colloidal silver paint, colloidal silver paste, colloidal carbon paint, or double-stick tape. If tape is used, some conductive paint or paste must bridge from the stub to the surface of the specimen, since tape is an insulator. Coverslips are also insulators, so they must be treated in the same fashion. The various pastes and paints must be allowed to dry before being introduced into a vacuum chamber, otherwise outgassing of the vehicle for the conductive materials will prevent the development of adequate vacuum. If microanalysis is to be attempted, tape, coverslips, and metal pastes or paints are contraindicated.

E. Coating Specimens

As previously noted, biological tissues are generally good insulators and, if not coated with a conductive material, will build up charge, since the primary electron beam current will have no way to reach ground potential. If the net specimen current (primary current minus secondary current) is not conducted to ground, image distortion results because the specimen builds up a negative charge from the beam, and the beam is repelled by the negative specimen charge.

In the early days of SEM, vacuum evaporators were used to coat specimens. Since vacuum evaporators are directional coaters, specimens were placed on motorized holders within the vacuum system in order to coat surfaces of variable topography evenly. These holders spun a disc containing the specimens at an angle while the whole disc was quickly rotated around a central axis, thus causing the specimens to go up and down in a clockwise direction while being rotated away from and toward the source (Figure 131). This procedure frequently heated the specimen to an unacceptable level while not producing a truly even metal coating on specimens with much topography.

Sputter coaters (see Chapter 8) were introduced to solve these problems. The earliest units still heated the specimens, but the ion cloud of argon and the metal atoms dislodged to coat the specimen produced well-coated surfaces for any topography. Later sputter coaters with a triode design (permanent magnet installed in the center of the metal target) reduced specimen heating to negligible levels.

Specimens are usually coated with gold or gold palladium, because the fine grain size necessary for TEM resolution is not required with conventional SEMs. However, the introduction of 0.7-nm FEG SEMs has recently led to the design and production of sputter coaters, such as the Denton Hi-Res 100 (Denton Vacuum, Inc.), capable of depositing 1.0-nm chromium metal coatings (in contrast to the usual 20 nm gold-palladium coatings used with conventional SEMs), as described by Apkarian (1989).

Specimens being prepared for microanalysis (EDS) should not be coated with metals, because the metal may cause overlapping emissions with areas of interest within the specimen. Carbon coating can be used in place of metal, and as mentioned above, the

FIGURE 131. Specimen holder for coating SEM samples in a vacuum evaporator (from a Ladd Vacuum Evaporator).

specimen is normally placed on pure carbon stubs with colloidal carbon, rather than on coverslips or aluminum stubs, to reduce the chance for spurious readings.

VII. ARTIFACTS AND THEIR CORRECTION

Some artifacts are inherent in specimens of a given type but can be minimized by instrument adjustments. Other artifacts are caused by improper instrument handling. Hayat (1978), Postek *et al.* (1980), and Crang and Klomparens (1988) show a number of examples of SEM artifacts, along with explanations concerning their derivation.

Charging occurs any time a negative charge builds up in the specimen. It can result in areas of strikingly different brightness in the specimen, movement of the specimen (particularly if the specimen has long, thin processes, such as fungal filaments extending above the specimen mount surface), or blurred images in photographs of the subject. To minimize charging, specimens should be freshly coated with a good conductive material such as gold-palladium (about 20 nm thick), the coating should be grounded to the specimen mount, and the specimen mount should be grounded to the SEM chamber. If charging is still severe, the specimen can be moved to the farthest working distance and/or the beam voltage can be reduced. A flat specimen produces less signal (and less potential charging) than a tilted specimen. If none of these suggestions helps, reduce the size and height of the specimen and consider increasing the bulk conductivity in future preparations of the particular specimen type by incorporating osmium during fixation, if possible.

Specimen burning can be a problem in situations that produce charging, as just described. In addition, nonconductive materials, such as wax coatings on leaf epidermal surfaces, can heat up sufficiently during bombardment with the beam to actually melt. In these cases, a heavier coating of conductive metal may help, as well as a decrease in the high voltage. Vascular casts made of nonconductive methacrylate plastic resins are also prone to beam damage. If the magnification is increased significantly above $2,000\times$ (during high-magnification work, the beam is concentrated over a smaller amount of the specimen, leading to greater specimen heating) at 15 kV, the plastic casts will actually begin cracking. Lower accelerating voltage or better conductive coatings can decrease the problem, but taking photographs at lower magnifications will also alleviate the problem. For some reason, thin layers of corneal tissues often are damaged by the beam, particularly when a line scan is used to saturate the SEM filament. If filament adjustment is not done quickly enough, the line scan will actually be burned into the specimen surface, even at low magnification. Another caution is to take low-magnification pictures of such samples prior to taking high-magnification pictures. As the magnification is increased, the specimen area scanned decreases. It is very disappointing to take a high-magnification photograph and then to decrease magnification in order to take a survey photograph, only to find that the specimen has a square burned into it where the high-magnification area was scanned by the beam.

Edge effects occur frequently. If villi of small intestinal surfaces are being observed, the pointed tips will tend to produce excess signal. If a cut surface of tissue is examined, the edges of lacunae will tend to be much brighter than surrounding flat areas. Decreasing high voltage, spreading the beam (condenser setting decreased), moving the specimen farther from the collector (increasing working distance), and making sure that the specimen is not tilted can decrease these effects. Preparing the specimen with osmium may also help. Working with brightness and contrast controls on the collector may also allow the operator to decrease an edge effect by adjusting the brightness and contrast of the edge to desired levels, disregarding the concomitant decrease in illumination and contrast of the surrounding area, if it is not the area of true interest.

Some specimens consisting of long filamentous structures (algal or fungal filaments or fruiting structures) extending well beyond the specimen stub surface will tend to move when exposed to the electron beam, making them unphotographable. One solution to this problem is to lay such structures down on a substrate, such as double-stick tape, so that they are closer to the specimen stub surface and, thus, better grounded.

VIII. SPECIALTY SEMS: FEG, LV, AND ESEM DEVICES

Hitachi's S-900 (Hitachi Scientific Instruments, Gaithersburg, MD) was the first of the contemporary SEMs to achieve high resolution, followed shortly thereafter by the JEOL 6400F. The use of a field emission gun (FEG) with its improved coherence and ability to work well at low accelerating voltage allows these instruments to provide 1.5- to 0.7-nm resolution with biological materials, permitting detailed examination of structures, such as the component parts of nuclear pores, previously possible only with a TEM (Ris, 1991). This technology is a major breakthrough for biologists because we can now examine structures with an SEM at the same level of resolution that has been standard for years with chemically fixed TEM samples (2 nm). Both types of microscopes have resolving capabilities that exceed our current conventional chemical fixation capabilities.

The Environmental Scanning Electron Microscope (ElectroScan, Danvers, MA) has already been mentioned as a major breakthrough in the SEM business because of its ability to examine hydrated specimens or specimens at almost atmospheric pressure (up to 50 Torr). This is possible because of a differential aperture just below the final lens, with high vacuum maintained in the beam-generating and -focusing part of the column, while low vacuum is tolerated in the specimen chamber. The collector is just below the differential aperture, and the specimen is located very near the collector (Fig. 132). Because of this physical arrangement, the scattering of the beam from any

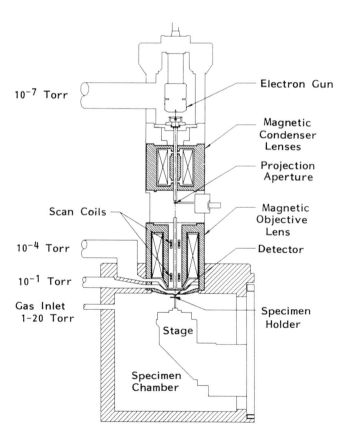

FIGURE 132. Diagram of the ESEM column. (Courtesy of Electro Scan.)

atmosphere present is minimized, and a large proportion of the signal produced from the specimen surface is collected.

JEOL has recently developed an instrument (JSM-5300LV) that can be used at specimen chamber pressures as high as 1 Torr. If a biological specimen is charging (which is not uncommon, as stated above), the chamber can be leaked slightly to introduce air that becomes ionized by the electron beam. The gas ions then can allow charge buildup in the specimen to pass to the grounded chamber, thus discharging the specimen. When the chamber is at a relatively high pressure (up to 1 Torr), the specimen must be examined with backscatter imaging, but when the specimen chamber is at high vacuum, the instrument performs just like any conventional SEI SEM. This instrument should also be of use with vacuum-sensitive specimens that get damaged by conventional high-vacuum chambers.

REFERENCES

Apkarian, R.P. 1989. Condenser/objective lens SE-I imaging of chromium coated biological specimens using a Schottky field emission source. In: *Proc. 47th Ann. Meeting Elec. Microsc. Soc. Amer.*, p. 68, San Francisco Press, Inc., San Francisco.

Claugher, D. (ed.). 1990. *Scanning electron microscopy in taxonomy and functional morphology.* Clarendon Press, Oxford.

Crang, R.F.E., and Klomparens, K.L. 1988. *Artifacts in biological electron microscopy.* Plenum, New York.

Edwards, D.F., Patton, C.S., Bemis, D.A., Kennedy, J.R., and Selcer, B.A. 1983. Immotile cilia syndrome in three dogs from a litter. *J. Am. Vet. Med. Assoc.* 183:667.

Haggis, G.H. 1972. Freeze-fracture for scanning electron microscopy. In: *Proceedings of the Fifth European Congress on Electron Microscopy*, p. 250, The Institute of Physics, London.

Hainfeld, J.F. 1977. Understanding and using field emission sources. *Scann. Electr. Microsc.* 1:591.

Hanstede, J.G., and Gerrits, P.O. 1982. A new plastic for morphometric investigation of blood vessels, especially in large organs such as the human liver. *Anat. Rec.* 203:307.

Hayat, M.A. 1978. *Introduction to biological scanning electron microscopy.* University Park Press, Baltimore.

Humphreys, W.J., Spurlock, B.O., and Johnson, J.S., 1975. Transmission electron microscopy of tissue prepared for scanning electron microscopy by ethanol-cryofracturing. *Stain Tech.* 50:119.

Kan, F.W.K. 1990. Use of Peldri II as a sublimation dehydrant in place of critical-point drying in fracture-label cytochemistry and in backscattered electron imaging fracture-label. *J. Elect. Microsc. Tech.* 14:21.

Kelley, R.O., Dekker, R.A.F., and Bluemink, J.G. 1975. Thiocarbohydrazide-mediated osmium binding: A technique for protecting soft biological specimens in the scanning electron microscope. In: *Principles and techniques of scanning electron microscopy: Biological applications, Vol. IV*, Hayat, M.A. (ed.), Van Nostrand-Reinhold Co., New York.

King, E.J., and Brown, M.F. 1983. A technique for preserving aerial fungal structures for scanning electron microscopy. *Can. J. Microbiol.* 29:653.

Mazia, D., Schatten, G., and Sale, W. 1975. Adhesion of cells to surfaces coated with polylysine. Applications to electron microscopy. *J. Cell Biol.* 66:198.

Nation, J.L. 1983. A new method using hexamethyldisilazane for preparation of soft insect tissues for scanning electron microscopy. *Stain Tech.* 58:347.

Pawley, J.B. 1988. Low voltage scanning electron microscopy. *EMSA Bull.* 18:61.

Pease, R.F.W. 1963. High resolution scanning electron microscopy. Ph.D. Dissertation, Cambridge University, England.

Postek, M.T., Howard, K.S., Johnson, A.H., and McMichael, K.L. 1980. *Scanning electron microscopy. A student's handbook*. Ladd Research Industries, Burlington, VT.

Ris, H. 1991. The three-dimensional structure of the nuclear pore complex as seen by high voltage electron microscopy and high resolution low voltage scanning electron microscopy. *Elect. Microsc. Soc. Am. Bull.* 21:54.

Suzuki, F. 1982. Microvasculature of the mouse testis and excurrent duct system. *Am. J. Anat.* 163:309.

Von Ardenne, M. 1938. Das elektronene-rastermikroskop, praktische ausfuhrung. *Z. Tech. Phys.* 19:407.

Zworykin, V.K., Hillier, J., and Snyder, R.L. 1942. A scanning electron microscope. *American Society for Testing Materials Bull.* 117:15.

CHAPTER 10

Cryotechniques

The first question that arises when considering cryotechniques in general is, what advantages are there to cryofixation? As we discussed in Chapter 1, fixation ideally instantaneously stops biological activity, immobilizes cellular components, and enables them to withstand any further processing procedures. In principle, biological materials can be fixed with either chemical or physical (cold, heat) techniques. Good chemical fixation requires the fast diffusion of chemical agents through membranes and cytoplasmic components. Cryofixation, on the other hand, requires the rapid diffusion of heat out of the specimen. The reason cryofixation can be superior to chemical fixation is that the rate of chemical diffusion into the specimen is much slower than the rate of heat diffusion out of the specimen. However, most cryofixation techniques adequately freeze samples to no more than 15 μm, in contrast to chemical fixation, which consistently fixes samples to a depth of 0.5 mm. Cryofixation is often inferior for structural studies because of the limited sample size available, but is vastly superior to chemical fixation for many microanalytical or immunocytochemical procedures.

The second question that requires investigation concerns the freezing process itself. To understand the final product of cryofixation, it is necessary to examine the effect of freezing biological samples. When water turns to crystalline ice, the fine structure of membranes vanishes and samples become cloudy. The improper freezing of samples can result in the growth of ice crystals by recruitment of water molecules from surrounding cytoplasmic areas, thus separating water from other nonaqueous molecules. As the crystals continue growing, they may draw extracellular water into cells. At a certain point, the attraction of water molecules to the growing ice crystals is overcome by the strength of the attraction of the remaining free water to other cellular constituents, thereby bringing a halt to further ice crystal growth. If a frozen cell is examined closely, ice crystals will be seen that have displaced normal cellular architecture and have actually penetrated structures such as membranes. The proteinaceous components of the cytoplasm and nucleoplasm become more concentrated than they were before freezing because they have been dehydrated during the growth of adjacent ice crystals. Thus, after improper freezing of cells physical damage will have been caused by crystal formation, water will have been gathered into crystals at the expense of dehydrating the areas where the water was located originally, and various cellular structures will have been physically displaced by the growth of ice crystals.

Modern cryotechniques attempt to avoid the formation of crystalline ice by freezing the sample so rapidly that migration of water molecules, followed by nucleation and crystal formation, cannot occur. If frozen quickly enough, samples will have vitreous (amorphous) ice formed without a crystal lattice. Unfortunately, technical difficulties usually limit us to samples that have microcrystalline ice whose crystals are not readily visualized by electron microscopy. Careful examination with a cold stage and electron diffraction will still reveal diffraction patterns consistent with the crystalline structure of frozen water.

This chapter is devoted to exploring the history of cryotechniques, the methods for fixing cells with cryotechniques that produce minimal crystallization, and the use to which these cryopreserved materials may be put. Three texts that are devoted to cryotechniques are those by Robards and Sleytr (1985), Roos and Morgan (1990), and Steinbrecht and Zierold (1987)

I. HISTORY

A. Organismal Period

This earliest period in cryobiology was devoted primarily to determining what organisms and life stages of organisms could withstand freezing. In the mid-1660s, Henry Power froze eel worms (nematodes) found in vinegar and discovered that they were still alive after thawing. In the late 1700s, Spalanzani froze rotifers, embryos, eggs, and adult forms of various organisms, and found that the embryos and eggs survived freezing more successfully than adults, which he attributed to the oily fluids that the embryos and eggs contained.

One of the major problems during this organismal period of cryobiology was the poor cryogens available. A saltwater and ice slush was usually the best cryogen available, which caused very slow freezing and, thus, a high probability of ice crystal formation. Toward the end of the 19th century, cryotechnology had advanced to the point where −200°C temperatures were possible and the list of organisms and tissues frozen and thawed had grown considerably. In addition, freeze-drying had been attempted on some tissues.

B. Mechanistic Period

Beginning in the 1900s, investigators began looking at the mechanisms involved in successful freezing of organisms as defined by the process producing minimal tissue damage.

In the early 1900s, the rate of freezing was determined to be a critical issue. By the 1930s, studies on the actual formation of ice crystals in protoplasm at the microscopic level demonstrated the cytoplasmic displacements that took place. By the 1940s it was determined that it was necessary to form vitreous ice within tissue for survival. During the same period, the cryoprotectant glycerol was used during cryofixation of avian sperm and other cells and tissues to decrease the formation of crystalline ice. Later workers have subsequently demonstrated that some overwintering insects naturally contain as much as 20% glycerol in their cells (Lee and Denlinger, 1991), the percentage usually used as a cryoprotectant for freeze-fracture work.

C. Cytological Period

This period was characterized by the successful use of cryotechniques for ultrastructural studies, a period that was ushered in by the development of freeze-fracture equipment to fracture membranes and cells for subsequent replica production and TEM examination. The first device was designed by Steere (1957) and became the commercially available Denton freeze-fracture apparatus. This was followed by the production of a freeze-fracture apparatus by Moor et al. (1961), which was much easier to use and gained wide acceptance when it was marketed by Balzers.

Sjostrand tried freeze-drying samples for electron microscopy in the 1940s with poor results but achieved reasonable preservation methods by 1958 (Sjostrand and Baker, 1958). Linner et al. (1986) described a highly effective method for cryofixing specimens and drying them in a molecular distillation device that was superior to previous freeze-drying equipment.

Fernandez-Moran, who was so instrumental in the development of diamond knives for ultramicrotomy, began exploring methods for freeze-substitution of samples wherein cryofixed samples were subsequently dehydrated at low temperatures and embedded for sectioning at room temperature (Fernandez-Moran, 1957).

By the 1980s, a number of different specimen-freezing methods had been developed that produced samples with vitreous or microcrystalline ice, and the techniques for cryoultramicrotomy pioneered by Bernhard (1965) had become routine. Tokuyasu (1973) introduced a technique for freezing tissues for cryoultramicrotomy that stimulated the development of immunolabeling of frozen ultrathin sections, which continues to be the major application for cryoultramicrotomy to biological samples.

In addition, in the 1980s low-temperature acrylic resins were developed for embedding specimens prepared by cryotechniques (see section on resins in Chapter 1).

II. PURPOSE

The purpose of cryofixation is to stop biological activity, which happens in about 10 msec (Gilkey and Staehelin, 1986), to prevent nucleation and crystallization of water, and to preserve the spatial relationships and as much biological activity of cellular constituents as possible.

Ice crystal development begins with a nucleation point to which water from the surrounding medium migrates, resulting in the growth of an ice crystal. This causes the solute concentration in regions between the ice crystals to increase, eventually reaching a concentration sufficient to prevent further growth of the adjacent ice crystal. At this point a solid eutectic has been formed, consisting of water and solute in a vitrified (glassy) state. In living organisms, ice crystal formation is thought to occur between -2 and $-80°C$, although some workers (Linner et al., 1986) maintain that some crystallization can occur even below $-100°C$.

The lowest temperature in the range is the recrystallization point. If a rapidly frozen specimen is allowed to warm to a temperature above this point, ice crystals will form. It is important to remember that ice crystals can form from freezing a specimen too slowly or warming a specimen too slowly.

Water that is ultrarapidly frozen with a slam-freezing device held at $-196°C$ with liquid nitrogen will demonstrate a structure no different from that of liquid water when studied with diffraction techniques (Linner et al., 1986). This vitreous or amorphous ice goes through a transition if warmed to -135 to $-120°C$ and becomes micro-crystalline in structure by the formation of cubic ice crystals. As the temperature increases from -120 to $-40°C$, several types of ice are formed. By the time the temperature has warmed to $-35°C$, further crystallization results in large ice crystals, and some melting has also occurred. The size of ice crystals is dependent on the rate of freezing or thawing and the solute concentration (cytoplasmic solutes and/or cryoprotectants added). At the high freezing rates achieved with modern cryotechniques, most of the water in a specimen crystallizes homogeneously into 20-nm^3 crystals that are difficult to resolve ultrastructurally, and the specimen is commonly described as vitrified. If these samples are examined in the native frozen state with a cold stage-equipped TEM set up for diffraction, crystal structure can still be demonstrated, so the sample is not truly vitrified but actually has a microcrystalline structure.

The concept of cryoprotection arose out of the knowledge that certain cell types with low water content (bacterial spores, some seeds, fungal spores) may be subjected to freezing temperatures and then thawed without appreciable damage. Cryoprotectants have been added to other types of cells and tissues to improve their capacity to withstand the freezing process. Glycerol, ethylene glycol, dimethylsulfoxide (DMSO), and sucrose have been used extensively in this regard. A buffered solution of 20–25% glycerol is one of the most extensively used cryoprotectants and was originally developed for use with freeze-fracture methods. Cells and tissues are usually first fixed with aldehydes, since many unfixed cells will shrink in glycerol because of osmotic effects. The fixative is then rinsed out with an appropriate buffer, and the cells are immersed in the buffer/glycerol mixture for 30–60 min prior to freezing. The larger the piece of tissue, the higher the concentration of glycerol used. More than 2 hr in glycerol is considered undesirable. Some workers mix glycerol with sucrose (10% glycerol/20% sucrose) with good results. If the glycerol concentrations are above 35–40%, specimen etching (if desired) becomes impossible because the water contained in the specimen cannot sublime from the surface under normal etching conditions.

Tokuyasu (1973) demonstrated that a light aldehyde fixation permeabilizes cytoplasmic membranes sufficiently to eliminate the possiblity of cryoprotectant-induced osmotic effects (2.1 M sucrose induces severe osmotic damage in unfixed cells or tissues but causes no damage if the cells have been lightly fixed in 2% buffered paraformaldehyde first).

III. CRYOGENS

Costello and Corless (1978) examined the actual temperatures and freezing rates of cryogens used for cryofixation (Table 8). Examination of this table reveals that the temperature of the cryogen is not the most important factor in determining its suitability for freezing specimens. Liquid nitrogen (−196°C) and nitrogen slush (−207°C) are the coldest cryogens in the list but have the slowest freezing rates (16,000 and 21,000°C/sec, respectively). To be effective for cryofixation, the cryogen must remain in contact with the specimen surface. Nitrogen has a low boiling point (−196°C) and quickly turns into a gas when confronted with a specimen at ambient temperature. The nitrogen gas formed is a significant thermal insulator, slowing the specimen-freezing process. The most effective cryogens are those that remain liquid when cooled almost to liquid nitrogen temperature, thus allowing them to wet the specimen for quickest cooling. In addition, they have a boiling point a number of degrees higher, so that they will remain liquid while the specimen is plunged (or sprayed) into them. Propane is the cryogen of choice for immersion, spray, or jet freezing, since it has a low temperature and a fast freezing rate.

With metal mirror (slam) freezing, a polished metal surface is cooled with liquid nitrogen to −190 to −196°C, at which time the specimen is brought into contact with the metal surface and held there until cryofixation is complete. With the specimen held in intimate contact with the nitrogen-cooled metal, the problem of nitrogen boiling is obviated.

IV. SAFETY PRECAUTIONS

All of the cryogens must be handled carefully to avoid injury to the user. Liquid nitrogen can produce serious freezing damage if trapped inside gloves, shoes, or other

TABLE 8. Temperature and Freezing Rate of Cryogens
Used for Cryofixation

Cryogen	Temperature at cryogen surface	Freezing rate (°C/sec)
Propane	−190°C	98,000
Freon 13	−185°C	78,000
Freon 22	−155°C	66,000
Freon 12	−152°C	47,000
Isopentane	−160°C	45,000
Nitrogen slush	−207°C	21,000
Liquid nitrogen	−196°C	16,000

articles of clothing. If liquid nitrogen is spilled on bare skin, its rapid evaporation usually results in minimal damage. On the other hand, the other cryogens do not evaporate quickly and so can cause severe damage to exposed skin with which they come into contact. An added danger with propane is its flammability. Liquid propane, if exposed to the atmosphere, is cold enough to cause condensation of air (containing oxygen) on its surface or the surface of the vessel in which it is contained. This combination of cooled propane and oxygen is extremely flammable. For this reason, sample freezing should take place under a fume hood to avoid a dangerous level of gaseous propane. After specimen freezing is completed, the cryogen should be evaporated under the fume hood at a slow rate. Liquid propane and gaseous propane are both heavier than air. They will seek the lowest level in a room and can actually "run" out from under a fume hood down onto a laboratory floor if the hood is turned off. As the gaseous propane diffuses (or liquid propane vaporizes), any spark or flame in the vicinity can cause an explosion.

Finally, liquid nitrogen is generally treated as a relatively benign substance, since it rarely causes serious freezing unless trapped against body parts, but the gaseous form of nitrogen can also be extremely dangerous. If the typical 160-l liquid nitrogen tank is stored in a small room without adequate ventilation, the very act of transferring nitrogen to another container for transport to specimen preparation areas can be dangerous since large volumes of gaseous nitrogen can be produced. In a poorly ventilated space, this can result in air being displaced by nitrogen, leading to the real possiblity of suffocation.

V. FREEZING METHODS

Four of the five basic freezing methods (immersion, metal mirror, jet, and spray freezing) will produce well-frozen cells only within 10–15 μm of the specimen surface. Below that thin layer the thermal mass of the rest of the tissue prevents sufficient heat extraction to produce good freezing. This limits the overall mass of tissue that can be cryofixed adequately. Specimens over 0.2 mm in thickness are clearly unnecessary and counterproductive. Monolayers of cells will typically be more successful subjects for cryofixation than multicellular tissues. Various methods have been developed to try to increase this depth of cryofixation, but the only one that has shown promise is the high-pressure freezer marketed by Balzers that has been reported to produce samples that are well frozen to depths of 0.5–0.6 mm, equal to the normal depths for good chemical fixation (Gilkey and Staehelin, 1986; Moor, 1986).

A. High-Speed Plunging/Immersion

Cryoprotected material may be frozen by being quickly plunged by hand (held with forceps) directly into liquid nitrogen. If crystal damage to tissue is observed, the specimens can be plunged by hand into a moderate cryogen, such as Freon 13 or 22 cooled by liquid nitrogen, with improved results.

Specimens that have not been cryoprotected have more stringent requirements for proper cryofixation. Those specimens must be immersed more quickly into a more effective cryogen, such as propane cooled with liquid nitrogen. Robards and Crosby (1983) studied the effect of the entry velocity of specimens into liquid nitrogen-cooled propane and found that specimen cooling increased with increasing speed of entry up to about 3 m/sec^{-1}. Above that speed, cooling rates decreased, possibly because of trapped air pockets or induced turbulence insulating the specimen surface. The only exception to this velocity limitation is specimens that are prepared with the aerodynamic shape of blades or needles, which have been successfully frozen with immersion rates of 15 m/sec^{-1} (Handley et al., 1981). Immersion-freezing devices that are laboratory built or industrially manufactured (Fig. 133) typically use springs or gravity to achieve the desired specimen velocity. It is important to keep the cryogen surface near the top of the vessel into which the specimen is plunged. If the cryogen is at the bottom of a long tube, the air space above the cryogen can be cool enough to freeze the specimen as it passes through on its way to the cryogen, but will still be too warm to freeze the specimen adequately, resulting in ice damage.

Not only are the specimen velocity, the placement of the cryogen in the freezing chamber, and the nature of the cryogen of importance, but also the shape and size of the specimen are factors that must be considered. Handley et al. (1981) put their specimens (blood) behind a thin titanium foil with an edged profile to improve heat transfer.

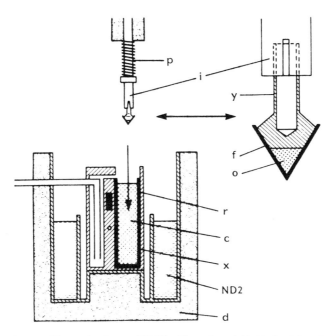

FIGURE 133. High-speed immersion freezer (KF-80) with specimen (o) behind metal conical foil (f) driven by spring (p) into liquid propane cryogen (c) cooled with liquid nitrogen (ND2) in surrounding chamber. (Redrawn from diagram, courtesy of Leica, Inc.)

Other workers have used foils of copper and gold.

Dubochet *et al.* (1983) described a technique using a 400-mesh grid to which a droplet of subcellular particles up to the size of bacteria or mitochondria (0.5-μm diameter) had been added. The grid was wicked almost to dryness with a piece of filter paper, thus leaving a liquid film about 100 nm thick. The grid with its thin film of specimen/liquid was immersed in liquid nitrogen-cooled propane, cryotransferred to a TEM cryostage, and observed with a low-dose unit to decrease heating of the frozen specimen. The immersion speed did not have to be particularly fast, since the heat-transfer characteristics of the metal grid were excellent.

Thin slices of tissue or suspensions of cells (20 μm thick or less) have been placed between thin metal sheets (titanium or gold, typically) about 50 μm thick, resulting in excellent heat transfer upon submersion in the cryogen (Gilkey and Staehelin, 1986). Some devices (Reichert KF-80) have a holder that grasps locking forceps holding the specimen. The KF-80 then plunges the specimen-holding forceps into a cylinder of nitrogen-cooled propane by spring-loaded pressure.

B. Spray Freezing

This technique (Bachmann and Schmitt-Fumian, 1973) is based on the principle that cells contained in droplets of fluid of about 10- to 30-μm diameter that are sprayed into liquid nitrogen-cooled propane will be quickly frozen because of the favorable surface-to-volume ratio. The frozen droplets are then transferred to a surface held at $-85°C$ and put under vacuum to extract the liquid propane. After the propane is removed, butylbenzene (f.p. $-95°C$) is mixed with the droplets to produce a paste that can be transferred to freeze-fracture planchets and dipped into liquid nitrogen before freeze-fracturing. Gilkey and Staehelin (1986) suggest that the high freezing rates possible with this technique ($>100,000°C/sec$) are outweighed by the tediousness of the processing and the difficulty in obtaining good fractured surfaces because of smearing of the butylbenzene by the fracturing blade.

C. Jet Freezing

Müller *et al.* (1980) introduced the double-jet propane freezer (Fig. 134). With this technique, a thin specimen is sandwiched between two metal planchets or sheets, which protect the surface of the sample from the high-velocity propane jet and ensure rapid heat transfer. The specimen is held between the two jets through which liquid nitrogen-cooled liquid propane is propelled by high pressure toward the sample. Since the specimen is cooled from both sides at once, Gilkey and Staehelin (1986) report that the specimen is, on average, frozen to a depth of 40 μm. Balzers produces a commercial jet-freezing device.

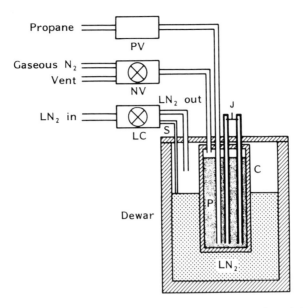

FIGURE 134. Propane jet-freezing device from Müller *et al.*, 1980. Propane (P) cooled by liquid nitrogen (LN$_2$) is forced by gaseous N$_2$ into two tubes with small orfices (J), producing a jet of cooled propane. (Redrawn from Gilkey and Staehelin, 1986, courtesy of Wiley-Liss, Inc.)

D. High-Pressure Freezing

This technique was developed by Moor and his colleagues (Moor and Riehle, 1968, Moor *et al.*, 1980). A commercially made unit produced by Balzers is shown in Fig. 135.

Specimens are placed between two metal planchets to help reduce the possibility of the samples being crushed. The specimen is placed within the high-pressure chamber, the chamber is pressurized, and then liquid nitrogen is forced into the chamber under pressure to freeze the sample.

The principle behind this technique is that high pressure causes a reduction in the cooling rate necessary to produce vitreous ice as water is frozen. In addition, the rate of ice crystal formation and growth is reduced by the application of high pressure during freezing.

It was found that pressurizing the chamber with liquid nitrogen alone resulted in a cooling rate that was too slow. The system currently available forces a small bolus of isopropyl alchohol (2 ml) into the chamber during initial pressurization so that the pressure rises to about 2000 bars before freezing begins when the nitrogen enters the chamber at about 2500 bars and is directed onto the specimen from jets in the chamber. The pressurization and freezing cycle, which lowers the specimen temperature to −150°C, takes about 0.5 sec. The specimen temperature remains below −100°C for

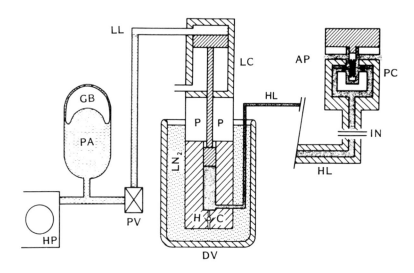

FIGURE 135. Diagrammatic representation of Balzers high-pressure freezer as designed by Moor *et al.* HC = high-pressure cylinder; DV = Dewar; PP = pressure piston; HL = high-pressure line; PA = pressure accumulator; GB = gas balloon; PV = pressure valve; IN = isopropyl alchohol insertion port; PC = pressure chamber containing specimen in metal sandwich; AP = exhaust aperture; LC = low-pressure chamber; LL = low-pressure line; HP = hydraulic pump; LN₂ = liquid nitrogen. (Redrawn from Gilkey and Staehelin, 1986, courtesy of Wiley-Liss, Inc.)

about 6–8 sec, allowing the operator to remove the sample for storage in liquid nitrogen before it thaws.

Moor (1986) points out several potential problems with high pressure freezing:

1. *Euglena* exposed to 2,000 bars for 0.1 sec have a 90% survival rate, while all those exposed for 4 sec die, indicating that freezing must quickly follow pressurization to ensure good results.

2. It is important to raise the chamber pressure to 2,100 bars before freezing takes place, which is the purpose of the isopropyl alchohol, thus delaying the entrance of the liquid nitrogen into the chamber until it has been sufficiently pressurized (this is critical since the liquid nitrogen is being used both to pressurize the chamber and to freeze the specimen). He reports that it takes 15 msec to pressurize the chamber sufficiently for the freezing process to work effectively.

3. The specimen should be cooled for a period following freezing so that there is no chance of the temperature rising above the crystallization point during specimen transfer to liquid nitrogen for storage.

4. Finally, as with all freezing techniques, the surface of the specimen freezes before the interior. In order to achieve the great depths of freezing (0.5–0.6 mm) reported for this technique, it is necessary to hold the pressure during freezing for at least 0.3 sec to ensure thorough freezing.

E. Metal Mirror/Slam Freezing

Van Harreveld and Crowell (1964) are credited with being the first to freeze samples for electron microscopy on cooled metal blocks. At the present time there are a variety of commercially produced slam freezers (Fig. 136) available from electron microscopy supply houses or directly from equipment manufacturers (Ted Pella, Inc.; RMC, Inc.; Leitz/Reichert; LifeCell Corp.; BioRad).

All of these devices consist of a polished metal surface, either of copper or gold-plated copper, which is cooled with liquid nitrogen or liquid helium. The specimen is attached to a plunger above the metal mirror and is subsequently impinged upon the mirror surface by gravity, electromagnetic force, gas pressure, or springs. Some of the devices rely on nitrogen gas rising up from the reservoir cooling the mirror to keep air from condensing on the mirror surface, while others have the mirror within a low vacuum chamber while it is being cooled and only open the chamber milliseconds prior to slamming the specimen onto the mirror. With most of the devices, the metal mirror is fabricated from copper. It is polished and cleaned prior to slamming so that the specimen will be slammed onto a blemish-free surface for best thermal conductivity. The mirror is then placed into the slamming chamber and cooled with liquid nitrogen. Liquid helium can also be used, but is much more expensive and does not appear to produce a significant difference in final specimen quality. The chamber containing the target mirror is covered during the cooling process to minimize the formation of

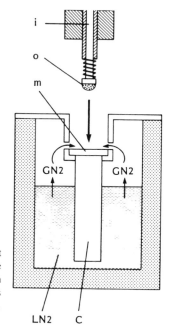

FIGURE 136. Detail of chamber and specimen plunger of Reichert KF-80 set up as a slam-freezer. Copper bar (C) with polished surface (m) is kept free of frost by gaseous nitrogen (GN2) produced from liquid nitrogen reservoir (LN2) cooling copper bar. Specimen (o) is driven onto mirror by a spring wrapped around the inner shaft (i). (Redrawn with permission from Leica, Inc.)

insulating frost on the mirror surface. Just as the specimen driver is being released, the chamber door is opened. After the specimen is slammed, it is held on the cooled metal for 20 or 30 sec to cool areas behind the superficial well-frozen 10–15 μm. After this cooling period, the specimen is transferred quickly to liquid nitrogen for storage until subsequent processing steps.

One problem that was noted during the development of slammers was that a specimen striking the target because of gravity (a free-fall device) tended to bounce after initial contact, briefly breaking thermal contact with the target. It was noted that this was sufficient to cause ice damage. To eliminate this problem, some of the devices have used shock absorbers to prevent rebound (Gentleman Jim®), and others have used gas-driven plungers or electromagnetic drivers to slam the specimen holder against the target while not allowing any recoil.

Another problem concerns the target surface. All targets develop frost after removing the frozen specimen. The frost must be removed prior to the next slamming cycle by thawing the target. In the process of heating and then recooling a copper target, the surface tends to oxidize, thus providing poorer thermal contact surfaces for the specimen. With most devices, it is necessary to remove the target and to polish it between slams for this reason. The LifeCell CF-100 has a novel approach to this problem in that it has a gold-plated copper target that does not oxidize. Because of this feature, 30–50 slams can be done without having to polish the target.

VI. USES OF FROZEN SPECIMENS

Once a specimen is frozen by one of the five techniques described above, further choices must be made between different sample handling procedures determined by the eventual sample fate. Frozen samples may be prepared for freeze-fracture, cryosubstitution, molecular distillation, immunolabeling, or microanalytical approaches.

Another parameter of cryotechniques concerns the quality of the final product examined. Most cryotechniques produce images that appear significantly different from conventional chemically fixed and epoxide-embedded materials (Fig. 137). Thus, when determining what techniques to employ, it is necessary to focus on the question being asked. If straight structural studies are desired, cryosubstitution or conventional chemical fixation will yield the most cellular detail. On the other hand, if immunolabeling is the primary objective, cytological preservation is sacrificed for antigen preservation and accessibility. Some of the immunolabeling procedures sacrifice membrane preservation, but compartments can still be identified, so antibody labeling can be described as occurring inside or outside of various cellular compartments despite the poor preservation of membranes. Frozen sections or samples prepared by molecular distillation also look significantly different from conventionally prepared samples, but both techniques can minimize exposure of samples to fixatives, solvents, heat, and strongly cross-linked resins. Thus, structural detail is often sacrificed for the preservation of other features being sought with cryotechniques.

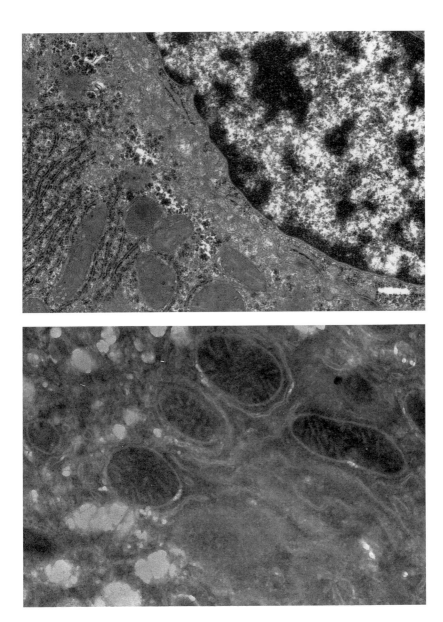

FIGURE 137. Top: Liver, conventional chemical fixation and embedment, post-stained with uranyl acetate and lead citrate. 19,200×. Bottom: Liver, Tokuyasu (1973) technique. Light aldehyde fixation followed by cryoprotection in 2.3 M sucrose, cryofixation in Freon cooled with liquid nitrogen, cryoultramicrotomy, and staining with ammonium molybdate. 27,000×.

A. Cryoultramicrotomy

Microtomes used for cryoultramicrotomy can have thermal advance (LKB Cryonova) or mechanical advance mechanisms (Dupont/Sorvall MT-5000 and 6000; RMC MT-6000 and 7000; Reichert Ultracut E, S). Since most workers cut ultrathin frozen sections on dry knives, eliminating the possibility of evaluating section thickness by the normal interference colors produced by sections floating on water when illuminated with diffuse light, mechanical specimen advance is preferred so that section thickness can be accurately determined. Thermal advance units routinely produce more variable section thickness, and so are less suitable for frozen-section work.

Cryoultramicrotomes equipped for fast arm return speeds, along with adjustability to very slow cutting speeds (< 0.1 mm/sec) and small cutting windows (< 1.0 mm), will give the most satisfying results for native frozen materials (those cryofixed without any chemical treatment) destined for microanalysis with or without cryotransfer to a cryostage in a TEM. These materials are generally sectioned at $-150°C$ after ultrarapid freezing without cryoprotection. If the cutting speed is too fast, section melting can occur, as revealed by the presence of knife marks in sections.

Block-face trimming is critical for success. The block should be reduced to 0.2–0.5 mm length, with the upper and lower edges exactly parallel, and then the face should be trimmed smooth with a glass knife. Glass knives cooled to $-105°C$ to produce sections by the Tokuyasu (1973) technique for immunolabeling are considerably harder than when used at room temperature and will cut many good frozen sections. Diamond knives are considered superior when cutting 10- to 50-nm sections of native frozen materials for microanalysis.

1. Immunolabeling (Tokuyasu, 1973)

Samples are typically fixed for 1 hr in 2–4% phosphate-buffered (pH 7.0) formaldehyde made up freshly from paraformaldehyde powder. After rinsing the samples in the same buffer, they are immersed in 2.3 M sucrose for 1 hr before being cryofixed. Because of the cryoprotection provided by the sucrose, they can usually be manually immersed in liquid nitrogen, producing adequate cryofixation. If excessive freezing artifacts (Fig. 138) are observed, immersion fixation in Freon cooled with liquid nitrogen may be necessary. The knife temperature ($-105°C$) should be slightly cooler than that of the specimen ($-100°C$). Sections for immunolabeling are picked up with a drop of 2.3-M sucrose just before the sucrose freezes (Fig. 139), and are transferred to the surface of a grid, which is then floated on distilled water in a small Petri dish to remove the sucrose. After blotting the grid dry, it can be immunolabeled and then negatively stained with ammonium molybdate.

2. Frozen Sections for Microanalysis

Native frozen tissues that have had no chemical fixation or cryoprotection can be cryosectioned, cryotransferred, and examined on a cryostage-equipped TEM. Native

FIGURE 138. Sample with freezing artifacts in the form of a generalized moth-eaten appearance of slam-frozen and cryosubstituted sample caused by the formation of numerous small ice crystals in cytoplasmic areas away from the sample surface (S) of mouse skeletal muscle. 2500×.

frozen sections can yield information concerning the location of diffusible elements, such as sodium and potassium, as well as insoluble materials.

Since no chemical fixation or cryoprotectants are employed, native frozen sections must be frozen by the most exacting methods. Most workers use jet-, slam-, or high-pressure freezing to produce materials suitable for cryosectioning. Samples are usually sectioned at −120 to −150°C. Dry sections are moved to film-coated grids

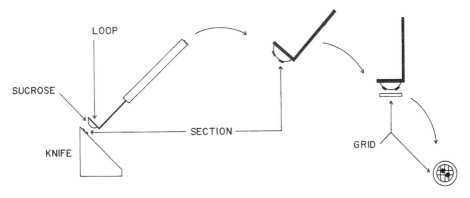

FIGURE 139. Tokuyasu (1973) procedure for transferring frozen sections to a grid with a loop containing a drop of sucrose.

within the cryosectioning chamber with a pre-cooled eyelash and then pressed down onto the grid surface with a cooled nylon-tipped rod. Static electricity is often a problem when working at low temperatures with native frozen sections, so many workers use devices such as a Zerostat® to decrease electrostatic charges in the ultra-microtome's cryochamber.

The grids with frozen sections are then cryotransferred to a cryostage-equipped TEM for microanalysis (see Steinbrecht and Zierold, 1987 for further applications).

3. Structural Analysis

Examination of sections prepared from native frozen materials or samples prepared by Tokuyasu (1973) techniques can provide structural information about tissues exposed to minimal preparatory techniques. Native frozen sections have been exposed to no chemical treatments whatsoever, while Tokuyasu sections have been only lightly fixed, rinsed in buffers, put into sucrose, and frozen, thereby eliminating osmium, dehydration agents, resins, and the heat used to polymerize most conventional resins.

In most cases, these frozen sections must be stained to produce contrast (e.g., with ammonium molybdate), although examining these materials with a Zeiss 902 TEM, which has the ability to eliminate scattered electrons from the imaging process, should produce good images, even without staining. Conventional chemical fixation and cryosubstitution produce different images (Fig. 140). Cryosubstitution often reveals materials not preserved by conventional chemical fixation, but freezing damage during initial cryofixation is a frequent artifactual element found in photographs of these materials.

B. Cryosubstitution

Cryosubstitution techniques have grown in popularity over the last 10 years for purely morphological investigations. With conventional chemical fixation, plasmalemmal profiles are often wavy, while cryosubstitution produces smooth profiles that are considered to be more representative of the living state. Studies of various fungi have shown that the populations of vesicles involved in tip growth of fungal hyphae, which appear homogeneous with conventional chemical fixation, show varied contents with cryosubstitution methods (Howard and Aist, 1979).

Cryosubstitution starts with cryofixation followed by transfer of the sample to a cryochamber containing a substitution agent that replaces water in the sample. After all the water has been replaced by the substitution agent, more of the substitution agent in which various fixatives have been dissolved is added to the chamber containing the samples. After a suitable time during which the fixative-laden substitution agent has diffused throughout the sample, the chamber temperature is raised slowly, allowing the chemical fixatives to interact with the samples.

Cryosubstitution ensures rapid cessation of biological activity as a result of cryofixation followed by dehydration and then infusion of chemical fixatives at temperatures too low (usually −80°C) for chemical activity. As the chamber temperature is

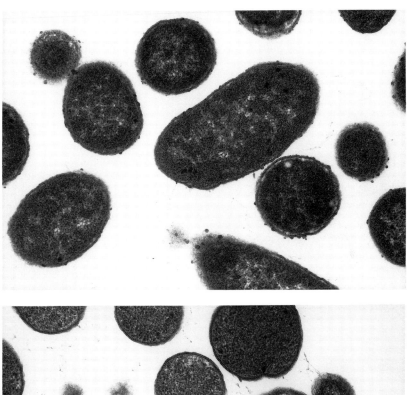

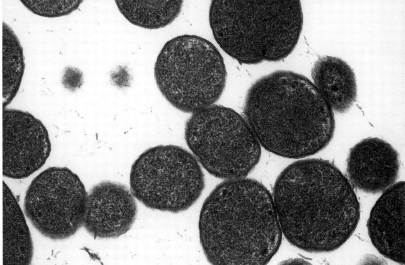

FIGURE 140. Top: *Escherichia coli* fixed with conventional chemical methods. 33,600×. Bottom: *Escherichia coli* fixed by slam-freezing, followed by cryosubstitution. 33,600×.

raised, the chemicals contained in the substitution agent simultaneously fix all areas of the sample once they reach their particular reactive temperature (e.g., −40°C for uranyl acetate and −20°C for osmium).

This is markedly different from conventional chemical fixation where the fixative must diffuse from the sample surface to the interior. During the process, the fixative becomes bound to tissue constituents, thus diluting the fixative solution reaching the interior of the sample. In addition, as the sample becomes fixed from the outside in, further diffusion of the fixation solution becomes impeded by cross-linking and chemical alteration of cellular constituents.

Some workers (Harvey, 1982) have used apolar substitution agents, such as diethylether, to preserve the original distribution of water-soluble substances for subsequent microanalysis, which would presumably have been mobilized by the use of polar substitution agents. Methods employing apolar agents take up to 14 days for cryosubstitution of the sample because of their poor miscibility with water.

Most cryosubstitution is done using polar solvents (see Steinbrecht and Müller, 1987), such as acetone, which is the most widely used agent. Methanol has the advantage that it substitutes more quickly at low temperatures than acetone and is more suitable for dissolving fixatives such as glutaraldehyde.

Polar agents such as acetone or methanol placed over cryofixed samples held at $-80°C$ typically dissolve frozen water and replace the frozen water within 6–12 hr. During this time the substition agent can be changed several times. The final change typically contains 2–4% osmium, although some investigators have added acrolein, glutaraldehyde, and uranyl acetate (see Steinbrecht and Müller, 1987).

C. Freeze-Fracture

Steere (1957) demonstrated virus particles within cells with freeze-fracture techniques, which ushered in the first widely applied cryotechnique applied to biological materials. Most early workers felt that freeze-fracture techniques could eliminate the need for chemical fixation and would provide a more realistic picture of cellular structures than chemical fixation. Unfortunately, initial freezing of tissues and cells without cryoprotection showed freezing damage, so most laboratories resorted to light aldehyde fixation followed by treatment with 20–25% buffered glycerol for cryoprotection during subsequent manual immersion freezing in Freon 13 cooled with liquid nitrogen.

Various freeze-fracture devices have been manufactured over the years, but the units sold by Balzers have gained the widest acceptance for a variety of reasons. These units have several specimen stages capable of holding small (3-mm diameter) gold planchets with favorable heat-transfer characteristics (Fig. 141). The cryoprotected sample is loaded into the planchets, quickly frozen, and stored under liquid nitrogen until use. After the knife arm and specimen stage have been cooled under vacuum to $-150°$ and $-100°C$, respectively, the bell jar is brought to atmospheric pressure, and the specimen is quickly loaded onto the specimen stage. The bell jar is then closed and evacuated to about 10^{-5} Torr. At that time, the knife-holding arm, which is loaded with a standard injector razor blade, is passed above the specimen and slowly lowered in 1-μm increments until the specimen surface is fractured. It is important not to make too many passes, as the deeper part of the sample will generally not be frozen as well as superficial regions. After the surface of the specimen is fractured, the knife is moved out of the way and a carbon/platinum coat is evaporated onto the specimen surface.

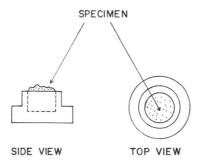

FIGURE 141. Gold specimen planchets used with Balzers freeze-fracture units.

Next, carbon is evaporated onto the specimen for added replica stability. The chamber is then brought to atmospheric pressure and the specimen is removed and thawed, and the replica is floated off the planchet on distilled water. The replica is then transferred with a loop to successive baths of acid and sodium hypochlorite (bleach) with intervening distilled-water washes and finally rinsed several times in distilled water. The replica, which has now had all biological material corroded from its surface, is picked up with a grid, dried, and examined with the TEM. The replica will reveal cellular structures (Fig. 142), but is mostly used to examine intramembrane domains (Fig. 143), demonstrating intramembrane protein particle distribution.

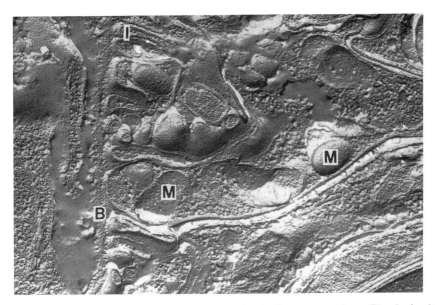

FIGURE 142. Freeze-fracture preparation of a rat kidney showing basement membrane (B), mitochondria (M), and basal infolding (I). 25,500×.

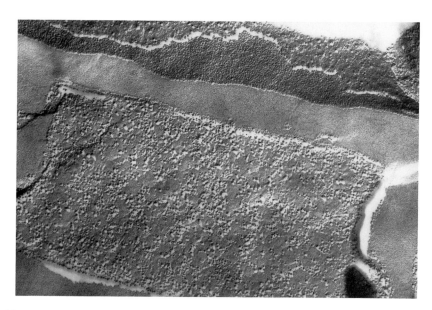

FIGURE 143. Freeze-fracture preparation of *Eschericia coli* cells showing intramembrane protein particles. 45,900×.

An alternative preparative procedure utilizes matched holders held in a hinged device (Fig. 144). The sample is placed between two planchets and then frozen. The frozen planchets with sample sandwiched between them are mounted on the cooled stage of the freeze-fracture device, the bell jar is closed, and the system is evacuated. The knife-holding arm is used to pop open the double-recovery device, exposing two complementary fractured specimen faces that are subsequently shadowed, producing complementary replicas that are cleaned as above. The replicas can be carefully examined, and complementary faces can be located to examine the insertion of membrane proteins into the lipid bilayer.

Freeze-etch samples are prepared the same way as standard freeze-fracture samples, except that after the specimen is fractured the cold knife holder is suspended over the specimen for a few minutes to allow frozen water from the sample to sublime and

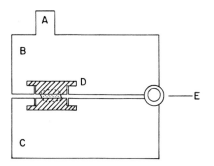

FIGURE 144. Freeze-fracture double-recovery device. Pin A is struck to open jaws (B,C) hinged at E. The sample (stippled) frozen between the two planchets (D) is broken in two by the process and is subsequently shadowed.

attach to the knife holder. After the etching period is finished, the knife is moved out of the way and a metal replica is made as described above. The purpose of etching is to make nonaqueous domains of the cell stand up in relief against the depressed areas from which water has sublimed.

Freeze-fracture techniques are not used as much today as in the past because of the development of more sensitive techniques utilizing gold-labeled antibodies, which can not only localize the protein components within a membrane, but also identify specific proteins, which is not possible with freeze-fracture techniques. Coupling gold-labeled antibody techniques with high-resolution FEG SEM allows the identification of several specific proteins simultaneously and also can use sample sizes much larger than with freeze-fracture methods. However, there is still a place for freeze-fracture techniques in conjunction with other methods, such as negative staining and polyacrylamide gel electrophoresis. This combination can answer basic cell biology questions, as was done in elucidating the structure and interaction of streptolysin with the lipid component of natural and artificial membranes (Bhakdi *et al.*, 1985).

D. Freeze-Drying as Typified by Molecular Distillation

Linner *et al.* (1986) described an improved freeze-drying device called a molecular distillation device (MDD). The MDD provided a means to reproducibly dry a cryofixed specimen, which could then either be vapor-fixed with osmium prior to infiltration with room-temperature Spurr resin for routine ultramicrotomy or directly infiltrated with a low-temperature acrylic resin such as Lowicryl. The former sample type proved to be good for structural studies, immunolabeling, and autoradiography, while the latter type was primarily intended for immunolabeling and microanalytical work.

The advantage of this specimen-preparation technique is that a native frozen specimen could be infiltrated with resin (with or without osmium vapor fixation) without ever having been exposed to aqueous fixatives, aqueous buffers, or dehydration agents. The dried sample was directly infiltrated with resin, which was either polymerized by heat (Spurr) or polymerized with ultraviolet light at low temperatures (Lowicryl). Thus, water-soluble enzymes and ions remain as they were in life, rather than being removed or rearranged during the aqueous processing phases employed during conventional fixation schedules. Naturally, the success of molecular distillation is highly dependent on good cryofixation, which led to the development of the CF-100 slam freezer to go along with the MDD (both available from LifeCell).

After the sample is slam frozen, it is put into a holder that is kept under liquid nitrogen (Fig. 145). The holder is then placed in a container filled with liquid nitrogen, which is attached to the bottom of the MDD device and, in turn, is suspended in a dewar of liquid nitrogen. The chamber containing the specimens is bolted to the MDD and then connected to a mechanical pump, which pumps the liquid nitrogen out of the specimen area (which is still held at liquid nitrogen temperatures by the external dewar of liquid nitrogen in which it is immersed). Following evacuation of the liquid nitrogen, the specimen chamber is disconnected from the mechanical pump, and a tur-

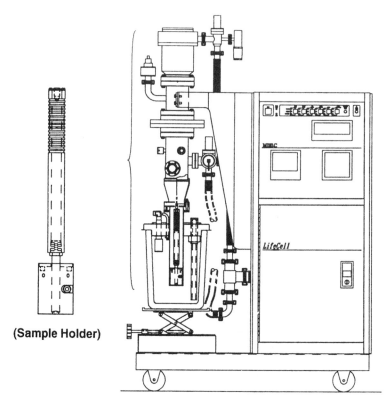

(Sample Holder)

FIGURE 145. LifeCell MDD-C molecular distillation dryer. (Courtesy of LifeCell Corporation)

bomolecular pump backed by the mechanical pump is connected to the specimen chamber. Water sublimes from the frozen specimens over a 5- to 7-day period and is removed from the system by a turbomolecular pump. The temperature of the specimen chamber is then brought up in steps specified by the MDD processing program until the chamber reaches $-40°C$, at which time Lowicryl resin is pulled into the chamber under slight vacuum to infiltrate the specimens. The resin can then be polymerized with ultraviolet radiation prior to bringing the chamber to room temperature and pressure for specimen removal.

Alternatively, the chamber containing dried specimens is brought to room temperature and low vacuum, a vial of osmium crystals is attached to a port in the chamber, and the low vacuum is used to pull osmium vapor into the chamber to vapor-fix the samples. Then the osmium vapor is evacuated from the chamber, and Spurr resin is pulled into the chamber under low vacuum. After the specimens are infiltrated, the chamber is opened, the specimens are removed and put into molds with fresh Spurr resin, and the resin is polymerized with heat (Fig. 146).

There are two major problems with this approach to sample preparation. The equipment is expensive, and many different types of tissue must have specific freezing methods designed for them. It is also important to realize that there are several other

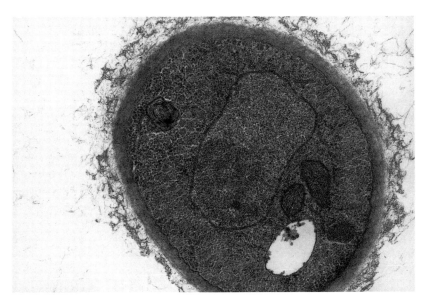

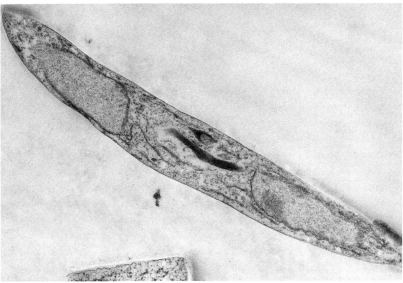

FIGURE 146. Top: *Aspergillus fumigatus* fixed with conventional chemical methods, embedded in Spurr resin, sectioned, and stained with uranyl acetate and lead citrate. 27,000×. Bottom: *Aspergillus fumigatus* slam-frozen, followed by MDD, osmium vapor treatment, Spurr resin infiltration, sectioning, and uranyl acetate staining. 12,000×.

techniques that are less expensive and that will answer the same questions. Unless the antigens that need to be localized are water soluble, Tokuyasu (1973) techniques employing cryoultramicrotomy for intracellular immunolabeling will probably work well, take less time, and cost considerably less. Native frozen sections cut on a cryoultramicrotome, cryotransferred, and examined with a cryostage on a TEM will usually give the same microanalytical possibilities as a section prepared by molecular distillation.

VII. ARTIFACTS AND THEIR CORRECTION

Materials prepared for electron microscopy utilizing cryotechniques suffer from many different types of artifacts. If chemical fixation precedes freezing, as with Tokuyasu (1973) techniques for cryoultramicrotomy, the chemical fixation artifacts described in Chapter 1 can occur.

Techniques that begin with cryofixation are consistently plagued with freezing artifacts. As mentioned previously, most freezing techniques are capable of optimal sample freezing of no more than 10 to 15 μm from the sample surface. Figure 147 clearly illustrates the tissue distortion caused by ice crystal formation and also shows that the farther from the sample surface, the larger the artifactual holes formed.

Another interesting feature associated with cryofixation is the insulating aspect of biological membranes. The freezing front passing through the tissue can be very irregular because of this aspect. Figure 147 shows cells of the bacterium *Escherichia coli* that were slam-frozen, cryosubstituted, and embedded in Spurr resin, followed by sectioning and normal post-staining with uranyl acetate and lead citrate. The cells showing dark surfaces with large open areas in the center were severely damaged during the freezing process. Note that relatively normal-looking cells are located back from the slammed surface, interspersed with damaged cells. This demonstrates that the freezing process does not proceed in a strictly linear direction from the surface of the sample inward, but skips around some cells.

Slam-freezing is one of the most successful, reproducible, and widely used methods for cryofixation, but it leads to a variety of artifacts. LifeCell Corporation has performed a number of studies on different types of cells and tissues (personal communication), which make it clear that each time a new type of sample is frozen, a series of trials to determine the best type of material to put behind the specimen must be undertaken. If the backing material is inappropriate for the sample, crushing artifact may occur, even at the sample surface. Figure 148 shows cells of the fungus *Aspergillus fumigatus* that were slam-frozen, freeze dried by molecular distillation, treated with osmium vapors, embedded in Spurr resin, sectioned, and stained with uranyl acetate. The nucleus in the right-hand cell has a discontinuous nuclear envelope because the cell was crushed during the cryofixation step.

Further discussions of cryofixation artifacts can be found in Crang and Klomparens (1988), Robards and Sleytr (1985), and Steinbrecht and Zierold (1987).

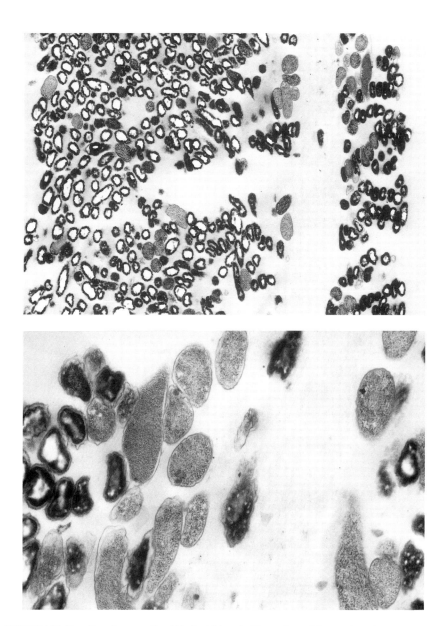

FIGURE 147. Top: Slam-frozen cells of *Escherichia coli*. The surface of the specimen is to the right. The entire thickness of the slammed sample is shown. 7,000×. Bottom: Slam-frozen cells of *E. coli* showing cells with relatively well-preserved cytoplasmic contents intermixed with cells severely damaged by the freezing process. 22,400×.

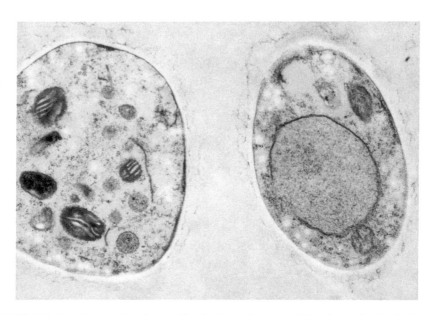

FIGURE 148. Slam-frozen cells of *Aspergillus fumigatus* that were subjected to molecular distillation, osmium vapor, and Spurr resin embedment before being sectioned and stained with uranyl acetate. Note that the cell on the right has a burst nuclear envelope and nucleoplasm spilled into the cytoplasm. The small electron-lucent areas in the cytoplasm are areas of ice nucleation during the freezing process. 24,800×.

REFERENCES

Bachmann, L., and Schmitt-Fumian, W.W. 1973. Spray-freezing and freeze-etching. In: *Freeze-etching, techniques and applications*, E.L. Benedetti and P. Favard (eds.), Societe Francaise de Microscopie Electronique, Paris.

Bernhard, W. 1965. Ultramicrotomie a basse temperature. *Ann. Biol.* 4:5.

Bhakdi, S., Tranum-Jensen, J., and Sziegoleit, A. 1985. Mechanism of membrane damage by streptolysin-O. *Infect. Immun.* 47:52.

Costello, M.J., and Corless, J.M. 1978. The direct measurement of temperature changes within freeze-fracture specimens during rapid quenching in liquid coolants. *J. Microsc.* 112:17.

Crang, R.F.E., and Klomparens, K.L. 1988. *Artifacts in biological electron microscopy.* Plenum Press, New York.

Dubochet, J., McDowall, A.W., Menge, B., Schmid, E.N., and Lickfield, K.G. 1983. Electron microscopy of frozen-hydrated bacteria. *J. Bacteriol.* 155:381.

Fernandez-Moran, H. 1957. Electron microscopy of nervous tissue. In: *Metabolism of the nervous system*, D. Richter (ed.), Pergamon Press, Oxford.

Gilkey, J.C., and Staehelin, L.A. 1986. Advances in ultrarapid freezing for the preservation of cellular ultrastructure. *J. Elect. Microsc. Tech.* 3:177.

Handley, D.A., Alexander, J.T., and Chien, S. 1981. The design of a simple device for rapid quench freezing of biological samples. *J. Microsc.* 121:273.

Harvey, D.M.R. 1982. Freeze-substitution. *J. Microsc.* (Oxford) 127:209.

Howard, R.J. and Aist, J.R. 1979. Hyphal tip cell ultrastructure of the fungus *Fusarium*: Improved preservation by freeze-substitution. *J. Ultrastruct. Res.* 66:224.

Lee, R.E., Jr., and Denlinger, D.L. (eds.). 1991. *Insects at low temperature.* Chapman and Hall, New York.

Linner, J.G., Livesey, S.A., Harrison, D.S., and Steiner, A.L. 1986. A new technique for removal of amorphous phase tissue water without ice crystal damage: A preparative method for ultrastructural analysis and immunoelectron microscopy. *J. Histochem. Cytochem.* 34:1123.

Moor, H. 1986. Theory and practice of high pressure freezing. In: *Cryotechniques in biological electron microscopy,* R.A. Steinbrecht and K. Zierold, eds.), Springer-Verlag, New York.

Moor, H. and Riehle, U. 1968. Snap-freezing under high pressure: A new fixation technique for freeze-etching. In: *Proceedings of the Fourth European Regional Conference on Electron Microscopy,* D.S. Bocciarelli (ed.), Tipografia Poliglotta Vaticana, Rome, 2:33.

Moor, H., Muhlethaler, K., Waldner, H., and Frey-Wyssling, A. 1961. A new freezing ultramicrotome. *J. Biophys. Biochem. Cytol.* 10:1.

Moor, H., Bellin, G., Sandri, C., and Akert, K. 1980. The influence of high pressure freezing on mammalian nerve tissue. *Cell Tissue Res.* 209:201.

Müller, M., Meister, N., and Moor, H. 1980. Freezing in a propane jet and its application in freeze-fracturing. *Mikroskopie* (Wein) 36:129.

Robards, A.W., and Crosby, P. 1983. Optimisation of plunge freezing: Linear relationship between cooling rate and entry velocity into liquid propane. *Cryo-Letters* 4:23.

Robards, A.W., and Sleytr, U.B. 1985. *Low temperature methods in biological electron microscopy.* Elsevier, New York.

Roos, N., and Morgan, A.J. 1990. *Cryopreparation of thin biological specimens for electron microscopy: Methods and applications.* RMS Microscopy Handbook No. 21. Oxford University Press, New York.

Sjostrand, F.S., and Baker, R.F. 1958. Fixation by freeze-drying for electron microscopy of tissue cells. *J. Ultrastruct. Res.* 1:239.

Steere, R.L. 1957. Electron microscopy of structural detail in frozen biological specimens. *J. Biophys. Biochem. Cytol.* 3:45.

Steinbrecht, R.A., and Muller, M. 1987. Freeze-substitution and freeze-drying. In: *Cryotechniques in biological electron microscopy.* R.A. Steinbrecht and K. Zierold (eds.), Springer-Verlag, New York, pp. 149–172.

Steinbrecht, R.A., and Zierold, K. (eds.). 1987. *Cryotechniques in biological electron microscopy.* Springer-Verlag, New York.

Tokuyasu, K.T. 1973. A technique for ultramicrotomy of cell suspensions and tissues. *J. Cell Biol.* 57:551.

Van Harreveld, A., and Crowell, J. 1964. Electron microscopy after rapid freezing on a metal surface and substitution fixation. *Anat. Rec.* 149:381.

CHAPTER 11

High-Voltage Electron Microscopy

I. HISTORY

The first high-voltage electron microscope (HVEM) capable of generating 1000 kV was put into operation in Toulouse, France, in the laboratory of Dr. G. Dupouy in 1960 (Dupouy, 1985). In 1969, Dupouy installed a 3000-kV instrument, which offered improved specimen penetration and reduced chromatic aberration, but did not offer significant advantages over the original 1000-kV instrument.

In the United States, federal funding helped several laboratories purchase and install HVEMs in the 1970s. In 1971, Dr. Hans Ris installed an AEI instrument at the University of Wisconsin; this was followed by the installation of a JEOL instrument at the University of Colorado in Dr. Keith Porter's laboratory and the installation of a HVEM instrument in Albany, New York, in 1978. All of the instruments listed above that are in the United States are available free of charge to American investigators because they are national resources and are subsidized by federal funds. To gain access, one should contact the laboratories about the availability of the instruments and proposal guidelines.

The original expectation was that the high accelerating voltage of these instruments would permit the examination of living materials in small, closed environmental chambers. However, Dupouy demonstrated that radiation damage was so severe that HVEMs have been mostly limited to the examination of fixed biological specimens and various types of nonbiological samples.

Intermediate-voltage electron microscopes (IVEM) capable of 200–400 kV were subsequently developed that could perform many of the same tasks of HVEMs but required less initial capital expenditure, less space, had greater stability, and were generally easier to operate. Recent developments in the design and application of IVEMs allow most of the work formerly done with HVEMs to be accomplished with less expensive, more versatile, and more stable IVEMs.

Buseck *et al.* (1988) edited a text with thorough discussions of the physics of high-resolution TEM and HVEM from a materials scientist's point of view. Johnson *et al.* (1986) edited another collection of chapters on HVEM applications to materials.

II. PURPOSE

After the initial attempts to examine living material in environmental chambers met with failure, investigators developed other uses for HVEMs. The superior beam-penetrating capability of these instruments permits examination of much thicker specimens (up to 10 μm in thickness) than is possible with conventional TEMs. When stereo pairs of these materials are prepared, the great depth of field possible from electron optics delivers impressive three-dimensional images of these specimens.

Chromatic aberration is reduced when greater accelerating voltages are employed because less electron-energy spectrum broadening caused by multiple inelastic and elastic scattering events within the specimen occurs. This improves X-ray signals that are measured point by point with energy-dispersive spectroscopy (EDS). Electron energy loss spectroscopy also can be performed on thicker specimens than in a conventional TEM because the mean free path of an electron between two inelastic scattering events is greater as a result of the higher accelerating voltage of the primary electrons.

The advantages to HVEM are all related to the increased accelerating voltage. There is greater specimen penetration, along with greater resolution, with thick specimens. The former leads to greater brightness for a given specimen, and the latter is mostly the result of reduced chromatic aberration because of reduced beam energy spreading. In addition, the specimens suffer less ionization damage and thus less heating.

The disadvantages of HVEMs are largely the result of physical parameters. The instruments are extremely costly (the original instruments in the 1970s cost millions of dollars, when a conventional 100-kV instrument could be purchased for less than $100,000); they require space specifically designed for them, since they are several stories tall and occupy more than the floor space of a conventional laboratory; and they are physically less stable than conventional TEMs because of their size.

As already mentioned, HVEMs caused considerably more specimen damage due to atomic displacement than originally expected, so living materials have eluded significant investigation with these instruments. These instruments are also not very useful for the examination of conventional specimens, since they suffer from reduced contrast and inferior image quality at low operating voltages.

III. FUNCTIONAL ASPECTS OF HVEMS

A. Resolution

Chromatic aberration is a major source of decreased resolution with all electron microscopes. This phenomenon can be initiated at the electron gun, resulting from the broad spectrum of energies exhibited by electrons emitted and accelerated from a tungsten filament. A smaller source provides less chromatic aberration (LaB_6, FEG), and the latest generation of FEGs provides the most focused electron beams with the least chromatic aberration available today. Field emission guns are found on only a few of the commercially available IVEMs currently available (e.g., JEOL JEM-2010F).

The most important source of chromatic aberration is energy losses of the incident electron beam within the specimen. As the image-forming electrons emerge from a specimen, some have become widely scattered from elastic collisions; they consequently strike the objective aperture and do not reach the imaging plane. Some electrons experience inelastic collisions within the specimen, with the result that electrons of greatly reduced energy levels emerge from the specimen and enter the projector lens system. Finally, some electrons emerge from the specimen having undergone neither elastic nor inelastic collisions. Chromatic aberrations result from the inelastic collisions occurring in the specimen, which produce electrons that cannot be focused in the same plane. As imaging electrons pass through the projector lens system, those with higher energies are refracted less than those with lower energies, producing several planes of focus for the image and resulting in decreased resolution. Since higher voltages produce electrons that pass through the specimen at higher velocities, less inelastic scattering occurs, leading to less chromatic aberration and higher resolution.

The practical side of this feature is that, for a given specimen, higher resolution is possible with higher accelerating voltage. Of course, chromatic aberration is directly proportional to section thickness. Since most investigators use HVEMs to examine specimens much thicker than those used for conventional TEM work, the final resolution achieved for very thick specimens does not approach that for ultrathin sections examined with a conventional TEM. For biological specimens, chromatic aberration will be reduced by at least 20 times in going from 100 kV to 1000 kV. In other words, it should be possible to achieve 0.5-nm resolution in a 1-μm-thick specimen imaged at 1000 kV.

B. Radiation Damage

Radiation damage arises from several different types of interaction between the electron beam and atoms within the specimen. Any structural changes that are induced in the sample can result in artifacts or, at the very least, a decrease in resolution.

At high accelerating voltages (over 100 kV), an accelerated electron can transfer enough energy to a specimen atom to actually displace the atom. This relatively rare event, however, is probably a negligible factor in terms of final image quality.

The most important source of biological specimen damage from an electron beam is ionization damage. This type of damage is of little consequence with metallic specimens, because any electrons ejected from the specimen are replaced by others flowing up from the normally grounded specimen holder. If, however, areas of a specimen are inadequately grounded (as is often encountered with poorly conductive biological specimens embedded in nonconductive plastic resins), charging can result, thus degrading the image. In addition, covalently bonded structures found in biological specimens may incur permanent changes after electron bombardment because of electron rearrangements (often accompanied by atomic rearrangements within molecules). The amount of specimen damage theoretically varies inversely to the square of the incident electron velocity. Thus, increasing the accelerating voltage from 100 to 1000 kV should reduce ionization damage by one third.

Beam heating can also damage specimens. For a specimen of a given thickness, beam-heating effects should be about three times less at 1000 kV than at 100 kV. However, in practice a HVEM is used to look at specimens at least 10 times thicker than those examined with conventional TEMs, so the amount of heat damage is still greater with HVEM than with conventional TEM.

C. Contrast

Contrast is essentially a signal to noise (S/N) problem. A thin specimen without added contrast-building materials (osmium, uranium, lead) produces a small amount of beam scattering and thus low contrast. For a given specimen thickness, scattering by the sample decreases as the incident electron energy increases. If a small feature is of low contrast, the S/N ratio may not be high enough to detect the feature. In this case, resolution is limited by a lack of contrast. As explained previously, with HVEM the

FIGURE 149. JEOL JEM-1000. The platform supporting the high-voltage generator is supported on steel pillars above the microscope column. Note the size of the operator chair for proper perspective. (Courtesy of JEOL USA, Inc.)

high accelerating voltages employed produce excellent specimen penetration and brightness, but also reduce contrast. One possible solution to this problem is dark-field imaging. By using an off-center aperture or electromagnetic deflection to eliminate the axial component of the beam following specimen interaction, the scattered components of the electron beam can be used to produce a dark-field image of materials with low contrast (Ottensmeyer, 1982).

IV. MICROSCOPE CONSTRUCTION

Microscopes designed for HVEM work are in all ways larger than conventional TEMs. They must have larger electromagnetic lenses to control the more energetic electron beam, the column is at least 3 m tall, and the high-voltage generator is itself larger than most conventional TEMs (Figs. 149 and 150).

The high-voltage generators and accelerators in most commercial HVEMs are gas insulated, occupying about 20 m³ of space. These large chambers are designed to prevent arcing of the high voltage.

Stages are larger than those found in conventional TEMs and thus suffer from more drift and vibration problems. The sheer height of the column makes all mechanical controls longer, bulkier, and less exact than those on conventional TEMs.

FIGURE 150. A close-up view of the JEOL Atomic Resolution Microscope (JEM-ARM1000). (Courtesy of JEOL USA, Inc.)

V. SAMPLE PREPARATION

Preparation of samples up to 1 μm thick can be done with conventional TEM techniques, but thicker samples often require selective staining methods to prevent confusion. As with all electron optics, such great depth of field exists that all materials within a section even 5 μm thick will be in focus at the same time. Being unable to optically section (by focusing up and down through the specimen, as one does with light microscopy) makes the flood of information presented simultaneously extremely difficult to interpret. Selective staining has been used by Franzini-Armstrong and Peachey (Franzini-Armstrong and Peachey, 1982; Peachey and Franzini-Armstrong, 1978) to stain the sarcoplasmic reticulum (SR) in muscle in thick sections so that its branching networks could be deciphered without the confusion caused by muscle proteins with which the SR is associated. Even with selective staining, however, the SR patterns within a thick section were difficult to decipher, so Peachey employed stereo pairs, allowing three-dimensional viewing of the SR.

Using HVEM, Wolosewick and Porter (1979) revealed a complicated network of cytoplasmic proteins, which they termed the *microtrabecular lattice*, in whole cells grown directly on coated grids. The cells were cultured directly on polymer-film-coated grids, allowing them to spread out so that they were thin enough to be penetrated easily by the HVEM electron beam.

Wolosewick (1980) also developed a technique utilizing polyethylene glycol (PEG) specimen embedment. Sections cut from fixed and PEG-embedded samples were placed on grids for subsequent removal of the PEG medium, followed by critical point drying. This allowed Wolosewick to examine thick sections of biological samples with conventional TEM and HVEM without having to deal with the electron-scattering (and contrast-robbing) aspects of nonbiological support media (embedding resins).

VI. APPLICATIONS

As mentioned above, sections of biological materials 1 to 10 μm thick can be examined with HVEM. Stereoscopy is usually applied to these materials to help reduce the confusion caused by all the cellular structures being in focus at the same time (Peachey, 1978). With selective staining of cellular components and/or stereoscopy methods, a true understanding of three-dimensional relationships between cellular structures is possible. With conventional TEM, the specimen is essentially two dimensional and extremely thin, so a large number of pictures of serial sections must be taken and analyzed to discern relationships revealed in one HVEM photograph (or stereo pair).

Another type of preparation used extensively to study cells as described above, is produced by growing cells directly on grids, which are then subjected to fixation and critical point drying prior to HVEM examination. This technique is limited to those cells that will adhere to a plastic film and spread out sufficiently to allow beam penetration. Large, multinucleated cells, such as myofibers, or thick-walled cells, such as those of plants, are not suitable for this technique.

Immunocytochemical techniques, particularly post-embedment localization in conjunction with cryosectioning, are another potential application of HVEM (see Chapter 14).

The dream of examining living cells with HVEM has only occasionally been realized because of the problems with radiation damage already mentioned. Over the years, various thin-film chambers and differentially pumped wet chambers were developed for these purposes, but little work continues in the area.

VII. INTERMEDIATE-VOLTAGE ELECTRON MICROSCOPY (IVEM)

Contemporary IVEMs, as typified by those from JEOL (JEM-2010, 3010) and Philips (CM-30), possess most of the advantages of beam penetration of HVEMs without most of the disadvantages associated with HVEM cost, stability, and space requirements. The original IVEM units did not provide images at low accelerating voltages (100–120 kV) that were comparable to conventional TEM images, but the current generation of IVEMs are considerably more flexible. It is now possible to achieve high resolution for both ultrathin sections and thick sections (up to 3 μm) with IVEMs.

Contemporary IVEMs designed for relatively low voltages (200 kV) offer several configurations that can be chosen for different applications. The JEOL JEM-2010 can be outfitted with one of two pole pieces, one permitting limited specimen tilting but offering high resolution (0.94 nm) and the other designed for high tilt capacity for analytical purposes, which has slightly less resolution (0.23 nm). In addition, the instrument can be provided with a conventional tungsten filament or a thermal field emission gun (FEG). The FEG would be chosen if microanalytical applications were desired without having to resort to scanning transmission electron microscopy, because it can produce a small probe diameter. As with all FEGs, high beam density and narrow-energy bandwidth are ideal for producing a fine probe diameter (1 nm with 1 μA of current) for microanalytical purposes.

The JEOL JEM-3010 (300 kV) possesses the advantages of a 300-kV instrument over one designed for 200 kV. The 3010 can be thought of as a 300 kV instrument with a 2010 built in. A 300 kV instrument such as the 3010 is probably most versatile for IVEM work with biological specimens, because it has better specimen-penetrating capabilities with the standard pole piece (allowing visualization of thicker sections) as well as having a high tilt stage (up to 40°). The 3010 thus obtains almost the same resolving capability (0.21 nm) available with the high-resolution pole piece available for the 2010 and simultaneously can do high tilt work.

As described by Glaeser (1982), when accelerating voltage is increased, the specimen exhibits reduced stopping power in regard to the electron beam. This means that less scattering occurs, resulting in reduced spherical aberration, reduced chromatic aberration, and increased brightness. The first two features are responsible for the increased resolving power of IVEMs.

As section thickness increases, higher accelerating voltages are required to main-

TABLE 9. Depth of Field in Micrometers
at Selected Accelerating Voltages
in Relation to Resolution[a]

	Depth of field in μm		
Resolution	100 kV	300 kV	1000 kV
1.5 nm	0.6	1.1	2.6
2.5 nm	1.7	3.2	7.2
4.0 nm	4.3	8.1	18.4

[a]From Glaeser (1982).

tain the same resolving capabilities. Thus, depth of field (section thickness) helps define the need for IVEM and HVEM. Table 9 (Glaeser, 1982) shows these relationships.

As mentioned previously for HVEM, using tilting stages (goniometers) to produce stereo pairs helps reduce the potential visual confusion caused by all structures in a thick section being in focus at the same time. Peachey (1986) discusses the relationships between section thickness and magnification when determining the proper tilting angles for stereo image production.

Lewis *et al.* (1988) discuss practical resolution as dependent on accelerating voltage, section thickness, magnification, and stereoscopy (Table 10).

Thus, as with HVEM, the major applications for IVEM involve examination of thick sections and whole mounts of cells. With three-dimensional imaging techniques, a wealth of information about the spatial organization of cellular materials is available. In addition, application of contemporary gold label antibody techniques to cell surfaces of intact cells can improve the knowledge of the spatial distributions of antigenic sites (Lewis *et al.*, 1988).

There are three IVEM facilities available to U.S. investigators (see Appendix B.V.): one devoted primarily to three-dimensional imaging of macromolecules, one to three-dimensional imaging and microanalysis of biomedical materials, and a third supporting IVEM in basic medical research.

TABLE 10. Factors Associated with Practical Resolution[a]

Accelerating voltage	Section thickness	Image quality	Useful magnification
100–120 kV	60–120 nm	Excellent	150,000+
	0.5–1.5 μm	Good with 3-D	5,000–15,000
300–400 kV	up to 1.5 μm	Excellent	15,000–40,000
	2.5–3.5 μm	Excellent with 3-D	25,000–35,000
1000 kV	3.5 μm+	Excellent with 3-D	25,000+

[a]From Lewis *et al.* (1988).

REFERENCES

Buseck, P., Cowley, J., and Eyring, L. (eds.). 1988. *High-resolution transmission electron microscopy and associated techniques.* Oxford University Press, New York.

Dupouy, G. 1985. Megavolt electron microscopy. In: *The beginnings of electron microscopy,* P.W. Hawkes (ed.), Academic Press, New York, pp. 103–165.

Franzini-Armstrong, C., and Peachey, L.D. 1982. A modified Golgi black reaction method for light and electron microscopy. *J. Histochem. Cytochem.* 30:99.

Glaeser, R.M. 1982. A critique of the theoretical basis for the use of HVEM in biology. *EMSA Proc.* 40:2.

Johnson, J.E., Jr., Hirsch, P., Fujita, H., Shimizu, R., and Thomas, G. 1986. *High resolution and high voltage electron microscopy.* Alan R. Liss, New York.

Lewis, J.C., Jones, N.L., O'Toole, E.T., Grant, K.W., and Jerome, W.G. 1988. Intermediate voltage electron microscopy in biomedical research. *EMSA Bull.* 18:2.

Ottensmeyer, F.P. 1982. Scattered electrons in microscopy and microanalysis. *Science* 215:461.

Peachey, L.D. 1978. Stereoscopic electron microscopy. Principles and methods. *EMSA Bull.* 8:15.

Peachey, L.D. 1986. The extraction of three-dimensional information from stereo micrographs of thick sections using computer graphics methods. *Ann. N.Y. Acad. Sci.* 483:161.

Peachey, L.D., and Franzini-Armstrong, C. 1978. Observations on the T system of rat skeletal muscle fibers in three-dimensions using high voltage electron microscopy and the Golgi stain. *Biophys. J.* 21:61a.

Wolosewick, J.J. 1980. The application of polyethylene glycol (PEG) to electron microscopy. *J. Cell Biol.* 86:675.

Wolosewick, J.J., and Porter, K.R. 1979. The microtrabecular lattice of the cytoplasmic ground substance: Artifact or reality? *J. Cell Biol.* 82:114.

CHAPTER 12

Microanalysis

In electron microscopy, the term *microanalysis* has historically referred to X-ray micro-analysis (energy-dispersive spectroscopy) and electron diffraction, with electron energy loss spectroscopy (EELS) added over the last decade. Materials scientists have utilized these techniques longer and more extensively than biologists, so their applications are most thoroughly discussed in texts on the subject (Chandler, 1978; Egerton, 1989; Goldstein *et al.*, 1981; Morgan, 1985; Newbury *et al.*, 1986; Russ, 1984).

Electron optical imaging and X-ray analysis were first combined in the late 1940s, and, since the early 1960s, X-ray microanalysis has been widely used in physics, electronics, metallurgy, mineralogy, and, more recently, biology.

I. MICROANALYSIS TECHNIQUES

A. Energy-Dispersive Spectroscopy (EDS)

X-ray microanalysis (EDS) provides a method to identify elements within thin and thick specimens with high sensitivity and precise location (particularly with thin specimens examined with a field-emission gun-equipped TEM). Electrons are arranged in a series of shells (K, L, M, etc.) around an atom's nucleus, with the shells nearest the nucleus having the least energy. If an electron is removed from the shell of an atom by an elastic collision of an electron from the primary beam, an ion is produced (Fig. 151). To stabilize the ion, an electron from a higher energy orbit (outer shell) must fill the gap formed when the electron is ejected from the atom. An amount of energy equal to the difference in energy between the outer shell and the inner shell is released in the form of an auger electron, a photon of light, or an X-ray photon. The X-ray energy is the potential energy difference between the two shells. In an atom with a large number of electrons, the initial electron loss can lead to a cascade of orbital electron shifts and the potential release of a large spectrum of different X-rays. A heavy atom such as

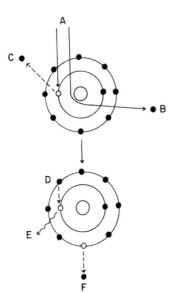

FIGURE 151. Diagram of the possible energetic events following primary electron beam interaction with a specimen atom. A = primary electrons; B = backscattered electron; C = secondary electron; D = electron shift to fill vacancy following release of secondary electron C; E = photon of light or X-ray photon; F = release of Auger electron after shift D.

uranium (atomic number 92) has a large number of possible spectral emissions, while a smaller atom such as sodium (atomic number 11) will exhibit far fewer. However, quantum theory principles predict that a number of the expected transitions cannot occur.

With EDS, the X-rays generated by the interaction of the primary beam with the specimen are sorted electronically. This technique is usually sensitive to elements with a Z number of 11 or above and to elemental concentrations typically less than 1% in a given sample (Vaughan, 1989). An X-ray photon that reaches the semiconductor detector first produces a current, which is then converted electronically into a voltage whose amplitude is directly proportional to the energy of the X-ray signal (Fig. 152). The voltage is then converted into a digital signal, which is, in turn, recorded as a count by a multichannel analyzer. The counts accumulated from a sample over time produce an X-ray spectrum of the specimen.

X-ray microanalysis units can be attached to SEMs, TEMs, and scanning transmission electron microscopes (STEMs). The advantage of STEMs over standard TEMs for this work is that STEMS are designed to produce a more finely focused beam that is scanned over a thin section, which gives more signal and higher resolution than with the less finely focused beam of a conventional TEM. Another comparatively recent development that produces a finer probe diameter is the application of field emission guns (FEG) to both TEMs and SEMs.

Vacuum systems of microscopes devoted to EDS need to have the capability of producing a vacuum of 5×10^{-5} Torr or better to prevent condensation of gas molecules on the detector, which is operated at cryogenic temperatures. If gases condense on the detector window, they can potentially absorb incoming X-rays, thus degrading the signal-detecting capabilities of the instrument. The accumulation of

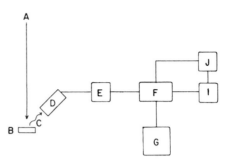

FIGURE 152. Diagrammatic representation of the signal-handling components of an EDS unit. The electron beam (A) interacts with the specimen (B), causing the release of characteristic X-rays (C), which reach the detector (D). The signal is amplified by the amplifier (E), passes through the multichannel analyzer (F), and then can be viewed on the display (G), stored in the computer (I), or can be fed to the data output device (J) by either the computer or the multichannel analyzer.

contaminants resulting from the interaction of the electron beam with the specimen under suboptimal vacuum conditions can also lead to the absorption of X-rays generated from beneath the contaminating material and the appearance of non-specimen-generated X-rays. Absorption of X-rays by surface contaminants can compromise quantitative studies. Low-energy X-rays generated by light elements are more likely to be absorbed by surface contaminants and thereby not reach the detector surface.

1. The Detector

The X-ray detector is a solid-state device consisting of a silicon-lithium crystal (Fig. 153) with a surface area in the range of 5–200 mm². It is sandwiched between two metal electrodes across which a bias voltage is applied. The detector crystal is kept under high vacuum and cooled with liquid nitrogen. The crystal is either part of a windowless detector or, more commonly, is covered with a thin (approximately 7-μm-thick) radiation-transparent beryllium window. In this latter configuration, elements below atomic number 11 are not detected, because their X-rays are absorbed by the beryllium. Each electron in the silicon portion of the collector can absorb 3.8 V. The negative charge on the detector is directly proportional to the energy of the incident X-ray. The X-ray is thus converted into an electron signal by the detector, which is then amplified and processed by a multichannel analyzer that separates the energy pulses in terms of amplitude and stores them in memory.

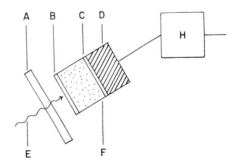

FIGURE 153. Diagrammatic sketch of an X-ray detector showing the characteristic X-ray (E) passing through the beryllium window (A), where it interacts with the solid-state detector composed of a positive layer (B), the silicon-lithium semiconductor (F), a negative layer (C), and the field effect transistor (D), which serves as a preamplifier for the signal that is then conveyed to the amplifier (H).

2. Emission Analysis

Emissions (characteristic X-ray lines) vary in intensity, and those that are the easiest to detect (those having the greatest intensity) are generally used for characterizing an element within a sample.

X-rays generated by transitions from the E_L to the E_K shell are called K lines. These transitions are the source of the most intense emissions, followed by E_M to E_L transitions (L lines) and so forth. Even within a given shell, there are a variety of electron jumps that can take place, designated by Greek letters along with numerical designations (α 1, α 2; β 1, β 2; γ 1, etc.). The higher its number and letter, the greater the intensity of an electron jump.

Mosely (1914) calculated the relationship between X-ray frequency (V or energy) and atomic number (Z) as follows:

$$V = 0.248 \ (Z-1)^2 \times 10^{16}$$

The X-ray frequency has been determined for all elements, which allows an element to be identified by the energy of the X-ray emissions generated by a given sample.

3. Electron Ionization

In order to produce an X-ray, the primary electron beam must have enough energy to remove an electron from an inner shell of a specimen atom. This energy is known as the *critical excitation potential* or the *absorption edge* and has a discrete value for each orbital electron energy level (K-shell electrons need more energy than L-shell electrons, etc.). The energy needed also tends to increase with increasing atomic number.

As mentioned previously, X-ray photon production does not always follow the ionization of atoms. Auger electrons or photons of light may also be released.

4. X-Ray Fluorescence

As was discussed in Chapter 9, the interaction of an electron beam with a specimen produces a myriad of events that take place simultaneously following the initial ionization event. We selectively record specific aspects of the signals generated, and the instruments are provided with several mechanisms to assist us with this selectivity. X-rays produced following ionization can cause other atoms within the sample to be secondarily ionized, leading to the generation of further X-rays. This phenomenon is known as X-ray fluorescence and may lead to quantitation errors, since the additional fluorescent X-rays can enhance or reduce the primary X-ray signal.

5. X-Ray Absorption

Absorption accompanies X-ray fluorescence. Some X-rays are stopped en route to the detector, or are scattered away from the detector, and are thus lost to the image-forming system. As previously mentioned, X-rays from higher Z-number atoms have

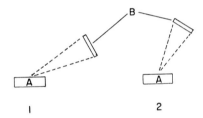

FIGURE 154. Location of an X-ray detector (B) in relationship to the specimen (A). If the detector is located at a low angle in relationship to the specimen (1), resolution will be lower than if it is located at a higher angle (2).

higher energy and are thus less likely to be absorbed within the specimen or within the detector window. Those with a Z number of 11 or lower have very low energy and will be absorbed by the specimen or by the beryllium detector window, so no signal can be recorded from these elements. To increase the likelihood of recording low-energy X-rays, the detector is usually placed at a high angle in relationship to the specimen (Fig. 154). This reduces the chance for X-rays from the sample to be self-absorbed.

6. Bremsstrahlung, Continuum, White Radiation

A primary electron that is scattered inelastically by the Coulomb field of an atomic nucleus can lose some or all of its energy. The amount of energy lost may vary from zero up to the initial energy of the primary electron beam. This lost energy may be released in the form of X-rays called *bremsstrahlung* (braking radiation), *continuum*, or *white radiation*. The amount of bremsstrahlung is directly related to the total number of atoms of all kinds within the sample, unlike the characteristic X-rays that are specific to one type of atom. Thus, bremsstrahlung is a source of specimen noise during qualitative analysis (elemental analysis). On the other hand, bremsstrahlung is a source of information concerning the mass thickness of a specimen, an important factor if quantitative analysis of thin sections is being performed.

7. Signal Analysis

Our discussion so far should make it clear that the X-ray spectrum of a given element is a complex signal with a number of different characteristic X-ray lines of different intensities (energies) against a background of bremsstrahlung. It is thus necessary to utilize a computer attached to the multichannel analyzer to determine which parts of the broad spectrum of signals is meaningful. In addition, a knowledgeable operator can select which part of the spectrum to evaluate, depending on the information being sought by the investigator.

With heavier elements, it is not always possible to produce K lines; thus, the L lines generated would need to be evaluated. Another factor that is of concern is the sensitivity of the X-ray detector. Various detectors are available, many of which will detect one line (M) more easily than another (L), even if the intensity of the L line is greater. Overlapping spectral lines are produced when a number of elements are present in the specimen. If this overlapping prevents clear analysis, another part of the spectral range can be selected in which the lines of the originally overlapping elements are further apart.

8. Sensitivity and Resolution

Sensitivity can be increased by placing the detector closer to the specimen (increasing count rate). Moving the detector closer to the specimen also reduces specimen damage because less beam exposure time is necessary to produce a usable spectrum. If too much signal is generated, however, more background will be recorded as well, leading to the potential for reduced resolution (the system can become overloaded with signals).

Higher accelerating voltages penetrate more deeply into the specimen, causing more beam spreading and, thus, less resolution, even though there is a simultaneous increase in excitation of the specimen. The high-density, finely focused (3–5 nm) beam generated by field emission guns is well suited to high-resolution EDS work, since the larger the probe diameter, the more beam spreading takes place within a sample, and the less resolution is possible.

9. Specimen Preparation

The best instrumentation for EDS work is generally found associated with engineering or materials science laboratories. The technicians who maintain and operate these instruments typically are very knowledgeable about the spectra of various elements and their overlaps, and about how to adjust the multichannel analyzer and computer analysis to separate elements with similar spectral characteristics for commonly read lines. At the same time, they generally have no knowlege of fixation, embedment, staining, and general preparative methods employed by biologists, as well as the interpretation of morphological features of biological materials. These matters have been discussed in detail by Coleman (1975) and Hall *et al.* (1974).

To work effectively with materials laboratories, it is necessary to communicate what chemicals with Z numbers over 11 were used in specimen preparation (osmium, lead, uranium, etc.) and what substrates (stubs, grids, coverslips, polymer films) were used as specimen supports.

If a TEM or STEM unit is employed for EDS, it is necessary to choose a grid material that will not interfere with the analysis desired. Grids are manufactured from copper, nickel, gold, molybdenum, titanium, chromium, platinum, beryllium, carbon-coated nylon, etc. If the biological sample is not expected to contain materials with spectra overlapping those of copper, copper grids will be adequate. Beryllium grids are generally considered to be good substrates for EDS work because of their electron transparency, but they are extremely hydrophobic, making section pick-up very difficult. In addition, they should never be cleaned with any solvents, since beryllium is extremely toxic and must not be solubilized in any fluids that might come into contact with living things. Grids coated with polymer films, such as Formvar or collodion, can have significant silicon, chlorine, and sulfur content, so many microanalytical laboratories use grids coated exclusively with carbon.

If there is any danger of solubilizing elemental components of cells or tissues during processing for conventional chemical fixation and embedment, cryotechniques can be applied. Native frozen sections cryotransferred to a microscope equipped with a cryostage can preserve soluble electrolytes, such as sodium, potassium, and chlorine, for subsequent microanalysis. Cryosubstitution also has advantages over conventional chemical fixation procedures, because substitution of methanol or acetone for water at low temperatures results in less extraction of water-soluble cellular components. Cryofixation followed by freeze-drying or molecular distillation can also preserve small soluble ions that otherwise would be leached from the specimen during conventional chemical processing, and thus would be eliminated from possible detection by EDS.

With preparations intended for SEM EDS work, carbon stubs are usually used to reduce spurious signals that would be generated from metal stubs. For the same reason, specimens are typically coated with carbon rather than gold-palladium or other metals. Carbon will still allow excess electrical current to be conducted to ground potential and will decrease specimen heating without producing conflicting signals characteristic of the heavy metals used for coating specimens for conventional secondary electron imaging. In addition, it should be remembered that any conductive coating will have some capacity to absorb X-rays generated from the specimen. If samples are mounted on glass coverslips, there may be further problems with spurious readings, since glass can contain magnesium, copper, and of course, silicon.

Coleman (1975) offers a brief discussion of sample preparation for biological fluids containing soluble ions. He also discusses the problems of drying samples obtained from buffered media. In the latter case, if a sample soaked in buffer is dried, the buffering salts tend to precipitate as the sample is dried, thus obscuring detail for SEI examinations, as well as producing large domains of crystals capable of being recorded by EDS.

10. Qualitative vs. Quantitative Work

Most EDS work performed on biological specimens is qualitative in nature and determines the presence of different elements within a sample. In some cases, it is desirable to ascertain the amount of a specific element within a specimen. In these cases, quantitative analysis is necessary. As mentioned above, this is made possible by utilizing the bremsstrahlung to determine the mass of the sample based on the non-specific radiation arising from the beam interactions with all the different atoms in the specimen. Once the mass of the specimen is recorded by the computer, the relative quantity of a given element of interest can be determined, and then the computer can make a quantitative analysis for that element. This method is discussed more thoroughly in the texts cited at the beginning of this chapter.

B. Electron Energy Loss Spectroscopy (EELS)

When the electron beam in a TEM interacts with a thin section of a sample, three events occur. Most of the primary beam electrons pass through the specimen without significant interaction, eventually striking the phosphor-coated viewing screen or photographic film. Primary electrons that experience elastic scattering within the specimen are swept through large angles, are scattered behind the objective aperture opening, and are prevented from reaching the viewing screen. Primary electrons experiencing inelastic collisions within the specimen suffer an energy loss (Δ E) but have only small changes in direction. Those electrons that suffer an energy loss because of one or more inelastic collisions within the specimen but that still arrive at the viewing chamber may be imaged in a different plane from the undeviated primary beam and become a source of chromatic aberration in the specimen image.

The Zeiss EM902 TEM incorporates an imaging electron spectrometer into the lens system, which can eliminate chromatic aberration. The unscattered part of the primary electron beam is bent through a 90° angle in the magnetic field of the prism (Fig. 155), reflected back through the electrostatic field of the mirror, and then deflected again by the prism back onto the optical axis. The slower electrons produced by inelastic collisions are less energetic and thus are more easily deflected as they pass through the magnetic field of the prism. As the electrons are refracted, they strike the

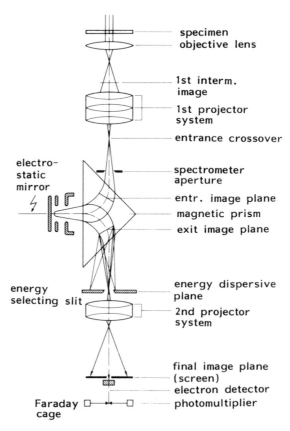

FIGURE 155. Diagram of the electron paths within a Zeiss EM-902. (Courtesy of Carl Zeiss, Inc.)

spectrometer slit and thus are prevented from reaching the viewing screen. The thicker a specimen, the greater the fraction of inelastically scattered electrons caused by multiple scattering within the specimen. Thus, a thicker specimen will generate a broader electron energy spectrum. In a Zeiss 902, when viewing a 0.75-μm-thick section, an 80-kV image is sharper than that produced from the same sample in a 200-kV instrument and is virtually as sharp as that produced from a 1000-kV unit. Of course, since all deviated electrons are kept from reaching the screen/film, image brightness is significantly diminished in this mode (if the slit filter is not turned on, the 902 works like a conventional TEM).

A procedure called *contrast tuning* can change the contrast of the specimen image by selecting electrons for imaging that have experienced various energy losses. If the spectrum is swept across the spectrometer slit, multiple scattered electrons with specific energy losses can be imaged.

A mode called electron spectroscopic imaging (ESI) images through an electron energy spectrometer, always using electrons of defined energy and a narrow bandwidth. Each element within a specimen has its own characteristic capabilities to produce scattering of the electron beam (both inelastic and elastic), largely due to the element's Z number.

Various manufacturers offer the option of electron energy loss spectroscopy (EELS) units similar to that supplied with the Zeiss 902. These identify elements within the specimen on the basis of their effect on the number and energy of electrons emerging from the specimen as compared to the number of electrons and energy of the primary beam. Egerton (1989), Hezel (1988), and Zaluzec (1986) discuss the instrumentation and theory behind EELS. Specimen analysis with EELS is more sensitive for elements at the lighter end of the spectrum, such as sodium and potassium, compared to EDS.

REFERENCES

Chandler, J.A. 1978. X-ray microanalysis in the electron microscope. In: *Practical methods in electron microscopy*. Vol. 5, Pt. 2, A.M. Glauert (ed.), North-Holland, New York.

Coleman, J.R. 1975. Biological applications: Sample preparation and quantitation. In: *Practical scanning electron microscopy. Electron and ion microprobe analysis*, J.I. Goldstein and H. Yakowitz (eds.), Plenum Press, New York, pp. 491–527.

Egerton, R.F. 1989. *Electron energy-loss spectroscopy in the electron microscope*. Plenum Press, New York.

Goldstein, J.I., Newbury, D.E., Echlin, P., Joy, D.C., Fiori, C., and Lifshin, E. 1981. Scanning electron microscopy and X-ray microanalysis. Plenum Press, New York.

Hall, T., Echlin, P., and Kaufmann, R. (eds.). 1974. *Microprobe analysis as applied to cells and tissues*. Academic Press, New York.

Hezel, U.B. 1988. Electron spectroscopy for imaging and analysis in the transmission electron microscope. *Am. Lab.* September, p. 51.

Morgan, A.J. 1985. X-ray microanalysis in electron microscopy for biologists. In: *Royal Microscopical Society Handbook* 5, Oxford University Press, New York.

Mosley, H.G. 1914. The high frequency spectra of the elements, part II. *Phil. Mag.* 27:703.

Newbury, D.E., Joy, D.C., Echlin, P., Fiori, C.E., and Goldstein, J.I. 1986. *Advanced scanning electron microscopy and X-ray microanalysis*. Plenum Press, New York.

Russ, J.C. 1984. *Fundamentals of energy dispersive X-ray analysis.* Butterworths, London.
Vaughan, D. (ed.). 1989. *Energy-dispersive X-ray microanalysis. An introduction.* Kevex Instruments, Inc., San Carlos, CA.
Zaluzec, N.J. 1986. *A beginner's guide to electron energy loss spectroscopy Part II—Electron spectrometers. EMSA Bull.* 16:58.

CHAPTER 13

Cytochemistry

Cytochemistry at the ultrastructural level is a broad subject, encompassing considerations of fixation and its effect on enzyme activity, problems with reagent and label
penetration of tissues and cells, and problems distinguishing reaction products from
other cellular constituents. The aim of this chapter is to introduce some of the different
general types of staining procedures frequently employed to demonstrate specific
chemical entities associated with cellular surfaces and cytoplasmic contents.

As with many of the specialized techniques in electron microscopy, it is always
advisable to examine the current literature concerning the group of organisms being
studied and the particular cytochemical procedures that are of interest. There are also
various routinely used techniques, which are described in Hayat (1981, 1989). Even
though now seriously dated, some of the classical cytochemical procedures (many of
them still widely used), as well as in-depth discussions of their mechanisms, are
provided in the five-volume series entitled *Electron Microscopy of Enzymes*, edited by
Hayat (1973–1977). Immunocytochemistry has evolved into a major specialized area
of cytochemistry that warrants the separate treatment provided in Chapter 14.

I. PROBLEMS

As was mentioned in Chapter 1, the goal of fixation is to stabilize the colloidal
suspension of cellular constituents so that the individual components are immobilized
and made resistant to the further steps of dehydration, embedment, and polymerization
that are employed with standard processing procedures. We have seen that cryotechniques avoid some of the stresses associated with the processing steps but introduce
other problems, primarily the shallow depth of fixation. If a cytochemical procedure
yields positive results after chemical fixation, that approach is recommended because
the sample size is generally larger than is possible with cryotechniques, and the
photographic image that results is generally more pleasing. On the other hand, if a

procedure does not work with conventional chemical fixation, cryotechniques offer a solution.

For enzyme cytochemistry, we have the twin problems of not only needing to stabilize cellular components in an attempt to prevent their extraction during processing, but also needing to maintain an active three-dimensional protein. Since enzyme specificity is determined by reactive sites defined by the complex tertiary folding of the protein molecules comprising enzymes, any of the cross-linking fixatives used in electron microscopy has the capacity to inactivate enzymes. This result may occur because the cross-linking process changes the spatial relationship between different areas within an enzyme's active site, or it may occur because the act of cross-linking the enzyme to another adjacent protein makes the enzyme's active site inaccessible for enzymatic activity. In addition, inactivation may result when the fixative becomes attached to the enzyme, since the fixative molecules themselves may provide steric hindrance to the binding site for the enzymatic reaction.

Each aldehyde fixative inactivates various enzymes to a different extent. Aldehydes such as glutaraldehyde and acrolein that more strongly cross-link are usually less desirable for enzyme cytochemistry than formaldehyde, which cross-links weakly. Since each aldehyde can inactivate a spectrum of enzymes, it is generally a poor idea to use a primary fixative containing a mixture of aldehydes, such as formaldehyde and glutaraldehyde (Karnovsky's, McDowell's, and Trump's), because more enzymes will become inactivated than if only one aldehyde is used. At the same time, all aldehydes produce membrane leakage, and thus increase the permeability of cells and organelles to reagents, which can be helpful. Unfortunately, the process also allows enzymes and other materials originally contained within organelles, such as lysosomes, to leak into the cytoplasm. This latter phenomenon presumably accounts for some of the leaching of cytoplasmic contents encountered with long-term storage of tissues in aldehyde solutions. As the cells become fixed, their membranes become permeable, their lysosomes leak some of their hydrolytic enzymes into the cytoplasm, where the enzymes then degrade cytoplasmic materials, which are subsequently washed out during further processing steps.

Most of the literature on cytochemical procedures for electron microscopy suggests that biological-grade 37–40% formaldehyde should not be used to formulate fixatives because of the small amounts of methanol added during manufacture to prevent polymerization. Various procedures manuals and textbooks for electron microscopy even state that structural work will be compromised unless formaldehyde is made from freshly depolymerized paraformaldehyde containing no additives. However, it is clear from all published sources using McDowell's and Trump's 4F:1G (see Chapter 1) that biological-grade formaldehyde is adequate for preparing fixatives for straightforward structural work.

A number of enzyme cytochemical procedures have been modified from techniques originally developed for light microscopy utilizing tissues fixed in 4% buffered neutral formalin formulated from biological-grade formaldehyde. This indicates that it is not absolutely necessary to go to the trouble of making up solutions of formaldehyde from paraformaldehyde, which have a relatively short shelf life. Nonetheless, most recipes in the literature that use formaldehyde still specify making it up fresh, and so it is customary to follow that procedure.

Other aspects of fixation procedures may cause problems. Osmium is a strong oxidant and will damage the reactivity of most enzymatically active sites in cells. The fact that osmium is a heavy metal may also interfere with enzymatic activity.

Even with the nonenzymatic cytochemistry used to localize carbohydrates or various types of charged sites associated with cells, processing chemicals may introduce problems. If a procedure for demonstrating negatively charged groups is employed, phosphate buffers will usually decrease or eliminate the stain's effectiveness. The negative charges of phosphate buffers will exhibit reactivity with some stains, possibly resulting in coprecipitation of the stain and the buffer, preventing the desired interaction of the stain with cellular components.

Cytochemical procedures are designed to leave a recognizable product at the site of a specific cellular component that is being identified. The reaction product should ideally be electron dense and hard to confuse with other normally encountered cellular components. It also needs to be large enough to be easily visualized with an electron microscope. Producing a reaction product of sufficient size to be visualized proves to be one of the most significant problems encountered in designing cytochemical protocols. Many products that are large enough to be easily visualized will not cross membranes, even after aldehyde fixation. One of the reasons for the popularity of peroxidase procedures is that the reagents have the capability of penetrating aldehyde-fixed membranes. However, the typical product of these reactions is an osmicated area within a cell, which can be confused with other normally osmicated materials within the cell.

Various procedures have been employed to improve reagent accessibility to cells and tissues. Conventional samples 1 mm thick can be exposed to cytochemical reagents, embedded, and sectioned, and then only the surface cells that have been easily penetrated can be examined. This procedure is fairly inefficient and produces limited usable sample size, but will often work. Another method frequently used involves aldehyde fixation followed by chopping or slicing of the tissue into 40- to 50-μm-thick slices (utilizing a Smith-Farquahar [Dupont TC-2] or MacIlwaine tissue chopper in the former case or a Vibratome® in the latter case) prior to cytochemical procedures. The thin slices of tissue produced provide superior reagent accessiblity. Some tissues are sliced or chopped in the living state, exposed to cytochemical procedures, and then fixed and embedded. After the tissue slices have been fixed and the cytochemical procedures have been completed, the slices may be put into molten (45–50°C) agar and centrifuged gently before the agar solidifies to produce a concentrated sample that is easier to handle in subsequent processing steps.

II. SPECIFIC REACTION PRODUCTS

To be an effective stain for electron microscopy, a material incorporated into tissues must be capable of scattering the electron beam, thus producing dark areas in the viewed image because of subtractive contrast. Some of the stain products are also visible at the light microscopic level, which can be helpful for the analysis of staining success. If possible, it is useful to ascertain that some staining is occurring and to examine a comparatively large amount of sample at the light level before embarking on the lengthy procedures usually required for the preparation of samples for electron microscopy. Since conventional chemical fixation and embedment procedures profoundly inactivate most enzymes, most cytochemistry is done *en bloc*.

A. Peroxidase Procedures

The enzyme peroxidase is easily demonstrated at the light microscopic level by exposing it to hydrogen peroxide, which is reduced to water. To produce an electron-dense stain for electron microscopy, 3,3′-diaminobenzidine · HCl (DAB) is provided as an electron donor.

It should be noted that DAB is a confirmed carcinogen and must be handled with great care. If used in powdered form, it can easily be scattered during weighing by electrostatic forces, so it is usually better to use sealed vials into which diluent is injected prior to removal.

The highly reactive DAB molecule becomes localized at the site of the peroxidase and can, in turn, reduce osmium applied as a post-fixative, thereby amplifying the peroxidase and producing an osmium black. The product is seen as an amorphous electron-dense area within the tissue that has no other characteristics, and so must be interpreted with care to prevent confusion with naturally osmicated regions. Other reactive chemical entities, such as Hatchett's brown (described below), can be localized within cells by amplification with DAB and osmium.

B. Lead Capture

Another commonly employed reaction product borrowed from enzyme histochemical techniques is lead. Various lead-capture techniques developed for the demonstration of enzymes at the light microscopic level are easily transformed into electron microscopy procedures because of the electron density of the product. These techniques may confuse naturally occurring lead products with products found in tissues post-stained with lead. If lead-capture procedures are used, it is advisable to prepare two grids for examination, one with post-staining and one without. If the grid that was not post-stained is examined first, the lead products will stand out clearly, allowing easier interpretation of the post-stained grid, which will have the additional lead deposits associated with post-staining.

C. Ferritin

The iron-containing protein ferritin was used extensively in the early days of immunocytochemistry and to demonstrate ionic sites on cell surfaces (Danon *et al.*,

1972) because of the electron density of the ferritin molecule. Since ferritin occurs naturally in some tissues, it must be used with care. In addition, it is not so dense as lead or so discrete as gold, so it is not used much in contemporary procedures.

D. Colloidal Gold

Colloidal gold, which will be discussed more extensively in Chapter 14, has been conjugated to antibodies for immunocytochemical procedures and to lectins to localize saccharides during the last 10 years. It also can be conjugated to enzymes such as phospholipase A_2 (Coulombe et al., 1988) and used to probe cells for the location of specific reactive sites. Despite its size (1 nm or larger), which prevents passage across membranes, colloidal gold is a clearly nonbiological entity and cannot be confused with any other stain products, characteristics that enhance its usefulness.

E. Ruthenium Red, Alcian Blue, Pyroantimonate

There are also a variety of naturally electron-dense chemicals, such as ruthenium red, alcian blue, and pyroantimonate, that can be easily visualized with electron microscopy following binding to specific cellular constituents. Again, these chemicals produce amorphous electron-dense areas within cells, so these areas must be interpreted with care.

III. EXAMPLES OF ENZYME CYTOCHEMISTRY (see Powell, 1986)

A. Peroxidase Methods

Naturally occurring endogenous peroxidases (Novikoff and Goldfischer, 1969) located in microbodies are easily demonstrated by briefly fixing tissues or cells with formaldehyde freshly prepared from paraformaldehyde. After quickly washing with buffer, samples are exposed to hydrogen peroxide and DAB, rinsed, then post-fixed in osmium, dehydrated, and embedded. As stated above, the product is an amorphous, electron-dense osmium black (Fig. 156). To interpret the results, various controls should be employed, such as eliminating the initial substrate (H_2O_2) or briefly boiling the tissue (for 5 min) following the rinse after initial aldehyde fixation.

Exogenously applied peroxidase techniques introduced by Graham and Karnovsky (1966) have also been utilized to demonstrate leakage between normally tightly bound cell populations, such as endothelial cells of vascular tissues or epithelial cells lining various organ systems. In this application, horseradish peroxidase is injected into a living animal as a tracer, tissues are removed and fixed, and then amplification of the peroxidase is produced by interaction with DAB and osmium.

Finally, peroxidases have been conjugated to other products that are bound by enzymes (see Sternberger, 1973), which are then localized by the interaction of the peroxidase with reaction mixtures of hydrogen peroxide and DAB, followed by osmium.

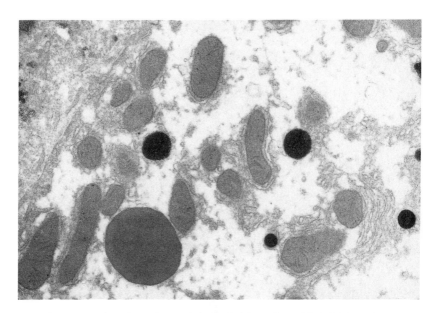

FIGURE 156. Demonstration of peroxisomes (microbodies) in rat liver with the DAB/osmium technique of Novikoff and Goldfischer (1969). The round, extremely electron-dense structures in the cytoplasm are microbodies. 18,400×. (Micrograph courtesy of Philip L. Sannes.)

B. Hatchett's Brown Methods

Several methods have been developed that utilize a reaction originally developed for light microscopy, which yields cupric ferrocyanide at the site of enzyme activity. This product is visible as a brown deposit with light microscopy, and it can be further amplified by DAB with subsequent osmication to produce an osmium black visible with electron microscopy.

This technique has been used to demonstrate the activity of aryl sulfatase on the substrate 4-nitro-1,2-benzenediol mono(hydrogen sulfate), as originally described by Hanker et al. (1975). Another application of this procedure is used to localize the enzyme involved in membrane turnover, acyl transferase, as shown in Fig. 157 (Dykstra, 1976).

C. Lead-Capture Methods

Lead-capture methods appropriated from light microscopy techniques were some of the first methods used to demonstrate hydrolytic enzymes associated with lysosomes such as acid phosphatase, alkaline phosphatase, and aryl sulfatase (see Essner, 1973).

Barka and Anderson (1962) introduced a still widely used modification of previous lead-capture methods for the demonstration of acid phosphatase. It was formulated with less lead than used with previous procedures and employed a maleate

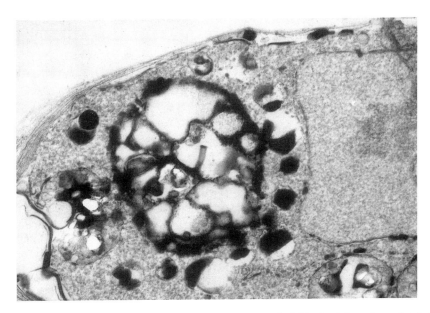

FIGURE 157. Acyl transferase demonstrated by a copper-capture method (Hatchett's brown) associated with membranes of *Sorodiplophrys stercorea*, a terrestrial protozoan. The product appears as electron-dense deposits within the cell, which was not post-stained. 32,400×.

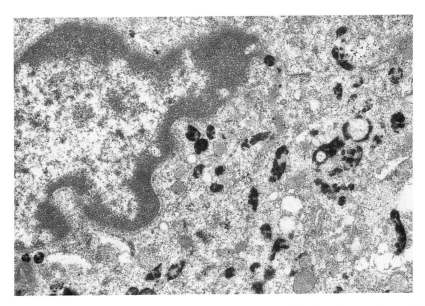

FIGURE 158. Demonstration of acid phosphatase located in lysosomes of rat alveolar macrophages with the lead-capture method of Barka and Anderson (1962). The lead phosphate product is visualized as the extremely electron-dense areas within the cell. 22,400×. (Micrograph courtesy of Philip L. Sannes.)

buffer, which was said to keep lead ions in solution, thus decreasing random deposition of lead in tissues and cells. As with the previous procedures, Na-β-glycerophosphate is the substrate for the enzyme, Tris-maleate buffer is used to maintain a pH of 5.0 for best enzymatic activity, and lead nitrate is the source of lead. Acid phosphatase causes the release of phosphate ions from the substrate, which replace the nitrate molecules attached to lead nitrate, resulting in the formation of insoluble lead phosphate at the site of enzymatic activity (Fig. 158).

IV. EXAMPLES OF NONENZYMATIC CYTOCHEMISTRY

A. Cationic Dyes

Three major cationic materials have been used extensively for the demonstration of the glycocalyx associated with most cell surfaces. Initially, only plants, fungi, and many bacteria were known to have walls composed of complex assemblages of carbohydrate moieties and other chemical entities, but during the 1970s closer examination demonstrated the presence of a polysaccharide cell coat on the surface of virtually all cell types. It became evident that these carbohydrate-loaded, typically negatively charged coats were frequently involved in cellular recognition, so cell biologists became more interested in being able to delineate them clearly. This was achieved by applying cationic ferritin, alcian blue, or ruthenium red. In some cases (Dykstra and Aldrich, 1978), it was necessary to use both alcian blue (0.5% in the primary fixative) and ruthenium red (0.05% in the osmium post-fixative) to demonstrate evanescent glycocalyx materials (Fig. 159). When using these dyes, it is imperative that cacodylate buffers be used rather than phosphate buffers, since the latter will coprecipitate with the dyes, rendering them useless. None of these dyes reveal the exact chemical nature of glycocalyx materials, only that they are present and possess some negative charges.

B. Polysaccharide Stains

Several widely used semispecific polysaccharide stains have been developed (Erdos, 1986), which primarily reveal hydroxyl groups of polysaccharides opened by oxidation with periodic acid. These techniques are essentially equivalent to periodic acid-Schiff's reagent reactions employed in light microscopy. The three most commonly employed stains are silver methenamine (Pickett-Heaps, 1967; Edgar and Pickett-Heaps, 1982), which is equivalent to Gomori methenamine-silver staining used for light microscopy (Fig. 160), thiocarbohydrazide (Thiery and Rambourg, 1975); and silver protein (Thiery, 1967). All three are used to stain carbohydrate molecules associated with cells and tissues that have been fixed, embedded in commonly used resins, and then sectioned before staining. Osmium will often react with these stains, so it must either be removed from the sections by fairly strong oxidants prior to the polysaccharide reactions or not be used in the initial fixation steps. If osmium is not used as a

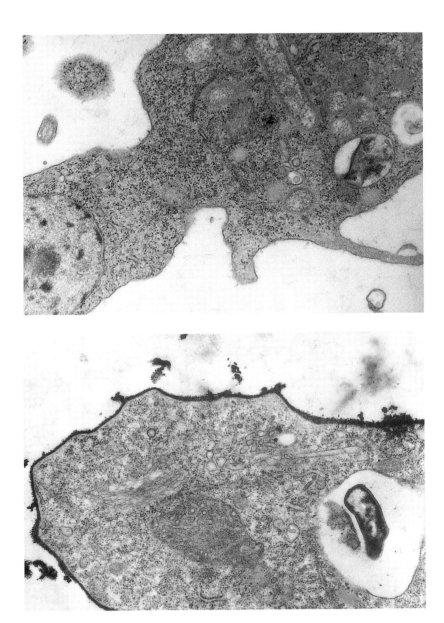

FIGURE 159. Amoebae of the slime mold, *Fuligo septica*. Top: A cell with no cytochemical staining. 20,500×. Bottom: A cell stained with 0.5% alcian blue in the glutaraldehyde primary fixative and 0.05% ruthenium red in the osmium post-fixation step. The glycocalyx can be seen as an electron-dense, granular deposit on the surface of the plasma membrane. No post-stain. 26,200×.

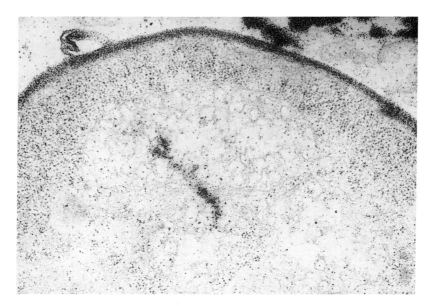

FIGURE 160. Demonstration of polysaccharide plates comprising the cell wall of the terrestrial protozoan *Sorodiplophrys stercorea* with silver methenamine staining. Large quantities of silver grains are deposited on the polysaccharide cell wall. Nonspecific deposits are located throughout the cell, presumably because of interaction with osmium, which was used as a post-fixative. 34,500×.

post-fixative, cell structure is somewhat compromised. But if optimal polysaccharide localization is desired with minimal interpretational difficulties, osmium should be eliminated.

All three techniques stain similar molecules, and all three should be used on sections picked up on inert grids made of stainless steel or nickel. Copper grids are chemically reactive and can interfere with the stains. The grids can be inserted into small segments of Tygon tubing with small slits in the surface (Fig. 161) to allow easy transfer from one staining solution to another.

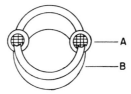

FIGURE 161. Tygon tubing (B) with slits holding grids (A) for cytochemical procedures.

TABLE 11. A Selection of Lectins and the Sugar Residues
with Which They React

Lectin	Source	Sugar specificity
Abrin	*Abrus precatorius* (jequirity bean)	D-galactose
Concanavalin A (Con A)	*Canavalia ensiformis* (jack bean)	α-D-mannose α-D-glucose
Limulus polyphenus (LPA)	*Limulus polyphenus* (horseshoe crab)	N-acetylneuraminic acid (sialic acid) Glucuronic acid Phosphorylcholine analogs
Ricin	*Ricinis communis* (castor bean)	N-acetyl-D-galactosamine β-D-galactosyl residues
Ulex	*Ulex europaeus* (gorse or furze)	α-L-fucose N,N′-diacetylchitobiose
Wheat germ agglutinin	*Triticum vulgaris* (wheat)	N-acetyl-β-D-glucosaminyl residues N-acetyl-β-D-glucosamine oligomers

C. Monosaccharide and Disaccharide Stains

The polysaccharide stains described above are still fairly nonspecific, staining charged entities such as osmium, in addition to polysaccharides. In addition, even when sufficient controls have been employed to unequivocally demonstrate polysaccharides, the specific sugars involved are not identified. Various workers (Sharon and Lis, 1989; Erdos, 1986) have found, however, that a number of plant proteins, called *lectins*, are capable of binding to specific sugar residues in or on cells. Initially, lectins were conjugated with fluorochromes, such as fluoroisothiocyanate (FITC) or peroxidase, attached to sugars associated with cells, and demonstrated with fluorescence or bright-field microscopy, respectively. Cell biologists then used lectins conjugated with compounds, such as ferritin or peroxidase, so that the lectins could be identified with the electron microscope. When colloidal gold became widely used as an electron-dense marker, investigators conjugated it with lectins to produce a superior marker for the presence of specific sugars in tissues and cells. Lectin binding of surface sugars is most easily done *en bloc*. If sugars inside cells are to be demonstrated, cryoultramicrotomy is indicated, since the lectin molecule with its associated label is too large to pass through a plasma membrane.

Lectin binding is particularly useful because of its specificity. A variety of sugars can be identified with the various lectins available (Table 11).

D. Calcium Staining

Calcium localization is of great interest to muscle researchers, membrane biologists, and others, since it is critically involved in numerous cellular functions. Po-

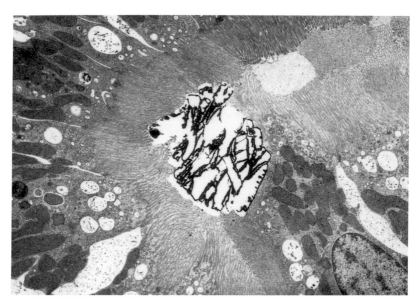

FIGURE 162. Proximal convoluted tubule of a rat kidney showing a large crystalline array of calcium oxalate stained with potassium pyroantimonate within the tubule lumen. Finely granular, black deposits of pyroantimonate-stained calcium are also located in apical vesicles of tubule epithelial cells. 2,150×.

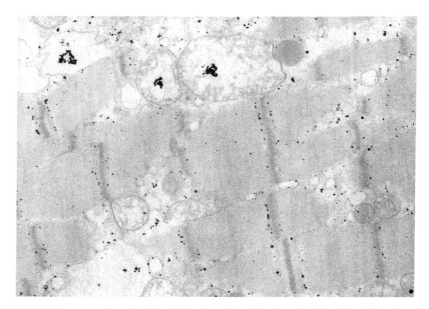

FIGURE 163. Demonstration of soluble calcium in canine skeletal muscle. Note that calcium is preferentially associated with the Z bands of sarcomeres. Larger granular deposits of calcium stained with pyroantimonate are located in the matrix area of mitochondria. If pyroantimonate staining is done during postfixation in osmium, rather than prior to aldehyde fixation, none of these deposits remain. 14,000×.

tassium pyroantimonate or sodium pyroantimonate has been used extensively (see review by Wick and Hepler, 1982) to demonstrate calcium. Pyroantimonate is not strictly calcium specific, because it also binds to sodium, potassium, and magnesium. With proper controls, however, pyroantimonate staining has been recognized as a relatively reliable and simple procedure to localize calcium. Pyroantimonate staining during post-fixation with osmium (Dykstra and Hackett, 1979), as shown in Fig. 162, will reveal calcium deposits as electron-dense material. Any soluble calcium in the tissue will have been washed out during fixation and processing up to the post-fixation step, however. To demonstrate soluble calcium, pyroantimonate must be applied to tissue during primary fixation. The techniques described by Oberc and Engel (1977) demonstrate such soluble calcium entities (Fig. 163) but do not work well with tissues other than muscle. Muscle tissue is relatively stable compared with other tissues. Even if muscle tissue is stored at 4°C on saline-dampened gauze for a number of hours, it will still look normal when fixed. The pyroantimonate method developed by Oberc and Engel uses a solution of pyroantimonate to stabilize calcium in the tissues prior to aldehyde fixation. Again, this method works for muscles, which are relatively stable, but works poorly for tissues, such as kidney, that suffer severe fixation damage during the procedure.

REFERENCES

Barka, T., and Anderson, P.J. 1962. Histochemical methods for acid phosphatase using hexazonium pararosanilin as a coupler. *J. Histochem. Cytochem.* 10:741.

Coulombe, P.A., Kan, F.W.K., and Bendayan, M. 1988. Introduction of a high-resolution cytochemical method for studying the distribution of phospholipids in biological tissues. *Eur. J. Cell Biol.* 46:564.

Danon, D., Goldstein, L., Marikovsky, Y., and Skutelsky, E. 1972. Use of cationized ferritin as a label of negative charges on cell surfaces. *J. Ultrastruct Res.* 38:500.

Dykstra, M.J. 1976. Wall and membrane biogenesis in the unusual labyrinthulid-like organism *Sorodiplophrys stercorea. Protoplasma* 87:329.

Dykstra, M.J., and Aldrich, H.C. 1978. Successful demonstration of an elusive cell coat in amoebae. *J. Protozool.* 25:38.

Dykstra, M.J., and Hackett, R.L. 1979. Ultrastructural events in early calcium oxalate crystal formation in rats. *Kidney Int.* 15:640.

Edgar, L.A., and Pickett-Heaps, J.D. 1982. Ultrastructural localization of polysaccharides in the motile diatom *Navicula cuspidata. Protoplasma* 113:10.

Erdos, G.W. 1986. Localization of carbohydrate-containing molecules. In: *Ultrastructure techniques for microorganisms*, H.C. Aldrich and W.J. Todd (eds.), Plenum Press, New York, Chap. 14.

Essner, E. 1973. Phosphatases. In: *Electron Microscopy of Enzymes: Principles and Methods*, Vol. 1, M.A. Hayat (ed.), Van Nostrand Reinhold, New York.

Graham, R.C., Jr., and Karnovsky, M.J. 1966. The early stages of absorption of injected horseradish peroxidase in the proximal convoluted tubules of mouse kidney: Ultrastructural cytochemistry by a new technique. *J. Histochem. Cytochem.* 14:291.

Hanker, J.S., Thornburg, L.P., Yates, P.E., and Romanovicz, D.K. 1975. The demonstration of aryl sulfatases with 4-nitro-1,2-benzenediol mono (hydrogen sulfate) by the formation of osmium blacks at the sites of copper capture. *Histochemistry* 41:207.

Hayat, M.A. (ed.). 1973–1977. *Electron microscopy of enzymes: Principles and methods*, Vols. 1–5. Van Nostrand Reinhold, New York.

Hayat, M.A. 1981. *Fixation for electron microscopy.* Academic Press, New York.

Hayat, M.A. 1989. *Principles and techniques of electron microscopy. Biological applications*, 3rd ed. CRC Press, Boca Raton, FL.

Novikoff, A.B., and Goldfischer, S. 1969. Visualization of peroxisomes (microbodies) and mitochondria with diaminobenzidine. *J. Histochem. Cytochem.* 17:675.

Oberc, M.A., and Engel, W.K. 1977. Ultrastructural localization of calcium in normal and abnormal skeletal muscle. *Lab. Invest.* 36:566.

Pickett-Heaps, J.D. 1967. Preliminary attempts at ultrastructural polysaccharide localization in root tip cells. *J. Histochem. Cytochem.* 15:442.

Powell, M.J. 1986. Cytochemical techniques for the subcellular localization of enzymes in microorganisms. In: *Ultrastructure techniques for microorganisms*, H.C. Aldrich and W.C. Todd (eds.), Plenum Press, New York, Chap. 15.

Sharon, N., and Lis, H. 1989. Lectins as cell recognition molecules. *Science* 246:227.

Sternberger, L.A. 1973. Enzyme immunocytochemistry. In: *Electron microscopy of enzymes. Principles and methods, Vol. 1*, M.A. Hayat (ed.), Van Nostrand Reinhold, New York, Chap. 7.

Thiery, J.P. 1967. Mise en evidence des polysaccharides sur coupes fines en microscopie electronique. *J. Microsc.* 6:987.

Thiery, J.P., and Rambourg, A. 1975. Polysaccharides cytochemistry. *J. Microsc.* 21:225.

Wick, S.M., and Hepler, P.K. 1982. Selective localization of intracellular Ca^{2+} with potassium antimonate. *J. Histochem. Cytochem.* 11:1190.

CHAPTER 14

Immunocytochemistry

I. PURPOSE

A good historical account of the beginnings of immunolabeling can be found in Romano and Romano (1984). This history includes the work of Coons *et al.* (1941), who introduced fluorescent-labeled antibodies to light microscopy, followed by the introduction of antibodies conjugated to ferritin (Singer, 1959), peroxidase (Nakane and Pierce, 1966), and gold (Faulk and Taylor, 1971; Romano *et al.*, 1974). The ability to easily detect antibodies at both the light and electron microscopic level has made it possible to localize a variety of antigenic substances associated with the surfaces and interiors of cells. Modifications of these techniques have been used to demonstrate the presence of nucleic acids (Raap *et al.*, 1989; Silva *et al.*, 1989), carbohydrates (Horisberger, 1985), and enzymes (Bendayan and Zollinger, 1983; Bendayan, 1985) as well. There are a number of excellent books dealing with immunolabeling at the ultrastructural level (Beesley, 1989a; Bullock and Petrusz, 1982, 1983, 1985, 1989; Hayat, 1989, 1990; Polak and Van Noorden, 1983; Polak and Varndell, 1984; Sternberger, 1986; Williams, 1977).

As with the techniques developed for ultrastructural localization of enzymes, fixation and embedment procedures must be critically evaluated prior to embarking on new immunocytochemical procedures. Willingham (1980) has described intracellular antigen localization with ferritin-labeled antibodies, using a technique to make membranes permeable to large molecules with saponin. The technique has been supplanted by cryosectioning procedures, but the stepwise approach to determination of the ideal fixation concentration and the design of labeling protocols may be applied to other systems, providing an excellent example of experimental design.

Ideally, the preparative procedures chosen will maintain good ultrastructural morphology without denaturing the antigenic determinants associated with cells and tissues that are to be demonstrated immunocytochemically. These two goals tend to be mutually exclusive, however. Typical chemical fixation, dehydration, and embedment pro-

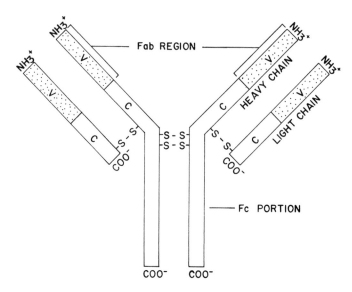

FIGURE 164. Diagram of an antibody molecule, showing the two potentially reactive sites, the Fab region, usually employed to react with specific antigens, and the Fc portion, to which protein A or G binds to demonstrate the presence of any immunoglobulin. V = variable region, C = constant region.

cedures utilizing epoxide resins have a strong tendency to either change protein structure or to make the active regions of proteins, such as enzymes and antibodies, inaccessible, as discussed in Chapter 13. Techniques developed during the last decade have applied approaches such as cryofixation, cryosubstitution, cryoultramicrotomy, molecular distillation, and acrylic resin embedding to solve some of the traditional problems associated with immunocytochemistry.

Some techniques use the Fab region of an antibody to recognize an antigen, while other techniques use another molecule (protein A or G) to recognize the Fc portion of an antibody molecule present in cells or tissues (Fig. 164). Direct and indirect labeling techniques (Falini and Taylor, 1983) have been developed, the latter primarily to amplify the resulting product for easier visualization.

II. PREPARATIVE TECHNIQUES

A. Pre-Embedding Labeling

If the antigen to be identified is associated with the surface of cells grown in culture, unfixed, unembedded samples can be labeled and subsequently negatively stained. Alternatively, living cells can be labeled, fixed, embedded, and then sectioned for transmission electron microscopy or prepared for scanning electron microscopy (Priestley, 1984). Since the cells are labeled prior to fixation, antigenicity is not diminished or destroyed, as often happens during fixation procedures. In addition, once the labeling procedures are completed, fixation by typical chemical procedures provides excellent ultrastructural morphology.

As mentioned in the previous chapter on cytochemical techniques, tissue choppers or slicers may yield thin samples of tissues that will be penetrated more easily by immunolabeling reagents, which otherwise would react only with the most superficial layers of cells within a sample. However, during the last 8–10 years cryosectioning techniques have become routine in laboratories performing ultrastructural immunolabeling and have tended to supplant tissue-chopping and -slicing techniques.

B. Post-Embedding Procedures

As previously stated, it is preferable to expose antigens to antibodies before the antigens are denatured or possibly eluted by processing and embedding, but it is difficult to guarantee adequate antibody penetration into fresh or fixed tissues. On the other hand, fixation before immunolabeling allows sample storage before further processing, and proper fixation has been shown to inhibit the binding of antibodies to Fc receptors on the surface of certain types of white blood cells (Leenen *et al.*, 1985).

Tissues can be dehydrated and rehydrated, freeze-thawed, and treated with dimethylsulfoxide (DMSO) or with detergents such as saponin (Willingham, 1980) or Triton X-100 to disrupt membranes sufficiently to allow antibodies to pass into cells. Most of these techniques compromise good ultrastructural preservation.

The study by Leenan *et al.* (1985) evaluated the effect of fixatives made from four different types of chemical stocks on the subsequent immunolabeling of 11 different antigens associated with surfaces of white blood cells. Glutaraldehyde fixatives were made from distilled stocks (8%), with a single UV absorbance peak of 235 nm (indicating a solution containing only monomeric glutaraldehyde) or 25% biological-grade glutaraldehyde, which contained monomeric and dimeric glutaraldehyde, as well as glutaric acid. Pure formaldehyde prepared from paraformaldehyde powder was compared to biological-grade formaldehyde (35%) containing methanol and traces of other compounds. All fixations were for 30 min at 4°C. The cells were then washed with Dulbecco's phosphate-buffered saline (DPBS) at pH 7.2 and then with DPBS supplemented with 0.02% gelatin prior to antibody labeling.

Beesley (1989b) explained the common use of gelatin or bovine serum albumin to reduce non-specific antibody binding in immunolabeling procedures:

> Bovine serum albumin in the buffer competes with the antibody proteins for non-specific 'sticky' sites on the preparation, thereby reducing non-specific background labeling. A short incubation of the tissue in 1% gelatin in phosphate buffered saline will also reduce the number of non-specific sticky sites. If the tissue has been fixed with an aldehyde it is useful to pre-treat the sample with 0.02 M glycine in phosphate-buffered saline in order to quench any free aldehyde groups which may promote 'non-immunological' protein sticking to the tissue.

Evaluation of antibody binding in the study by Leenen *et al.* (1985) was by ELISA. In some cases, fixation decreased antibody binding, in some cases it increased binding, and in some cases it had no effect on antibody binding. This demonstrates the difficulties experienced in predicting the outcome for antibody labeling procedures used for the first time on new systems.

Formaldehyde fixation, whether from biological-grade stocks or prepared from paraformaldehyde powder, had similar effects on antigen binding in most cases. Leenen *et al.* (1985) found that two of their antibodies (out of 11) were unaffected by paraformaldehyde or had increased binding compared to unfixed controls. Formaldehyde resulted in a maximum loss of antibody binding of 25–30%. It was suggested that this latter effect might be due to the presence of methanol in formaldehyde stocks.

Both types of glutaraldehyde stocks examined by Leenen *et al.* (1985) diminished antibody binding to a greater extent than formaldehyde. High concentrations of glutaraldehyde left only 55–70% of certain antigenic determinants intact. One antibody was hardly affected by distilled glutaraldehyde but had seriously diminished binding if fixation was performed with biological-grade glutaraldehyde. In some instances, crude glutaraldehyde had a denaturing effect at lower concentrations than distilled glutaraldehyde.

Leenen *et al.* further state that aldehyde groups are known to react mostly with free amino groups, present mainly on lysine residues. Formaldehyde is also known to react with arginine and asparagine side chains. Both of these aldehydes form intermolecular and intramolecular cross-links, but glutaraldehyde is more effective in this regard because of the two aldehyde groups per molecule. Primary and tertiary structures of antigens, thus, can be changed by aldehyde fixation. However, fixation has minimal effect on antibody binding when the antigenic determinant has a primary structure without lysine residues. Antibody binding may also be inhibited because of steric hindrance from neighboring groups.

Leenen *et al.* also suggest that if glutaraldehyde is used for fixation, pretreatment with sodium borohydride prior to immunolabeling might restore some of the decreased immunoreactivity, since borohydride reduces free aldehyde groups and/or double chemical bonds induced by the fixation procedure.

Bendayan and Zollinger (1983) clearly demonstrated that some antigens retain antigenicity, even after conventional fixation with aldehydes and osmium, dehydration, infiltration with epoxide resins, and heat polymerization. They also showed that the problem of penetrating the surface of strongly cross-linked, hydrophobic epoxide resins with immunoprobes can be overcome by etching sections with sodium metaperiodate. Unfortunately, many antigens are completely or partially inactivated by conventional processing methods.

Various workers have realized the efficacy of developing techniques by which samples can be fixed and embedded in resins, and then repeatedly sectioned. If numerous grids of sections can be prepared, it is easy to test various controls or to examine a variety of antibody concentrations (Newman and Jasani, 1984).

Resin components not only react with each other during polymerization, but also react with substituents and end groups on the biological material (Carlemalm and Villiger, 1989). This may make sectioning easier, since the tissue becomes integrated with the resin matrix. On the other hand, it may interfere with post-embedding labeling, since reactive sites may be modified or occupied by the resin. Epoxides such as

Epon are reactive to a number of groups (carboxyl, amino, indol, amide). To minimize reductions in antigenicity, low-temperature acrylic embedding media, such as the LR White and Lowicryl resins, were developed (Carlemalm and Villiger, 1989). These resins are much more selective in their reactivity than epoxides, resulting in poorer retention of cellular materials in general but in less antigen modification. Acrylics are also more water miscible, allowing processing schedules with transitions from lower alcohol concentrations (70%) directly into the resins. The higher hydrophilicity of these resins when compared to most epoxides also allows easier access to antigens within sectioned cells and tissues. Acrylic resin-embedded material prepared for immunolabeling frequently is not osmicated to avoid the potential for antigen oxidation or heavy-metal denaturation from the osmium treatment. The dehydration process and the lipid-extractive capabilities of the acrylic resins themselves result in images of cells that have indistinct membrane domains, since most of the lipids are extracted. Other cellular features are not as crisp and distinct as with conventionally fixed and epoxide-embedded tissues, but all the cellular compartments (nuclei, mitochondria, endoplasmic reticulum, lysosomes, etc.) are still identifiable. Thus, determinations of the location of antigens within or outside of these compartments can be made effectively with these techniques.

Molecular distillation (see Chapter 10) of cryofixed specimens followed by osmium vapor exposure and Spurr resin embedment produces plastic blocks that often can be subsequently sectioned and immunolabeled with excellent success. This procedure also works well with dried specimens that have been directly infiltrated with acrylic resins (Lowicryl K4M) without osmication.

C. Cryoultramicrotomy Technique

An alternative procedure for immunolabeling that involves no resin embedment was developed by Tokuyasu (1973, 1984) and employs brief fixation of cells and tissues in relatively weak aldehyde solutions, usually 2–4% phosphate-buffered formaldehyde prepared from paraformaldehyde with or without a trace (0.01–0.1%) of glutaraldehyde.

Unbound fixative is washed from the sample with buffer, and the specimen is infiltrated with 2.1–2.3 M sucrose in buffer. The sample is then frozen by plunging into liquid nitrogen, nitrogen slush, or liquid nitrogen-cooled cryogens such as Freon 22. More elaborate freezing techniques are rarely necessary because ice crystal formation is minimized by the presence of sugar. Once frozen, the sample is placed in the cryochamber of a cryoultramicrotome, and ultrathin sections are cut. After the sections are picked up on coated grids, they are removed from the chamber and thawed. Floating the grids section-side down on a small dish of distilled water allows the sucrose to diffuse out of the specimen, which is then allowed to air dry. The air-dried, unembedded sections are then immunolabeled and examined (usually after negative staining with 2% ammonium molybdate) with the transmission electron microscope. The advantage of Tokuyasu's technique is that antigens that are normally difficult to label within intact cells become accessible to antibodies and labels too large to pass through membranes because the cells have been physically cut open during cryoultramicrotomy. This technique is also useful for improving surface labeling of cells within dense tissues. In the intact specimen, cells are so densely packed or actually interconnected by junctional complexes that access to nonsuperficial cells by large stain molecules is blocked.

D. Negative Staining Procedures

If viruses, bacterial flagella, and pili, or the surfaces of isolated membranes or subcellular components are to be probed by antibodies, visualization of products is quite simple. The sample is usually attached to a film-coated grid, as with normal negative-staining procedures (see Chapter 3). The grid then can be floated on drops of the various reagents used for immunolabeling, preferably in a moist chamber on a gently rocking platform. Inert grids made of nickel or stainless steel are usually used to avoid the possibility of copper interacting with the antibodies. Gold-labeled antibodies are the label of choice, since the gold particles are so discrete and have such great electron density. After the final immunolabeling step, the grid is rinsed and then negatively stained with routine procedures (see Chapter 8). Since no fixation is employed prior to negative staining, the antigens are unperturbed by chemicals and reactivity remains high (Fig. 165).

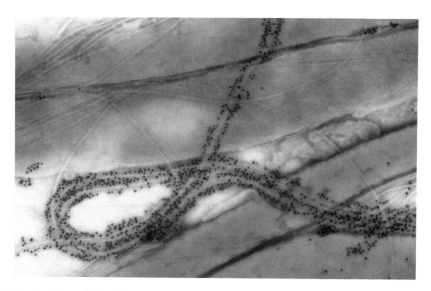

FIGURE 165. Transmission electron micrograph of *Borrelia burgdorferi* (Lyme disease spirochete) reacted with a monoclonal antibody to a flagellar protein, followed by a gold-labeled (5-nm) antibody to the primary antibody. Negatively stained with 2% PTA. 65,100×.

III. IMMUNOGLOBULINS

A. Protein A and Protein G Techniques

Antibodies can be localized by immunocytochemical techniques. Evaluation of autoimmune disease problems often involves probing tissues for the presence of immune deposits, typically IgG, in inappropriate locations. In this case, the precise identification of the antigen is not necessary, and a technique such as the protein A technique would be applicable.

Protein A has a molecular weight of 42,000 and is obtained from the bacterium *Staphylococcus aureus*. It binds to the Fc portion of IgG of many animal species without interacting at the antigen binding site. It binds strongly to IgG of humans, rabbits, guinea pigs, swine, and dogs. Binding to IgG of cows, mice, and horses is weaker, and binding to IgG of goats, rats, sheep, and chickens is very weak.

Protein G is produced by Group C *Streptococcus* sp. and binds more efficiently to the Fc region of IgG of sheep, goats, and bovines. Otherwise, it is used similarly to protein A.

B. Polyclonal and Monoclonal Antibodies

In many cases it is important to precisely identify the specific antigen present, usually against a background of other specific antigens. This requires the use of a polyclonal or monoclonal antibody developed against the specific antigen. Polyclonal antibodies are more cross-reactive than monoclonals. Beesley (1989b) points out that monoclonal antibodies are more specific than polyclonal antibodies, because they react with only a single antigenic determinant consisting of a few amino acids. He suggests that light fixation is required to prevent damage to these important amino acids. A fixative containing 4% formaldehyde with 0.05% glutaraldehyde in either phosphate or cacodylate buffer is suggested. If antigenicity is not optimal, he suggests deleting the glutaraldehyde. Beesley further states that since polyclonal antibodies are reactive with a number of different epitopes, fixation with 1% glutaraldehyde is usually permissible.

IV. COMMON IMMUNOLABELING TECHNIQUES FOR ELECTRON MICROSCOPY

A. Immunoferritin

Ferritin is a naturally occurring biological protein with a molecular weight of 750,000, an electron-dense iron core, and a diameter of 7.0 nm. Its apoprotein coat permits conjugation to other proteins, such as antibodies, by means of bivalent reagents, typically glutaraldehyde. When antibodies are coupled to ferritin by reaction with glutaraldehyde, there may be a diminution of antibody activity, primarily resulting from the bound glutaraldehyde decreasing access to the characteristically reactive regions of antibodies. Other problems include the heterogeneity of products produced during the coupling procedures, which necessitates the isolation of active ferritin/antibody conjugates from other moieties in the reaction mixture. The general inefficiency of the coupling procedure also causes problems. Not only will many of the antibody molecules remain unlabeled and thus unviewable because of insufficient inherent electron density, but unlabeled antibody in the reaction mixtures can compete with the labeled antibody for antigens within cells and tissues, denying access to the labeled antibody molecules. This, of course, is one of the standard controls to demonstrate the specificity of labeling procedures. A pretreatment of cells or tissues with unlabeled antibody, followed by incubation with labeled antibody to the same antigen, should prevent all labeling. The use of ferritin as a cationic stain as described in the section on cytochemistry explains why ferritin can bind nonspecifically to some charged sites. It even binds nonspecifically to some epoxide resins during post-embedding immunolabeling techniques.

Since ferritin is a naturally occuring molecule in certain tissues, a careful evaluation of the system to which it is to be applied must be undertaken. The size of the molecule prevents passage through intact membranes, so it is appropriate only for pre-embedment labeling of cellular surfaces (Fig. 166) or post-embedding labeling of sectioned materials. If ferritin is to be used in post-embedding procedures, acrylic resins should be used for sample embedment because of the potential reactivity of ferritin with epoxide resins.

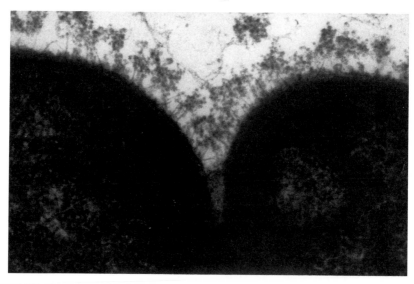

FIGURE 166. Transmission electron micrograph of pre-embedding labeling of pili on the surface of the causative agent of bovine pinkeye, *Moraxella bovis*. A ferritin-labeled antibody was used to demonstrate the primary antibody raised against the pili. 49,200×.

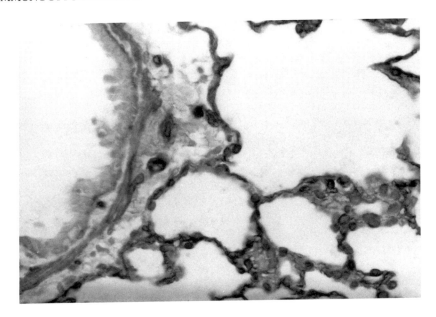

FIGURE 167. Light and electron microscopic demonstration of peroxidase/avidin/biotin immunolabeling. (Courtesy of Philip L. Sannes.) Top: Light micrograph of an adult rat lung immunolabeled with an avidin-biotin complex/peroxidase procedure for the localization of chondroitin sulfate proteoglycan. Alveolar basement membranes, capillary basement membranes, and external laminae of smooth muscle are stained. 660×. Bottom: Electron micrograph of an adult rat lung immunolabeled with an avidin-biotin complex/peroxidase procedure for the localization of heparan sulfate proteoglycan. The alveolar basement membrane profile is divided into the lamina rara externa (LRE) adjacent to a type I epithelial cell (vertical line), the lamina rara interna (LRI) adjacent to a capillary endothelial cell (E), and the lamina densa (LD). Reaction product, indicating the heparan sulfate proteoglycan, is darkly stained. 160,000×.

B. Immunoperoxidase Techniques

All of the commonly used immunoperoxidase techniques are described in the review article by Falini and Taylor (1983). The peroxidase molecule is one-tenth smaller than ferritin, with a molecular weight of 40,000. It is easily conjugated to antibodies with glutaraldehyde. Because of its relatively small size, it can penetrate membranes of cells fixed with aldehydes. It was first utilized for light microscopy techniques, and its easy visualization at the light microscope level allows relatively quick assessment of the efficacy of localization techniques to be utilized for electron microscopy (Fig. 167).

The immunoperoxidase technique has the disadvantage of all enzymatic procedures, in that the catalytic activity of enzymes can result in the accumulation of products capable of diffusion away from the site of the enzymatic activity and, in this case, the site of the antigen being localized. Another potential problem that must be considered is the presence of endogenous peroxidase activity within cells. The endogenous peroxidases may make interpretation of immunoperoxidase techniques difficult.

Another problem with immunoperoxidase techniques is that if a single antibody molecule with peroxidase bound to it attaches to an antigen, and then the peroxidase activity with the hydrogen peroxide substrate is amplified with DAB and osmium, the product may not be discernible. As a result, different amplification techniques have been developed. Falini and Taylor (1983) provide illustrations of direct and indirect conjugate procedures, labeled antigen methods, enzyme bridge methods, and the most commonly used peroxidase procedure, the peroxidase-antiperoxidase procedure (PAP) illustrated in Figure 168.

C. Immunogold Techniques

The strength of immunogold techniques is that the gold particles that are bound to antibodies are totally opaque and discrete. They are not diffuse products like the

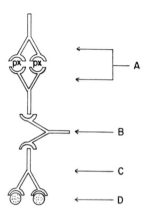

FIGURE 168. PAP procedure (after Falini and Taylor, 1983). A = mouse anti-px (peroxidase); B = goat anti-mouse; C = mouse antibody to D; D = antigen D.

osmium blacks produced by immunoperoxidase procedures and, unlike peroxidase and ferritin, which occur naturally in some tissues, gold is a nonbiological element. In addition, gold cannot diffuse from the site of binding, as can happen with enzymatic procedures. Gold can be bound to immunoglobulins (De Mey and Moeremans, 1986), to protein A and protein G (Bendayan and Zollinger, 1983), to lectins (Horisberger, 1985), and to enzymes (Bendayan, 1985). The process by which gold is bound to these molecules is poorly understood but is a simple adsorptive process that involves no chemical conjugation. Binding only requires proper pH, reagent concentration, and ionic strength of the reaction mixture. Gold itself has relatively low nonspecific adsorption to specimen surfaces (unlike ferritin, which binds readily to many epoxides). Gold probes may be prepared in a variety of sizes (from 1 nm to 150 nm), determined by buffer and pH (see De Mey and Moeremans, 1986, for specific recipes). For electron microscopy, gold probes over 20 nm are rarely employed, since larger particles are thought to produce steric hindrance for immunolabeling procedures. In addition, the density of labeling generally increases with decreasing gold particle diameter. Five-nanometer probes are usually the smallest used, even though 1-nm probes are now available, because anything smaller becomes harder to locate in low concentrations. The introduction of high-resolution, low-voltage scanning electron microscopes (see Chapter 9) recently has made it possible to easily visualize all gold labels with scanning as well as TEM techniques.

After macromolecules are adsorbed to gold particles, most still have their full biological activity (tertiary structures are preserved). During the preparation of gold probes, it is usually necessary to coat them with the active macromolecules desired (antibodies, lectins, protein A) fairly quickly, since uncoated gold particles often self-assemble into aggregates.

REFERENCES

Beesley, J.E. 1989a. *Colloidal gold: A new perspective for cytochemical marking.* Oxford University Press, New York.

Beesley, J.E. 1989b. Immunocytochemistry of microbiological organisms: A survey of techniques and applications. In: *Techniques in immunocytochemistry*, Vol. 4, G.R. Bullock and P. Petrusz (eds.), Academic Press, New York, pp. 67–93.

Bendayan, M. 1985. The enzyme-gold technique: A new cytochemical approach for the ultrastructural localization of macromolecules. In: *Techniques in immunocytochemistry*, Vol. 3, G.R. Bullock and P. Petrusz (eds.), Academic Press, pp. 179–201.

Bendayan, M., and Zollinger, M. 1983. Ultrastructural localization of antigenic sites on osmium-fixed tissues applying the protein A-gold technique. *J. Histochem. Cytochem.* 31:101.

Bullock, G.R., and Petrusz, P. (eds.). 1982, 1983, 1985, 1989. *Techniques in immunocytochemistry*, Vols. 1–4. Academic Press, New York.

Carlemalm, E., and Villiger, W. 1989. Low temperature embedding. In: *Techniques in immunocytochemistry*, Vol. 4, G.R. Bullock and P. Petrusz (eds.), Academic Press, New York, pp. 29–45.

Coons, A.H., Creech, H.J., and Jones, R.N. 1941. Immunological properties of an antibody containing a fluorescent group. *Proc. Soc. Exp. Biol.* 47:200.

De Mey, J., and Moeremans, M. 1986. The preparation of colloidal gold probes and their use as marker in electron microscopy. In: *Advanced techniques in biological electron microscopy III*, J.K. Koehler (ed.), Springer-Verglag, New York, pp. 229–271.

Falini, B., and Taylor, C.R. 1983. New developments in immunoperoxidase techniques and their application. *Arch. Pathol. Lab. Med.* 107:105.

Faulk, W.P., and Taylor, G.M. 1971. An immunocolloid method for the electron microscope. *Immunochemistry* 8:1081.

Hayat, M.A. (ed.), 1989, 1990. *Colloidal gold: Principles, methods and applications*, Vols. 1–3. Academic Press, New York.

Horisberger, M. 1985. The gold method as applied to lectin cytochemistry in transmission and scanning electron microscopy. In: *Techniques in immunocytochemistry*, Vol. 3, G.R. Bullock and P. Petrusz (eds.), Academic Press, New York, pp. 155–178.

Leenen, P.J.M., Jansen, A.M.A.C., and Ewijk, W.V. 1985. Fixation parameters for immunocytochemistry: The effect of glutaraldehyde or paraformaldehyde fixation on the preservation of mononuclear phagocyte differentiation antigens. In: *Techniques in immunocytochemistry*, Vol. 3. G.R. Bullock and P. Petrusz (eds.), Academic Press, New York, pp. 1–24.

Nakane, P.K., and Pierce, G.B. 1966. Enzyme-labeled antibodies: Preparation and application for the localization of antigens. *J. Histochem. Cytochem.* 14:929.

Newman, G.R., and Jasani, B. 1984. Post-embedding immunoenzyme techniques. In: *Immunolabeling for electron microscopy*, J.M. Polak and I.M. Varndell (eds.), Elsevier, New York, Chap. 5.

Polak, J.M., and Van Noorden, S. (eds.). 1983. *Immunocytochemistry. Practical applications in pathology and biology*. Wright PSG, London.

Polak, J.M., and Varndell, I.M. 1984. Immunolabeling for electron microscopy. Elsevier, New York.

Priestley, J.V. 1984. Pre-embedding ultrastructural immunocytochemistry: Immunoenzyme techniques. In: *Immunolabeling for electron microscopy*, J.M. Polak and I.M. Varndell (eds.), Elsevier, New York, Chap. 4.

Raap, A.K., Hopman, A.H.N., and Van Der Ploeg, M. 1989. Hapten labeling of nucleic acid probes for DNA *in situ* hybridization. In: *Techniques in immunocytochemistry*, Vol. 4. G.R. Bullock and P. Petrusz (eds.), Academic Press, New York, pp. 167–197.

Romano, E.L., Stolinski, C., and Hughes-Jones, N.C. 1974. An antiglobulin reagent labelled with colloidal gold for use in electron microscopy. *Immunochemistry* 11:521.

Romano, E.L., and Romano, M. 1984. Historical aspects. In: *Immunolabeling for electron microscopy*, J.M. Polak and I.M. Varndell (eds.), Elsevier, New York, Chap. 1.

Silva, F.G., Lawrence, J.B., and Singer, R.H. 1989. Progress toward ultrastructural identification of individual mRNAs in thin section: Myosin heavy-chain mRNA in developing myotubes. In: *Techniques in immunocytochemistry*, Vol. 4, G.R. Bullock and P. Petrusz (eds.), Academic Press, New York, pp. 147–165.

Singer, S.J. 1959. Preparation of and electron dense antibody conjugate. *Nature* 183:1523.

Sternberger, L.A. 1986. *Immunocytochemistry*. John Wiley and Sons, New York.

Tokuyasu, K.T. 1973. A technique for ultracryotomy of cell suspensions and tissues. *J. Cell Biol.* 57:551.

Tokuyasu. K.T. 1984. Immuno-cryoultramicrotomy. In: *Immunolabeling for electron microscopy*, J.M. Polak and I.M. Varndell (eds.), Elsevier, New York, Chap. 6.

Williams, M.A. 1977. *Autoradiography and immunocytochemistry*. Elsevier, New York.

Willingham, M.C. 1980. Electron microscopic immunocytochemical localization of intracellular antigens in cultured cells: The EGS and ferritin bridge procedures. *Histochem. J.* 12:419.

CHAPTER 15

Autoradiography

I. HISTORY

The historical record for autoradiography begins with the report published in 1867 by St. Victor, which described the blackening of silver chloride and silver iodide emulsions exposed to uranium nitrate and uranium tartrate, even when the silver halides and uranium compounds were separated by colored sheets of glass. He thought that the phenomenon was caused by luminescence of the uranium compounds.

In the 1890s, Henri Bequerel attempted to prove that fluorescence was the cause of photographic emulsion development by uranium compounds. He exposed uranyl sulfate to light and then placed a photographic plate in close contact with the uranium sulfate, with only two sheets of black paper between them. At one point, there was no sun for several days, so he left the uranium sulphate and film together in a drawer and still noted that an image was produced on the film. Two years later, the Curies described radioactivity for the first time.

In the 1920s, Lacassagne and his coworkers were the first to demonstrate radioactive compounds (polonium) within biological tissues. Physicists studied the phenomena of radioactivity quite actively for the next 20 years, but biologists did not examine radioactivity to any significant extent again until the 1940s. A seminal study by Leblond in 1943 demonstrated the distribution of iodine within the thyroid by autoradiography, followed by the application of liquid emulsions to samples to produce autoradiographs in his laboratory. Finally, in the 1950s Liquier-Milward showed the first autoradiographs produced from ultrastructural specimens. These were poor, but they facilitated the effusion of ultrastructural autoradiographic studies that were undertaken during the 1970s. By the late 1970s, the number of autoradiographic studies declined, particularly after improved immunocytochemical techniques were introduced in the late 1970s and early 1980s. As will become evident in this chapter, the twin problems of time investment and potential difficulties with resolution that made autoradiography unattractive at the ultrastructural level are eliminated if other types of probes, such as gold-labeled antibodies, enzymes, or lectins, can be used for compound localizations. Autoradiography is still used at times for ultrastructural studies but has its broadest application to light microscopy studies, particularly with plastic resin-embedded materials (epoxide semithin sections or JB-4 [methacrylate] sections of 1- to 2-μm thickness).

There are not a large number of texts covering autoradiography in great depth, but the ones by Rogers (1979), Bancroft and Stevens (1977), Baker (1989), and Williams (1977), along with an informative booklet from Kodak (Kodak Tech Bits, 1988), will give more than enough overview of autoradiography applications to biological materials at both the light- and electron-microscopic level.

II. PURPOSE

The purpose of autoradiographic techniques is to demonstrate the distribution of a radiolabeled compound within an organism or cell and to localize the site of incorporation of this compound into cells or tissues. Autoradiography can also reveal the location of the final product in which a radioactive precursor resides following anabolic reactions.

III. THEORETICAL ASPECTS

A. Detection of Radioactivity

There are three major classes of detectors for radioactivity. The first utilizes electrical measuring devices, such as Geiger tubes, ionization chambers, or gas flow counters, to record radioactive decay. In scintillation counters, which indirectly measure radioactive events, materials that have absorbed energy from incident radiation reemit some of the energy in the form of visible light (fluorescence), which is subsequently recorded. Autoradiography is the most sensitive of the three techniques for recording radioactivity, because each silver halide crystal within a photographic emulsion can be thought of as an independent detector that is insulated from other crystals by the gelatin of the emulsion. As with any photographic process, exposure of the silver halide crystal to energy (in this case, radioactivity) forms a latent image that is made permanent by photographic development. The record is cumulative and spatially accurate. The method is relatively slow compared to the other monitoring devices, but is notable because it offers the best resolution.

B. Types of Particles

Autoradiographic emulsions will record a variety of energetic particles including α and β particles, cosmic rays, gamma rays, X-rays, extranuclear electrons, and delta rays. For light- and electron-microscopic autoradiography, β particles are intended to be the main type of radiation recorded, although all the other types can be sources of noise (non-informational events) in the final product.

Uncharged particles (like neutrons) and electromagnetic radiations, such as X and gamma rays, only lose energy by direct collisions with electrons or nuclei. Since these emissions are relatively rare, they do not lend themselves to autoradiographic studies, although X-rays and gamma rays do make sporadic electron tracks through auto-radiographic emulsions.

Cosmic rays are highly energetic charged particles that bombard the upper atmosphere. When they penetrate the atmosphere and strike autoradiographic emulsions, cosmic rays produce straight tracks, with occasional delta tracks (described below) being generated.

Alpha particles are like the nucleus of a helium atom, consisting of two protons and two neutrons. They are emitted by isotopes of elements with high atomic numbers, such as uranium and thorium. The initial energy of an α particle is usually between 4 and 8 MeV. Since these particles are relatively energetic, they are not easily deflected. There is a mutual repulsion between α particles and nuclei, however. An α particle also attracts electrons, since it exhibits the positive charge of its two protons. Thus, an α particle dissipates its energy rapidly as it interacts with electrons in the specimen or emulsion. The range of an α particle is thus short despite its initial high energy. The particle creates a large number of latent images, affecting every silver halide crystal through which it passes. The path seen in a nuclear emulsion is very dense, straight, and short (from 15 to 40 μm long). The track is fairly wide, since the energized electrons resulting from particle collisions will travel short distances laterally through the emulsion. Autoradiographic studies can be undertaken using α emitters, but β emitters are considered to be far superior.

Beta particles have the same mass and charge as electrons but are generated from an atomic nucleus. Emission of a β particle is the most common way in which isotopes achieve stability. Beta particles show a broad spectrum of energies from one particular isotope. The spectrum shown varies between different isotopes as to the shape of the distribution curve of energies. Tritium has an E_{max} of 18 KeV, while other isotopes have an E_{max} up to 3 MeV. The mutual repulsion between β particles and electrons produces a random buffeting effect on the path of a β particle. Occasional collisions between β particles and electrons will cause the ejection of the latter from its orbit to become a delta ray (Fig. 169).

The track length of a 20-KeV β particle is about 3 μm long in Ilford G5 emulsion (Rogers, 1979). For a 6-MeV β particle, the track generated would be about 10 mm long. In contrast, an α particle of 6 MeV would generate a track 26 μm long. A β particle may pass through many crystals without producing a latent image. The track thus has large grains, small grains, and gaps. It is usually meandering and sometimes has an abrupt change in direction with a branch. The longer portion of the path is, by convention, the β-particle track, and the shorter is the Δ-ray track. The developed autoradiographic emulsion tends to show larger grains near the end of a β-particle track. If the energy range is above 500 KeV, the energy loss of the particle on its path through the emulsion is constant. As the energy drops below 500 KeV, the rate of energy loss increases rapidly until, near the end of the track, the rate of energy loss is

FIGURE 169. Diagram of a β-particle track (B) and an associated delta ray (A).

about eight times greater than at 500 KeV. This accounts for the increase in both number and size of latent images recorded near the end of a track. Beta-particle images must be recorded on more sensitive emulsions with more critical development than α-particle tracks.

In addition to β particles, autoradiographic emulsions can record extranuclear electrons. These are electrons produced from isotopes that release low-energy gamma photons as a result of rearrangements of neutrons and protons. The gamma photons often lose their energy to electrons within the inner shells of these isotopes, causing them to be ejected from their orbits. Once ejected, the electrons behave like a β particle emitted from the nuclei of isotopes. Extranuclear electrons have low initial energies, similar to the energy of a β particle from tritium, thus providing very good resolution with autoradiography.

C. Nuclear Emulsions

Photographic emulsions prepared for autoradiography (nuclear emulsions) have a much higher ratio of silver bromide to gelatin than normal photographic emulsions. The crystals are also more uniform in size. This confers a greater stopping power to the emulsion for radiation, thus producing relatively short tracks. Emulsions with silver halide crystals of 0.2–0.4 μm in diameter are typically employed in autoradiography, compared to those in X-ray films, which range from 0.2 to 3.0 μm in diameter. Emulsions with smaller crystals are less sensitive than those with larger ones, because a small crystal can contain only a small part of the trajectory of a charged particle. On the other hand, smaller grains will result in higher resolution. Nuclear emulsions exhibit extremely high recording efficiency for β particles and extranuclear electrons, particularly those with low energy. Thus, in order of efficiency for recording, tritium, carbon, sulfur, and iodine can be localized with varying degrees of success. The characteristics of these and other β emitters are shown in Table 12.

Nuclear emulsions can measure decay rates in a cell of one disintegration a day, which is considerably more sensitive than pulse counters (Geiger tubes), which measure about 10–20 counts/minute background. Different emulsions have silver grains of different size ranges, which helps determine the sensitivity of the specific emulsion and facilitates the selection of a particular emulsion type to use with a given emitter in an experimental setting (Table 13).

The gelatin of autoradiographic emulsions isolates individual silver halide crystals from each other to prevent neighbor catalysis. Gelatin also serves as an acceptor of bromine atoms, which migrate from the silver halide crystal when metallic silver is deposited (during the formation of the latent image and development).

TABLE 12. Characteristics of Some Common Radioisotopes

Radioisotope	Decay type	Half-life	Energy (MeV)
Tritium (^{3}H)	β	12.26 yr	0.018
Carbon (^{14}C)	β	5760 yr	0.159
Sulfur (^{35}S)	β	87.2 da	0.167
Iodine (^{125}I)	ENE*	60 da	
Iodine (^{131}I)	β	8.04 da	0.61
Phosphorus (^{32}P)	β	14.3 da	1.71

*ENE = extranuclear electron.

Nuclear emulsions have a shelf life (refrigerated) of only a few months, and they must never be frozen. Consequently, it is important to purchase stocks that have not been stored by vendors for any significant period of time. Eastman Kodak Company will ship fresh stocks directly when an order is called in, thus eliminating suppliers who may handle alternative products, such as those from Ilford. The Ilford products are generally diluted before specimens are coated, which must be done in nearly total darkness. The Kodak products, on the other hand, are not intended for dilution. They are heated to liquify them and then are used to coat specimens directly, which is procedurally somewhat simpler than working with Ilford emulsions.

D. Determination of Isotope Dose Level

Williams (1977) devotes six pages to determinations of radiochemical dose levels for autoradiography. He states that to prepare successful autoradiographs for ultrastructural examination, it is often necessary to have radiochemical dose levels 10 times those necessary for light-level examination and 10–100 times greater than those necessary for scintillation counting.

Radioisotopes are typically administered at 10–40 μCi/g body weight for *in vivo* studies. Thus, most studies of large animals are difficult because of the large amount of radioisotope necessary and because of further problems with disposal of the animal

TABLE 13. Characteristics of Several Nuclear Emulsions Available from Eastman Kodak

Emulsion	Silver halide crystal diameter (μm)	Use
NTB	0.28 μm	Most α emitters and β emitters with energies less than 0.025 MeV
NTB2	0.31 μm	Low/medium energy β emitters (^{3}H, ^{14}C, ^{35}S) and electrons with energies less than 2.0 MeV and all α emitters
NTB3	0.35 μm	Highest sensitivity of NTB products; will record all charged particles, but recommended for high-energy β emitters (^{32}P, ^{131}I, ^{125}I)

after the study (not to mention various wastes produced by the animal while being housed for the study).

Williams provided formulas for determining whole-body doses based on the type of study to be done. He calculated that a mouse with a body volume of 40 ml dosed with 250 μCi of tritiated leucine would produce a final tritium concentration within tissue (assuming uniform labeling) of 3.125 μCi/ml. Since the volume of an ultrathin section is on the order of 5×10^{-8} ml, the whole ultrathin section would produce 0.347 distintegrations/min. He then calculated that the number of silver grains/cell profile/min would be 17.2×10^{-6}, which means that a 121-day exposure period would be necessary to produce an average of three silver grains/cell profile. These calculations explain why whole-body dosing leading to autoradiography at the ultrastructural level is not encountered frequently in the literature.

Williams then points out the benefits of preparing light microscope autoradiographs of materials destined for electron microscopic autoradiography to estimate appropriate exposure times in the latter case and to assess the practicality of attempting ultrastructural localizations. He provides the following formula to make these determinations:

$$\text{Exp}_{EM} = \frac{\text{Exp}_{LM} \times T_{LM}}{T_{EM}} \times \frac{\text{Ef}_{LM}}{\text{Ef}_{EM}},$$

where Ef = efficiency %, Exp = exposure time in days, T = thickness in nm, LM = light microscopic, and EM = electron microscopic.

Williams calculated with this formula that if it took 7 days exposure with a 1.0-μm-thick section to produce more than 10 silver grains per cell profile at the light microscopic level, 70 days of exposure would be required to produce a similar grain density on an ultrathin (50-nm) section. This formula explains the basic rule of thumb: the exposure time for ultrathin sections is about 10 times that needed for light microscopy.

E. Rules for Autoradiography

As with any photographic emulsion, autoradiographic emulsions will record energy from various sources, including photons of light, nuclear radiation, heat, and pressure. Thus, it is necessary to avoid as many of the noise-producing sources as possible. Keeping the emulsion cool and dry during exposure and handling carefully during specimen coating to prevent mechanical exposure from overvigorous stirring or contact with metal stirring objects (use glass rods instead) are necessary. Specimens must be coated in total darkness or at extemely low levels of dark red light to minimize incidental photoexposure. The time period of emulsion exposure to the specimen should be kept to the minimum necessary to record sufficient signal. Oxygen can also result in spurious signal recording in some cases.

To produce the highest signal-to-noise ratios in final autoradiographs, specimens should be "hot" enough to produce the signal desired within 1 days to 3 weeks at the light microscopic level. The specimen should be coated with recently purchased, minimally heated emulsion that has had as little light exposure during the coating process as possible. After the specimens are coated and dried for 1–2 hr in the dark at room temperature, they are placed within a relatively small light-tight box provided with Drierite® or some other suitable drying agent. The light-tight containers should then be securely sealed with electrical tape and placed in a refrigerator for the desired exposure time. Whenever a specimen is removed for development, the container should be warmed to room temperature before being opened to prevent moisture accumulation on the specimens that will be exposed for longer periods.

As mentioned above, an exposure that provides proper signal levels for light microscopy will have to be extended about 10 times longer for electron microscopy, since there will be significantly fewer emitters in an ultrathin section compared to a typical specimen prepared for light microscopy. Another significant problem for autoradiography at the ultrastructural level is that the resolution of the microscope will frequently vastly exceed the resolution possible for autoradiography. The developed silver grain produced from exposure to radiation is comparatively large (see Table 13, which shows silver halide crystal size before development; the final developed grain will usually be larger), and radiation events can be recorded at oblique angles some distance from the source, depending on the energy of the emitter (Fig. 170). Thus, an emitter located within the endoplasmic reticulum can easily be recorded by areas in the emulsion not directly over the endoplasmic reticulum. In contrast, at the light microscopic level the resolution of the optical system does not exceed that of the autoradiographic preparation.

Rogers (1979) and Williams (1977) discuss the two major concepts for describing resolution with autoradiographic materials. The most commonly used is the HR value, which is the radius of a circle about a point source of radiation within which half the silver grains produced by the source may be found (Fig. 171). The second concept, the HD value, is derived by measuring bands of equal width parallel to a linear source. The half-distance (HD) is that distance containing half the silver grains produced from the linear source.

There are a number of factors that govern resolution in a given autoradiograph. The higher the energy of the isotope, the less resolution is possible, since emissions from a given source radiate out in all directions. Many of the emissions of low-energy sources are self-absorbed by the specimen and never get out of the specimen to interact

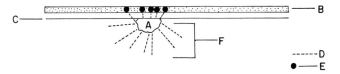

FIGURE 170. Problems with autoradiographic resolution at the electron microscopic level. A = source; B = emulsion; C = specimen surface; D = β emissions; E = exposed silver halide crystals; F = self-absorbed specimen emissions.

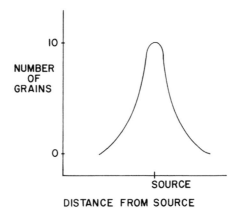

FIGURE 171. Distribution of silver grains around a point source in an autoradiograph.

with the emulsion. The higher the energy of the isotope, the farther away from the source that silver grains are produced (Fig. 172).

Another major consideration is the distance between the source and the emulsion. Silver halide crystals at the sides of the source are almost as likely to be struck by emissions as those directly over the source if the emulsion and source are separated. If the emulsion contacts the source, the probability of being hit falls off considerably even a short distance from the source.

The thickness of the source can significantly affect resolution. The first specimen layer beneath the emulsion has a characteristic curve of resolution, but the next deeper layer in the specimen will be separated from the emulsion by the first layer of speci- men. The deeper layer will actually have more resolution than the first layer because those emissions that do not emerge from the specimen vertically have an increased probability of being self-absorbed by the specimen (Fig. 173).

The emulsion itself determines resolution in several ways. First, the thicker the emulsion, the less resolution, because some oblique paths become recorded within the emulsion. Second, the size of the silver halide crystals within the emulsion has a direct

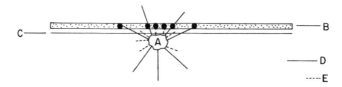

FIGURE 172. An illustration of a relatively high-energy isotope exposing silver halide crystals in the autoradiographic emulsion some distance from the source because of oblique particle paths. The resolution is reduced compared to a lower energy source, where most of the particles with oblique paths are self-absorbed prior to interaction with the emulsion. A = source; B = emulsion; C = specimen surface; D = high-energy emissions; oblique trajectories still reach the emulsion; E = low-energy emissions; only the most vertical emissions reach the emulsion.

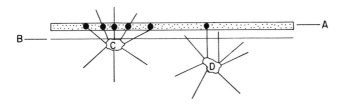

FIGURE 173. Diagram showing silver grain distribution from the first and second layer of a specimen (B) within which a radioisotope is bound. All path lengths are equal in this example. Oblique paths from the first source layer (C) reach the emulsion (A) and are recorded, while only some of the vertical emissions from the second source layer (D) can reach the emulsion.

bearing on resolution. The latent image of a β-particle path will lie at the preformed sensitivity speck of the exposed silver halide crystal, not necessarily where the particle path was within the crystal. Thus, the smaller the silver halide crystal, the closer the sensitivity speck that has been transformed into a latent image will be to the original particle path within the silver halide crystal (Fig. 174).

The length of exposure of an autoradiographic emulsion to a radioisotope source will also affect resolution. A short exposure time will give a sharp peak of silver grain distribution after development. Short exposures assure no double hits. A double hit of a single silver halide crystal still produces only one silver grain, just like a single hit. Long exposures give flatter, broader curves, with lower resolution.

With relatively insensitive emulsions, only low-energy portions of a particle path or the portions near the end of a track are recorded. With higher-sensitivity emulsions, a larger spectrum of particle energies can be recorded, including events nearer the source.

Final resolution is also affected by the actual development process. Different

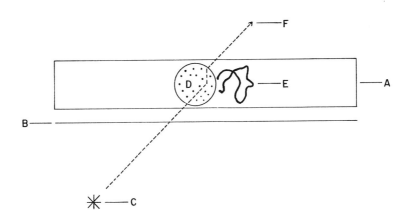

FIGURE 174. Diagram showing the relationship of the developed silver grain (E) in the emulsion (A) to the silver halide crystal (D) and the original path of the β particle (F) emitted from the source (C) below the specimen surface (B) producing the silver grain.

developers produce different sizes of developed silver grains, while higher developer temperatures and longer development times will produce larger silver grains.

In summary, separation of the source and emulsion is the chief source of decreased resolution. Isotope type (energy), specimen thickness, and emulsion thickness are also capable of affecting resolution. Silver halide crystal diameter, developed silver grain size, and emulsion sensitivity are relatively minor factors in determining potential resolution.

F. Light Microscopy Autoradiography

Light microscopy autoradiography utilizing 0.5- to 1-μm-thick sections of plastic-embedded cells and tissues is an extremely successful technique, particularly when using tritiated probes. The 18.5-KeV β particles from this source have a range seldom exceeding 2 μm in biological specimens and 1 μm in nuclear emulsions (Rogers, 1979). This assures high resolution at the light microscopic level. Naturally, specimens over 2 μm in thickness will have maximum loss of resolution, thereby assuring less resolution with paraffin-embedded or cryostat sections.

Another consideration for light microscopy autoradiography, particularly with frozen sections or paraffin sections, is the risk of diffusion of the product being localized. An interesting technical approach to this problem was described by Young and Kuhar (1979). They used tritiated ligands to demonstrate opioid receptors in lightly fixed tissue sections mounted on glass slides. To prevent tritiated ligands from diffusing into aqueous media (the liquified emulsion usually used to coat tissue sections on slides), coverslips were coated with emulsion and then dried before being affixed to the slides bearing tissue sections. The coverslips were attached by one end to the slides with cyanoacrylic glue and then held tightly to the tissue during exposure with paper clips. After an appropriate exposure time, the coverslip was held slightly off of the tissue surface during development of the emulsion by a Teflon® spacer. After rinsing, the spacer was removed, allowing the coverslip to resume its position in close proximity to the tissue sections, and the slide was then examined.

The effects of staining techniques used for light microscopy must be taken into account when preparing materials for light-level autoradiography (see Williams, 1977). In many cases, staining sections prior to exposure will produce unwanted silver grains because of interactions between stain molecules and silver halide crystals. In other cases, contact between tissues or cells and the autoradiographic emulsion can cause direct reduction of silver grains, a process known as chemography. At the ultrastructural level, some workers recommend coating radiolabeled sections with evaporated carbon (5 nm) prior to coating with liquid emulsion to avoid chemography.

Light-level autoradiography is procedurally quite simple in most cases, but it is important to read further about the general pitfalls of the technique (Rogers, 1979; Williams, 1977) to design appropriate protocols that avoid some of the simpler confounding factors presented in this brief discussion.

REFERENCES

Baker, J.R. 1989. *Autoradiography: A comprehensive overview.* Oxford University Press, New York.

Bancroft, J.D., and Stevens, A. 1977. *Theory and practice of histological techniques.* Churchill Livingstone, New York.

Kodak Tech Bits. *Autoradiography at the light microscope level.* Spring 1988.

Rogers, A.W. 1979. *Techniques of autoradiography.* Elsevier/North-Holland, New York.

Williams, M.A. 1977. *Autoradiography and immunocytochemistry.* Elsevier/North Holland, New York.

Young, S., III., and Kuhar, M.J. 1979. A new method for receptor autoradiography: [^{3}H]Opioid receptors in rat brain. *Brain Res.* 179:255–270.

CHAPTER 16

Computer-Assisted Imaging

Most biologists interested in morphometric analysis purchase programs to do the work, and a specific knowledge of all the algorithms and statistical calculations behind their operation is not critical. The programs are booted up, the menus are consulted to see what operations are possible, and the operator instructs the computer to perform certain operations from the menu. At the same time, if the operator does not possess knowledge concerning the types of questions that may be asked, the limitations inherent in the sampling techniques being used, and the general underlying assumptions necessary to computer-assisted image analysis, the equipment may not be used to its ultimate capacity. The purpose of this chapter is to give an overview of the capabilities and a few of the shortcomings of computer-assisted image analysis procedures used today. Three texts (Russ, 1986, 1990; Hader, 1991) cover most aspects of the mechanics behind image analysis and address specific applications in great detail. They are recommended reading prior to embarking on a project involving computer-assisted image analysis. The text by Misell (1978) was written prior to most of the computer-assisted analysis equipment currently available but deals extensively with principles and methods for image enhancement and analysis not confined to computers.

I. PURPOSE

Computer-assisted imaging encompasses techniques devoted to image acquistion (photography, digitizing, and storage of digitized images), image manipulation (changing gray scales, sharpening edges, smoothing edges, dilation, erosion, skeletonization, colorization), and subsequent image analysis (morphometry, stereology). Most of these techniques were originally developed without the benefit of computers, but with the development of the desk-top computer industry in the 1970s, the disciplines of morphometry and stereology became more focused on the digitized images, which produced information much more easily. While humans are best at making subjective comparisons between structures, computers are best at measuring objects and performing statistical analyses.

Typical images of use to cell biologists are acquired from transmission electron microscopes, scanning electron microscopes, or light microscopes, and may be either black-and-white images or color images. Measurements are usually made from real or optical sections and projected images wherein a once three-dimensional structure is viewed as a more-or-less two-dimensional structure.

The kind of information usually sought from these images encompasses object counting, measuring the percent area occupied by features within a field of view, and describing the position, shape, and size of objects within an image. Computers are particularly suited to these tasks because they can see all the details in an image and have no inherent interest in non-informational trivia within the image, unlike the human observer. Of course, humans have far better capabilities of understanding incomplete or unusual objects, of discriminating between contiguous objects, as well as being able to turn objects over mentally and to view them at different angles from those actually presented (a standard component of psychological testing).

II. RESOLUTION AND DISCRIMINATION

Computers typically convert an analogue image, such as a typical RS-170 video signal, into a digital signal composed of pixels (picture elements), which then can be stored easily and manipulated on a pixel-by-pixel basis. A relatively inexpensive computer display image consists of 512×512 pixels, and higher resolution units are available. The actual image resolution is usually less than the pixel count, because more than two pixels are typically required to discriminate a feature or boundary. A 512×512 pixel image consists of 262,144 pixels, which clearly discriminate less than the human eye with its 150 million individual receptors. On the other hand, the human eye can distinguish only 20–30 different levels of gray in a monochrome image, while a computer can discriminate among 256 brightness levels (gray levels).

Another aspect to be considered is the resolving capability of the original image-retrieving device. A typical inexpensive video camera (e.g., a Vidicon camera operating with RS-170 standard) has a spectral sensitivity similar to the human eye, consisting of 525 lines per image, seen at 30 frames/sec. Sixty interlaced 1/2 frames/sec are produced, each consisting of 262 lines. Some of the newer video cameras can produce 1,000–2,000 lines in a video image, but compared to photographic films, which can produce images with 100–200 lines/mm, video images have far less resolution than direct photographic images. In addition, video cameras can suffer from pincushion and barrel distortion (discussed in Chapter 4), uneven coatings, vignetting, and blooming (bright objects or edges will cause adjacent areas to develop brightness, resulting in bright objects appearing larger than they really are). Solid-state cameras consist of a receptive surface covered with transistors. Some have as many as one-third of a million pixels across their surface. With charge-coupled devices (CCD), signals are passed to the right on the screen on clock-based time and read out a line at a time. CCD units are smaller and more rugged than Vidicon units, and record dim images more easily, but their resolution is limited by the spacing of the individual solid-state detectors that make up their receptive surfaces, and, in addition, the individual sensors may not have identical sensitivity.

As with some of the techniques we have discussed previously (e.g., immu-nocytochemistry), it is important to identify the specific information being sought before choosing approaches. For example, in the case of scanning electron microscopy, in which where an image is typically recorded from a cathode ray tube (CRT), the original image already has resolution considerably inferior to the resolution capability of the film used to record the image from the CRT. For that reason, if the image is digitized directly from the SEM signal and stored as a digitized image, the image resolution is similar to that recorded on high-resolution photographic film from the CRT, and the digitized image may actually record more useful information for purposes of morphometric analysis than an original analogue image.

III. IMAGE PROCESSING

Image processing can be used to make an image more appealing or to extract meaning, data, or numbers from an image. There are several data, or numbers that can be extracted from an image. These data can then be manipulated in several ways, as outlined by Russ (1990) and shown in the chart below. As we go down the chart, there is a reduction in the amount of data remaining at each level. Any lateral movements in the chart change the form of the data but do not change the amount present.

Original image ↓	Usually analogue, and 2-D from the eye, film, or CRT, up to 150 million receptors [= pixels] in the human eye produce an analogue image
Digitization ↓	Finite pixels, finite gray scale, less than 10^6 pixels
Gray-scale image ↓	Image processing; does not reduce the volume of information; can correct for defects in image; can enhance or suppress data; can rearrange data
Discrimination ↓	Often reduces gray scale to binary scale; foreground and background discriminations can be made
Binary image ↓ *Segmentation* ↓	Allows easy image editing
Objects ↓	Feature selection made here
Measurement ↓	Area, etc.
Data ↓	Statistical analysis here
Stereology ↓	Reconstruction of 3-D information from 2-D image
Structure	Final understanding of structures originally captured in 2D image

Image processing can be used to correct defects in the image caused by uneven illumination, excess charging (SEM images), or non-informational noise. Averaging techniques can be used to reduce random noise (e.g., "snow" in a video image) by including many images, thus averaging out the noise that appears in different locations in individual successive images. This technique, of course, is applicable only to stationary images.

If only one image with excessive noise is available, the values of nine nearest neighbor pixels can be averaged, and then each pixel value can be replaced with the average value for the array to decrease the noise.

A set of nine pixels with brightness from low levels (1) to high levels (9) is as follows:

$$2 \quad 4 \quad 6$$
$$1 \quad 7 \quad 9 \qquad \text{averaged} = 5.88 \text{ (rounded to 6)}$$
$$8 \quad 8 \quad 8$$

The display after the "smoothing" operation is

$$6 \quad 6 \quad 6$$
$$6 \quad 6 \quad 6$$
$$6 \quad 6 \quad 6$$

The programs that perform this sort of operation are actually much more sophisticated than the example given, but they still tend to decrease the sharpness of real edges or discontinuities within the image. For this reason, another method, known as applying a median filter (Russ, 1990), is preferred. This procedure uses the same array of pixels shown above, consisting of a 3 × 3 square of pixels. The nine pixels are ranked in order of brightness, and then the median brightness is assigned to the central pixel (seven in the case cited above). An image can have median filtering applied to it repeatedly until no further changes are noted. Since the isolated random noise pixels will rarely have the median brightness value with which they are replaced, filtering will smooth out noisy, but uniform, areas. Surface edges or discrete object boundaries will not be moved, made broader, or have contrast reduced by this technique.

Image processing can be used to separate foreground information from background information, usually on the basis of brightness. Processing can also quickly reveal changes in the location of objects (one method for tracking motion) by subtracting a later image from an earlier one. Objects that remain in the same place are effectively subtracted from the combined image, leaving only those that have moved.

Other common operations are erosion, dilation, and skeletonization. Erosion removes any pixel with a preassigned character (i.e., black color) that touches more than a selected number of neighbors. Dilation is used to fill in spaces, essentially by a reverse of the process of erosion. Dilation can decrease noise in an image by filling in spaces in a broken-up image, thus increasing the size of features. Skeletonization is a form of severe erosion created by producing lines of pixels that define the midline of a feature. Skeletonization allows the operator to segregate touching objects that would normally be counted as one into two objects, which can be separately scored. Producing a skeleton can also be useful to more easily analyze branching patterns within a structure.

The methods, statistical bases, and applications of these and other image processing operations performed to improve images and to prepare them for analysis are thoroughly discussed by Russ (1986, 1990) and Hader (1991). The first two texts deal extensively with the statistical and mathematical models behind the operations upon which image analysis is based, but the second is more devoted to applications in the analysis of specific problems.

IV. MORPHOMETRIC ANALYSIS

Once a two-dimensional image is digitized and subjected to the variety of image-enhancement techniques briefly alluded to above, the image can then be analyzed. Analysis can yield information about the brightness or color of an object (e.g., protein gels). The specific position of objects also can be ascertained, as well as information about the number of objects/unit area, the amount of space occupied/unit area, any preferred orientation of the objects, and the distribution of the objects (are they evenly spaced or clustered?). Distance measurements can be made to determine the distance from one object to another, as well as the distance between a given object and a specific boundary. Size measurements can be taken that describe an object's area, its perimeter, the center of balance for the object (centroid feature), the equivalent diameter (a circle of the same area as the entire feature), and the longest and shortest chords of the feature.

Shape analyses are possible with most programs that are based on form factors, such as dimensional ratios, topological information (e.g., links, nodes, ends, branches), roundness, solidity, convexity (ratio of taut string to perimeter), and aspect ratio (length to breadth).

V. STEREOLOGY

An essential problem of quantification in biological studies is that most of the images with which we work are two dimensional, while most biological structures are three dimensional. We are thus faced with the problem of interpreting the true three-dimensional structure from a two-dimensional image.

Stereology is the science that derives three-dimensional information from the analysis of two-dimensional images. Morphometric techniques are first used to derive spatial information from the two-dimensional images, and then further mathematical manipulations project three-dimensional images from the two-dimensional constructs. All known practical counting rules (e.g., the rules for counting red blood cells with a hemocytometer) over a decade old are biased. Gundersen et al. (1988a) state that "counting on certain edges and corners of the counting frame . . . systematically over-estimates the number" of objects in the field of view. The fundamental principles employed in programs for computer-assisted stereology correct for this source of error (see Gundersen et al., 1988a, 1988b, for a discussion of contemporary stereological methods).

There are a number of quantitative questions that biologists attempt to answer with stereological techniques, such as what the volume occupied by a given type of structure is (e.g., how many cells are present per unit area of a liver as a measure of hypertrophy of the organ), how many structures are within a unit volume (e.g., how many mitochondria are present within a plant parenchymal cell), and how large the individual structures in the field of view are (volumetric analysis).

Stereological methods provide information about two types of geometrical properties of structures: topological and dimensional. Topological properties include the number of disconnected features in a set and their connectivity (the number of redundant connections in the structure of a network). Dimensional properties are more commonly studied and encompass the collective volume of the set of features, the area of the surfaces of the structures in the set, the length of any lineal feature of interest within the structure, and the total curvature of the structure surface.

To perform stereological analysis, certain fundamental assumptions are made: (1) The structure counted must be randomly oriented, (2) the structure must be randomly distributed, and (3) the sample must be representative. In short, the sample must be isotropic, uniform, and random (unbiased). Naturally, biological structures rarely possess all three of these attributes, so stereological techniques must be used with caution.

Stereological methods are based on further assumptions, namely, that the volume of an object within a given specimen volume is proportional to the area of a given object within a given specimen area, which is proportional to the number of points (from a grid) resting on an object out of the number of points over the entire field of view of the specimen, which is proportional to the length of an object in relation to the length of the field of view of the specimen. In other words, $V_V = L_L = P_P = A_A$, where V = volume fraction, L = lineal fraction, P = point fraction, and A = areal fraction.

When spherical objects in a field are either physically sectioned or optically sectioned (e.g., a high-magnification light microscope image of a thick specimen, with little depth of field), one sees a collection of circular profiles with different chord lengths (diameter) (Fig. 175). To determine the true size distribution of spherical or nearly spherical objects from two-dimensional images, the chord lengths (diameters) of the two-dimensional structures are measured or, if the objects are not circular, the major and minor axes are measured (a and b, respectively) and the diameter d of the structure is determined by the formula: $2 \times \sqrt{a \times b}$. The measured diameters are then divided into 10–15 equal groups, and the number of objects in each size category (N_A) is calculated. The results of this calculation are used to calculate the number per unit volume in each size class in the original three-dimensional sample.

As previously mentioned, to calculate the volume fraction V_V of objects within a specimen, the assumption is made that the volume fraction of the objects within a volume of sample is proportional to the areal fraction A_A of the objects and both of these are proportional to the fraction of points on a grid striking the objects compared to the number of points within the whole specimen volume (P_P). The volume of a particular object or set of objects within a selected specimen volume may be determined by two major methods: point counting and planimetry.

In planimetry, the calculated area of objects subjected to the application of algorithms in the computer program is used to calculate the volume fraction of the objects. Point-counting methods require several decisions by the computer operator. The general rule of thumb is to have no more than 0.6–1 point striking the object's

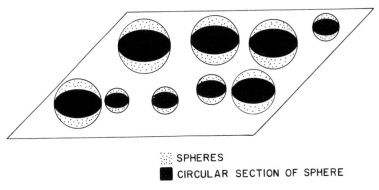

SPHERES
CIRCULAR SECTION OF SPHERE

FIGURE 175. Illustration of an optical section through a three-dimensional volume containing spherical structures.

profile. The correct grid spacing is thus dependent on the magnification of the image being analyzed and on the number of object profiles present within the image. It is also necessary to calculate how many "hits" are necessary to assure that P_P will predict V_V within reasonable limits of confidence. To achieve a confidence limit of 5%, 400 independent hits must be recorded. Finally, if an average of 40 independent hits are recorded per photograph, then 10 photographs will be necessary to estimate V_V within a 5% confidence limit. It is also necessary to remember that the point counting must be done on 10 different photographs, and not just by placing a grid randomly on a given photograph 10 different times. The latter case would not yield 400 independent hits.

The classic stereological methods discussed above, which are geometry based and are best used to evaluate cylinders, spheres, etc. that are isotropic and randomly oriented, relate surface area to unit volume and work, in many instances, but fall short with anisotropic objects. The newer methods of stereology (see Gundersen et al., 1988a, 1988b, for thorough discussions of these methods) were developed to accommodate objects typically encountered with biological specimens, which are not randomly oriented or uniform. The new methods are geometry independent, which allows nonrandomly oriented material to be evaluated as if it were isotropic, uniform, and random.

VI. COMPUTER-ASSISTED ANALYSIS OF MOVEMENT

A final area of general interest to biologists to which computer-assisted methods are applied is the analysis of object movements within a field of view. The analysis of sperm motility after the application of therapeutic drugs (of interest to pharmaceutical companies) and the chemotactic movement of bacteria, which involves changes in the direction of cell rotation, represent two typical applications. In both situations, the number of cells and the speed of movements make human analysis difficult. Live digital image acquisition followed by computer analysis easily delivers information concerning the rotational movements of individual cells and their track lengths over selected time periods. Takahashi (1991) discussed automated measurement of movements of halobacteria. He used computer programs to quantify the percent of reversing cells (cells changing directions from clockwise to counterclockwise and vice versa) in a population and to calculate the average number of reversals occurring in a population at a given time point. Finally, he calculated the latency period between the application of a chemotactic stimulus and the subsequent reversals exhibited by the cells.

The programs available for tracking cell motion can evaluate populations by measuring changes in population density resulting from cells migrating toward or away from chemical or environmental stimuli. Computers also can be used to track individual cells, which allows greater time resolution than population methods. Cells can therefore be exposed to stimuli for shorter time periods than with population analysis methods.

REFERENCES

Gundersen, H.J.G., Bendtsen, T.F., Korbo, L., Marcussen, N., Moller, A., Nielsen, K., Nyengaard, J.R., Pakkenberg, B., Sorensen, F.B., Vesterby, A., and West, M.J. 1988a. Some new, simple and efficient sterological methods an their use in pathological research and diagnosis. *Acta Pathol. Microbiol. Immunol. Scand.* 96:379.

Gundersen, H.J.G., Bagger, P., Bendtsen, T.F., Evans, S.M., Korbo, L., Marcussen, N., Moller, A., Nielsen, K., Nyengaard, J.R., Pakkenberg, B., Sorensen, F.B., Vesterby, A., and West, M.J. 1988b. The new stereological tools: Disector, fractionator, nucleator and point sampled intercepts and their use in pathological research and diagnosis. *Acta Pathol. Microbiol. Immunol. Scand.* 96:857.

Hader, D.-P. 1991. *Image analysis in biology.* CRC Press, Boca Raton, FL.

Misell, D.L. 1978. *Image analysis, enhancement and interpretation.* North-Holland, New York.

Russ, J.C. 1986. *Practical stereology.* Plenum Press, New York.

Russ, J.C. 1990. *Computer-assisted microscopy. The measurement and analysis of images.* Plenum Press, New York.

Takahashi, T. 1991. Automated measurement of movement responses in halobacteria. In: *Image analysis in biology,* P.-P. Hader (ed.), CRC Press, Boca Raton, FL, pp. 315–378.

CHAPTER 17

Scanning Tunneling Microscopy and Its Derivatives

The scanning tunneling microscope (STM) was first developed to examine the surface of solids. The properties of the surface of a solid sample are distinctly different from those within the solid, because the atoms on the surface are often arranged much differently from those in the interior of a sample. In the interior of a solid, atoms are surrounded by other atoms, whereas those on the surface interact only with those directly below them or adjacent to them, while being free to react with atoms above the surface of the solid (gases, etc.).

Binnig and Rohrer invented the STM in 1981 to allow a clearer picture of the atomic nature of the surface of solids and the interactions taking place on sample surfaces (Binnig and Rohrer, 1985). The silicon chip industry was particularly concerned with the surface of microelectronic devices because surface areas become more important as the surface-to-volume ratio increases with the reduction in overall size of these components.

Electron microscopes require high vacuum, electromagnetic lenses, and an electron source. The accelerated electrons used for imaging penetrate the sample surface to varying degrees. Thus, surface structures are difficult to visualize with the electron microscope at the molecular level. On the other hand, STMs do not image with free particles like the electrons used for electron microscopy, so vacuum systems, lenses, and illumination sources are unnecessary with the STM.

The underlying principle for STM imaging is that a cloud of electrons exists above the surface of a sample. The number of electrons above the sample surface decreases exponentially with the distance from the surface. The STM probe and the sample itself each have a cloud of electrons associated with their surfaces that protrude into the space between them. The probe is mechanically scanned over the sample surface while a voltage is applied between the probe tip and the specimen surface. The voltage causes a current to flow between the adjacent electron clouds of the probe and sample, which is called the *tunneling current*. This current falls off exponentially as the distance between the clouds increases. A distance as small as the diameter of a single atom can cause a 1,000-fold decrease in the tunneling current (Arscott, 1990; Binnig and Rohrer, 1985). Thus, the vertical position of atoms at the sample surface can be measured with great precision.

The STM probe is maintained at a fixed voltage potential in relationship to the sample as the probe is scanned across the specimen surface. A feedback loop maintains the tip of the probe at a set distance, ensuring a constant current flow between the probe and the specimen at the fixed voltage. Thus, as the topography of the sample changes, the feedback mechanism raises and lowers the probe to maintain a constant current flow. The movements of the probe are electronically amplified and analyzed by an associated computer that can project an image onto a cathode ray tube.

The gap between the probe and the sample is maintained at about 1.0 nm, and the stability and precision of the instrument allows the measurement of 0.001-nm distance variations. To achieve this level of precision, the STM must be in a vibration- and draft-free environment, the probe tip must be extremely small, and the probe drives must be extremely precise. Binnig and Rohrer (1985) show photographs of their STM and explain the mechanics and physics of its operation in significant detail.

Since the introduction of the STM, various "children" have been created (Pool, 1990), including the friction force microscope, atomic force microscope, magnetic force microscope, electronic force microscope, attractive mode force microscope, scanning thermal microscope, optical absorption microscope, scanning acoustic microscope, and the molecular dip-stick microscope. All of these instruments measure some aspect of force generated between a probe and a sample surface.

The atomic force microscope measures sample topography, even in nonconductive materials; the attractive mode force microscope measures the attractive force between molecules (van der Waals forces, etc.); and the magnetic force microscope maps out magnetic fields along a sample surface with a 100-nm resolving capability. For a brief description of these and other types of STM progeny, see Pool (1990).

Biological subjects for investigation with the STM and its progeny have included nucleic acids (Beebe *et al.*, 1989; Lindsay *et al.*, 1989), nucleic acids and their associated proteins (Amrein *et al.*, 1988, 1989), freeze-fracture replicas of biomembranes (Zasadzinski *et al.*, 1988), and isolated hepatic gap junctions (Hoh *et al.*, 1991).

The STM and its allies can be used to examine unfixed, hydrated biological materials that are not subjected to the high vacuum conditions used for electron microscopy samples (see Yao *et al.*, 1991, for some methodologies). Thus, within the limitations imposed by viewing only bulk sample surfaces at the molecular level, the degree of resolution with largely unprocessed materials is much greater than is possible with standard electron microscopy techniques. At a recent congress (Behm *et al.*, 1990), the instrumentation and applications of STM technology to specific problems were discussed.

REFERENCES

Amrein, M., Stasiak, A., Gross, H., Stoll, E., and Travaglini, G. 1988. Scanning tunneling microscopy of recA-DNA complexes coated with a conducting film. *Science* 240:514.

Amrein, M., Durr, R., Stasiak, A., Gross, H., and Travaglini, G. 1989. Scanning tunneling microscopy of uncoated recA-DNA complexes. *Science* 243:1708.

Arscott, P.G. 1990. Taking the measure of the molecule by scanning tunnelling microscopy. *Am. Soc. Microbiol. News* 56:136–138.

Beebe, T.P., Jr., Wilson, T.E., Ogletree, D.F., Katz, J.E., Balhorn, R., Salmeron, M.B., and Stiekhaus, W.J. 1989. Direct observation of native DNA structures with the scanning tunneling microscope. *Science* 243:370.

Behm, R.J., Garcia, N., and Rohrer, H. (eds.). 1990. *NATO Advanced Study Institute on basic concepts and applications of scanning tunneling microscopy.* Kluwer, Boston.

Binnig, G., and Rohrer, H. 1985. The scanning tunneling microscope. *Scient. Amer.* 253:50.

Hoh, J.H., Lal, R., John, S.A., Revel, J.-P., and Arnsdorf, M.F. 1991. Atomic force microscopy and dissection of gap junctions. *Science* 253:1405.

Lindsay, S.M., Thundat, T., Nagahara, L., Knipping, U., and Rill, R.L. 1989. Images of the DNA double helix in water. *Science* 244:1063.

Pool, R. 1990. The children of the STM. *Science* 247: 634.

Yao, J.E., He, J., Shang, G.Y., Kuang, Y.L., Wei, J., Zeng, K., Lin, K.C., Dai, J.W., and Su, Y.X. 1991. Scanning tunneling microscope and its biological applications. In: *International symposium on electron microscopy*, K. Kuo, and J. Yao (eds.), World Scientific, Singapore, pp. 81–95.

Zasadzinski, J.A.N., Schneir, J., Gurley, J., Elings, V., and Hansma, P.K. 1988. Scanning tunneling microscopy of freeze-fracture replicas of biomembranes. *Science* 239:1013.

APPENDIX A

Laboratory Safety

Laboratory safety has been addressed repeatedly in this text in the appropriate areas. It is important to remember that the laboratory setting provides exposure to a number of hazards, but if common-sense laboratory practices are followed, the laboratory should be no more dangerous in the short or long term than any other workplace.

Within a typical electron microscopy laboratory, sources of injury include various chemicals, cryogens, hot objects (e.g., pumps), high voltage, glass and razor blades, inflammable materials, and potential carcinogens.

All chemicals should be treated as if they are toxic unless Material Data Safety Sheets (MSDS) available from suppliers declare them harmless. Even though microscopists work with many toxic chemicals, they are generally handled in small amounts, and the chance of personal injury is extremely small if the chemicals are properly handled, as suggested in the MSDS instructions. Hands and eyes should be protected at all times, and sinks for washing should be readily available (as well as eye-wash stations and showers for spills). Clearly marked spill and first aid kits should be provided within the laboratory. The Occupational Safety and Health Administration (OSHA) has published specific guidelines for the safe operation of workplaces (available from institutional administrations) including scientific laboratories. The guidelines mandate what manner of safety equipment must be made available, what type of safety training needs to be administered, and to whom it should be offered. All government, educational, and private laboratories are required to have a safety plan in place that describes how to deal with emergency situations.

It would be redundant to go into the broad principles of safety in an electron microscopy laboratory because of the federal regulations mandating compliance with the OSHA training rules. Sources of heat, cold, and electricity necessitate the same precautions that would be exercised in any workplace.

Material safety data sheets are available from all chemical suppliers and should be requested whenever new chemicals are ordered. An overview of chemical handling and disposal is offered in the books listed below:

Anon. 1981. *Prudent practices for handling hazardous chemicals in laboratories*. National Academy Press, Washington, D.C.
Anon. 1983. *Prudent practices for disposal of chemicals from laboratories*. National Academy Press, Washington, D.C.

Literature Sources for Electron Microscopy

Throughout the text, various books and journal articles have been listed. There are atlases on ultrastructure for insects, protozoans, fungi, mammals, and other organisms. The following list of journals and general books is not exhaustive, but it offers a starting point to begin researching questions about biological ultrastructure.

I. ATLASES

Beckett, A., Heath, I.B., and McLaughlin, D.J. 1974. *An atlas of fungal ultrastructure*. Longman, London.
Fawcett, D.W. 1981. *The cell*, 2nd ed. W.B. Saunders, Philadelphia.
Gunning, B.E.S., and Steer, M.W. 1975. *Ultrastructure and the biology of plant cells*. Edward Arnold, London.
Kessel, R., and Kardon, R.H. 1979. *Tissues and organs, a text-atlas of scanning electron microscopy*. W.H. Freeman, San Francisco.
Lott, J.N.A., and Darley, J.J. 1976. *A scanning electron microscope study of green plants*. C.V. Mosby, St. Louis.
Palmer, E.L., and Martin, M.L. 1988. *Electron microscopy in viral diagnosis*. CRC Press, Boca Raton, FL.
Rhodin, J.A.G. 1974. *Histology, a text and atlas*. Oxford University Press, New York.
Threadgold, L.T. 1976. *The ultrastructure of the animal cell*, 2nd ed. Pergamon Press, New York.

II. JOURNALS

There are numerous journals dealing with cell and molecular biology, with specialized tissues and groups of organisms, and with the physics of electron microscopy, in addition to journals devoted to materials analysis. The ones listed below constitute a short list of the most readily available journals that have the most concentrated focus on biological ultrastructure questions.

Biology of the Cell (Societe Française de Microscopie)
Biotechnic and Histochemistry (*Stain*, before 1991)
Journal of Cell Biology
Journal of Cell Science
Journal of Cellular Biochemistry
Journal of Electron Microscopy (Japanese Society of Electron Microscopy)
Journal of Electron Microscopy Technique
Journal of Histochemistry and Cytochemistry
Journal of Structural Biology
Micron et Microscopica Acta
Microscopy, Microanalysis, and Microstructures (Societe Française de Microscopie)
Ultramicroscopy
Ultrastructural Pathology

III. SOCIETY PUBLICATIONS

In the United States, the Electron Microscopy Society of America (EMSA) is composed of electron microscopists in both the biological and materials sciences. Numerous organismally oriented or cyto/histochemically oriented societies also publish significant numbers of cytological papers. Membership in EMSA includes the *EMSA Bulletin*, which contains helpful technical hints and articles on ultrastructural techniques. The EMSA bulletin also lists affiliated regional and state electron microscopy societies.

IV. BOOK SERIES

Glauert, A.M. (ed.). 1972–1991. *Practical methods in electron microscopy.* 13 vols. North-Holland/Elsevier, New York.

Hayat, M.A. (ed.). 1973–1977. *Electron microscopy of enzymes: Principles and methods.* 5 vols. Van Nostrand Reinhold, New York.

Hayat, M.A. (ed.). 1970–1976. *Principles and techniques of electron microscopy.* 6 vols. Van Nostrand Reinhold, New York.

Hayat, M.A. (ed.). 1974–1978. *Principles and techniques of scanning electron microscopy.* 6 vols. Van Nostrand Reinhold, New York.

Royal microscopy society handbook. 17 vols. through 1989. Oxford University Press, Oxford. Covers both light- and electron-microscopy topics.

V. NIH RESOURCES FOR INTERMEDIATE VOLTAGE ELECTRON MICROSCOPY (IVEM) (From Lewis *et al.*, 1988)

Three-dimensional electron microscopy of macromolecules, University of Arizona, Department of Biochemistry, Tucson, Arizona, (602) 621-7524. *Contact W. Chiu, Ph.D.*

Biomedical three-dimensional imaging and analysis, University of Pennsylvania, Department of Biology, Philadelphia, Pennsylvania, (215) 898-5788. *Contact L.D. Peachey, Ph.D.*

MICROMED: IVEM in basic medical research, Bowman Gray School of Medicine, Department of Pathology, Winston-Salem, North Carolina, (919) 748-2675. *Contact J.C. Lewis, Ph.D.*

Electron Microscopy Equipment and Supplies

Listed below are some of the suppliers in the United States of the products necessary for electron microscopy work. Although not exhaustive, this list is fairly comprehensive and will provide a starting point for locating specific items.

I. EXPENDABLE SUPPLIES AND SMALL EQUIPMENT

Bal-Tec Products Inc.
984 Southford Road
P.O. Box 1221
Middlebury, CT 06762
(203) 598–3660

Electron Microscopy Sciences
321 Morris Road
Box 251
Fort Washington, PA 19034
(800) 523–5874

EMITECH, USA
5206 FM 1960 W.
Suite 100
Houston, TX 77069
(713) 893–8443

Energy Beam Sciences
P.O. Box 468
Agawam, MA 01001
(800) 992–9037

Ernest F. Fullam, Inc.
900 Albany Shaker Road
Latham, NY 12110–1491
(800) 883–4024

Ladd Research Industries, Inc.
P.O. Box 1005
Burlington, VT 05402
(800) 451–3406

Ted Pella, Inc.
P.O. Box 492477
Redding, CA 96049–2477
(800) 237–3526

Polysciences, Inc.
400 Valley Road
Warrington, PA 18976–2590
(215) 343–6484

Sigma Chemical Co.
P.O. Box 14508
St. Louis, MO 63178
(800) 325–3010

SPI Supplies
P.O. Box 656
569 East Gay Street
West Chester, PA 19381–0656
(800) 242–4774

Technical Products, Inc.
154 Haddon Avenue
Westmont, NJ 08108
(800) 858–3675

Tousimis Research Corporation
P.O. Box 2189
Rockville, MD 20847
(301) 881–2450

II. ELECTRON MICROSCOPES

Amray, Inc.
160 Middlesex Turnpike
Bedford, MA 01730
(617) 275–1400

Electro Scan Corporation
100 Rosewood Drive
Danvers, MA 01923
(508) 777–9280

Hitachi Scientific Instruments
Electron Microscope Department
25 West Watkins Mill Road
Gaithersburg, MD 20878
(800) 638–4087

JEOL USA, Inc.
11 Dearborn Road
Peabody, MA 01960
(508) 535–5900

Leica Inc.
Microscopy and Scientific Instruments Division
11 Deer Lake Road
Deerfield, IL 60015
(708) 405–7052

Philips Electronic Instruments, Inc.
Philips Analytical Division
85 McKee Drive
Mahwah, NJ 07430
(201) 529–3800

Carl Zeiss, Inc.
Electron Optics Division
One Zeiss Drive
Thornwood, NY 10594
(914) 681–7745

III. DIAMOND KNIVES

Delaware Diamond Knives, Inc.
3825 Lancaster Pike
Wilmington, DE 19805
(800) 222–5143

diaTECH, Inc.
6408 Clinton Highway, Suite 5
Knoxville, TN 37912
(615) 524–4541

DiATOME US
P.O. Box 125

321 Morris Road
Fort Washington, PA 19034
(215) 646–1478

Micro Engineering
Rt. 2, Rural Box 474
Huntsville, TX 77340
(800) 533–2509

IV. HIGH-VACUUM PUMPS

Alcatel Vacuum Products, Inc.
40 Pond Park Road
Hingham, MA 02043
(617) 749–8710

Balzers
8 Sagamore Park Road
Hudson, NH 03051
(800) 352–5012

Edwards High Vacuum, Int'l.
301 Ballardvale Street
Wilmington, MA 01887
(508) 658–5410

V. ULTRAMICROTOMES

Leica, Inc.
111 Deer Lake Road
Deerfield, IL 60015
(708) 405–7052
(LKB and Reichert Ultramicrotomes)

RMC, Inc.
Grant Road Industrial Center
Suite 122
1802 West Grant Road
Tucson, AZ 85745
(602) 882–7900

VI. EQUIPMENT FOR CRYOTECHNIQUES

Available from various EM supply houses and:

Bal-Tech Products, Inc.
984 Southford Road

P.O. Box 1221
Middlebury, CT 06762
(203) 598–3660

LifeCell Corporation
3606-A Research Forest Drive
The Woodlands, TX 77380
(713) 367–5368
(Equipment marketed through RMC, Inc.)

VII. SPUTTER COATERS AND VACUUM EVAPORATORS

Available from various EM supply houses and:

Anatech Ltd.
5510 Vine Street
Alexandria, VA 22310
(703) 971–9200

Bal-Tech Products, Inc.
984 Southford Road
P.O. Box 1221
Middlebury, CT 06762
(203) 598–3660

Denton Vacuum, Inc.
Cherry Hill Industrial Center
Cherry Hill, NJ 08003–4072
(609) 424–1012

Index

Acetone, 35, 37, 104, 157
  as a dehydrating agent, 32, 33, 34, 39, 238
  compatibility with acrylic resins, 35
  compatibility with epoxides, 32, 35
  as pump contaminant, 157
  use in cryosubstitution, 264, 291
Acid fuchsin, 169–170
Acid phosphatase, 22, 300, 301, 302
Acrolein, 264, 296
  fixation, 15–17
Acrylic resins, 34–35, 37, 167, 249, 267, 310, 313, 316
Acyl transferase, 300, 301
Agar embedment of cells and particulates, 38–39, 101
Alcian blue, 20, 299, 302, 303
Alcohol, 157, 172, 173
Aldehydes, 219, 250, 312, 313
  fixation for SEM, 8, 12, 15, 16, 19, 20, 21, 22, 39
  use in cryofixation, 237, 259, 264, 296
Alkaline phosphatase, 300
Ammonium molybdate, 259, 262, 313
  as a negative stain, 220
  for staining sections, 259, 260
Amyl acetate, 103, 105, 238
  usage with critical point drying (CPD), 238
Araldites, 35, 36
Argon, 152, 217, 241
Aryl sulfatase, 300
Astigmatism, 119, 126, 137, 146, 202
Atomic force microscopy (AFM), 114, 342
Attractive mode force microscopy, 342
Auger electrons, 230, 235–236, 286, 288

Auger electron imaging (AEI), 235
Autoradiography, 184, 267, 321–330
  electron microscopy, 326–330
  light microscopy, 330
Avidin-biotin complex, 317

Bacitracin, 221
Backscattered electron imaging (BEI), 234–235
Backstreaming, 157
Bacteria, 218, 302
  fixation, 73
  staining, 179
Basic fuchsin, 170, 171
BEEM capsules, 38, 104
Beryllium, 96, 287, 290
Block trimming, 87–91, 168
Braking radiation, 287
Bremsstrahlung, 289
Buffered neutral formalin (BNF), 12, 13, 15, 296
Buffers, 24–32
Butvar, 104, 105
Butylbenzene, 254

Cacodylate buffer, 21, 23, 31, 39, 41, 43, 46, 172, 173, 302, 315
Calcium, 23
  in fixatives, 21
  interaction with buffers, 23
  localization, 305–307
Carbohydrates, 297
  fixation, 10, 14, 16, 18
  immunolabeling, 309
  staining: see Polysaccharide staining

Carbon
  coating, 111, 221, 234, 241
  stubs, 240, 242, 291
  support films, 103, 104, 111, 112
Carbon rod electrodes, 212, 213, 214, 215
Carbon/platinum coating, 264
Carson's fixative, 25, 46
Catalase, 149
Cathodoluminescence, 230, 235, 237
Cellulose fibrils, 179, 180
Cellulose nitrate, 184
Cetylpyridinium chloride, 20
Chatter, 98, 99
Chemography, 330
Chlorofluorocarbons, 157
Chloroform, 91, 105
Chloroplasts, 12
Cholesterol, 19
Chromatic aberration, 113, 117, 125–126, 129,
    147, 180, 234, 275–277, 281, 292
Chromium, 217, 290
Cilia, 103
Cold fingers, 162
Collagen, 20
Collidine buffer, 31
Collodion, 103, 104, 105, 221, 290
Colloidal gold
  bound to enzymes, 299
  bound to lectins, 299
  immunocytochemical applications, 299, 305
Concanavalin A (Con A), 305
Contrast 278–279, 290
  tuning, 293
Copper block, use in cryofixation, 254, 257
Critical point drying, 32, 224, 234, 238–240, 280
Cryofixation, 6, 7, 247, 249–251, 252, 262, 267,
    270, 291, 310, 313
Cryoprotection, 250, 253, 264
Cryosectioning, 261, 281, 309, 311
Cryosubstitution, 7, 258, 261, 262–264, 291, 310
Cryotechniques, 291, 295, 296, 352
Cryoultramicrotomy, 7, 29, 80, 81, 83, 249, 259,
    260, 270, 305, 310, 313–314
Cytoplasm, 9, 13, 46, 71, 73, 272

Dark-field imaging, 180–181, 279
Dehydration agents, 32–34, 39, 309, 311
3,3′-diaminobenzidine HCl (DAB), 298, 299,
    300, 318
Diethylether, 264
Digital images, 333
Digitonin, 19
Dimethoxypropane (DMP), 33

Dimethylsulfoxide (DMSO), 250, 311
Dulbecco's phosphate-buffered saline (DPBS), 311

Electrodes, types of, 212, 213–215
Electron guns, 131–134, 228
Electron scattering, 129, 209
  elastic, 126, 128, 234, 232, 276
  inelastic, 126, 128
Electron energy loss spectroscopy (EELS), 114,
    141, 276, 285, 291–293
Electron lenses, 114, 117, 118–127
Electronic force microscopy, 342
ELISA, 311
Embedding resins, 34–38, 280
Embedment, 295, 309, 311
  media, 11, 32
Endoplasmic reticulum, 13, 20, 71, 73, 313
Energy-dispersive spectroscopy (EDS), 141, 223,
    229, 236–237, 241, 276, 285–291
Environmental Scanning electron microscope
    (ESEM), 223, 238, 244
Enzymes, 16, 17, 322
  gold-labeled, 309
  negative staining, 218
Eosin, 224
Epon, 180, 313
Epoxide resins, 32, 33, 35, 80, 167, 179, 310,
    312, 313
Epoxides, 38, 91, 319
Erosion, 333, 336
Esterases, 22
Ethanol
  as a dehydrating agent, 32, 33, 34, 35, 39, 68,
    224, 238, 239
  for grid cleaning, 104
Ethylene dichloride, 105
Ethylene glycol, 250

Ferritin, 218, 298–299, 302, 305, 309, 315, 316,
    318
Field emission guns (FEG), 267, 277, 281, 290
  in SEM, 223, 224, 229, 233, 234, 241, 243
  in TEM, 131, 132, 161, 164, 285, 286
Film-casting, 104–105
Fixation, 5–24
  chemical methods, 7–24
  physical methods, 6–7; see also Cryofixation
  simultaneous, 21, 39
  supplements, 19–20
Flagella, 209, 210, 218, 314
Fluorescent-labeled antibodies, 309
Fluoroisothiocyanate (FITC), 305
Fog, 184, 190

Formaldehyde, 12–15, 21, 31
  use in enzyme cytochemistry, 296, 297, 299, 311, 312
Formvar, 35, 37, 103–104, 105, 108, 111, 221
Freeze drying, 7, 240, 267, 291
Freeze-fracture, 249, 250, 258, 264–267, 342
Freezing
  high pressure, 255, 261
  immersion, 252, 253
  jet, 252, 254, 261
  metal mirror, 251, 252, 257–258; see also slam freezing
  spray, 252, 254
Freon, 238, 251, 252, 264, 313
Fresnel fringes, 119, 120, 122–124, 146
Friction force microscopy, 342

Glutaraldehyde, 171, 180, 264, 296, 303, 311, 312, 315, 316, 318
  fixation for TEM, 6, 14, 16, 17–19, 20, 21, 22, 29, 31, 39, 40–42, 45, 46, 74,
Glutaraldehyde-carbohydrazide (GACH), 37
Glutaraldehyde/osmium fixation, 46
Glycerol, 249, 250, 264
Glycocalyx stains, 302, 303
Glycogen, 10, 14, 24, 46, 175
Glycoproteins, 220
Gold, 211, 217, 299
  coating, 241
  film-casting, 104
  grids, 96, 97, 290
  use in cryofixation, 254, 257
Gold-labeled antibodies, 267, 282, 309, 314, 322
Gold-palladium coating, 149, 217, 241, 291
Golgi bodies, 210, 218
Gomori methenamine-silver staining, 302
Graphitized carbon, 149
Grids, 104, 106, 179, 180
  film coating, 37, 219, 261, 280, 313, 314
  types of, 96–97, 290, 304, 314

Hematoxylin, 224
Hemoglobin, 116, 218
HEPES, 32
Hexamethyldisilizane, 240
High-voltage electron microscopy (HVEM), 37, 128, 275–281
  advantages, 276
  construction, 279
  resolution, 276–277, 278
Histones, 10, 173
Holey films, 103, 111, 112
Horseradish peroxidase, 299

Hydrogen peroxide, 299, 318
Hydroquinone, 15, 16
Hypersensitivity, 38
Hysteresis, 119

Image digitization, 333, 335
Image processing, computer-assisted, 333, 335–337
Image resolution, 334–335
Image rotation, 120–122
Immunocytochemistry, 281, 295, 298, 335
  fixatives, 13, 14, 17, 18
  specimen embedding, 34
Immunoferritin, 315–316
Immunogold techniques, 318–319
Immunolabeling, 7, 249, 258, 260, 267, 309, 311
Immunoperoxidase techniques, 318, 319
Immunoprobes, 312
Intermediate-voltage electron microscopy (IVEM), 115, 128, 275, 281–282, 347–348
Ion getter pumps, 131, 161–162, 164, 165
Ionization damage, 276, 277
Isopentane, 251
Isopropyl alcohol, 256

JB-4 resin, 34

Karnovsky's fixative, 21, 22, 27, 28, 46, 296
Knives
  diamond, 82–83, 87, 98, 260, 351–352
  glass, 83–85, 86, 87, 98, 101, 260
  storage, 86–87
  ultramicrotomy, 79–87

Lanthanum hexaboride (LaB6), 131–133, 161, 164, 215,, 229, 277
Latex spheres, 149
Lead, 20, 299, 302
  contrast building with, 278, 290
  electron scattering by, 126
  post staining with, 171, 172, 175–179, 259, 269, 270
Lead-capture techniques, 298, 300–302
Lecithin, 36
Lectins, 224, 299,305, 319, 322
Lipids, 75
  fixation, 8, 9, 14, 16, 18, 19, 20
Liquid nitrogen, 162, 224, 251, 252, 253, 255, 257, 259, 267, 287, 313
Lowicryl resins, 35, 36, 37, 167, 267, 268, 313
Lowicryl K4M, 36, 313
LR Gold, 36, 167
LR White, 36, 167, 313

LR White resin, 35, 36, 37, 313
LX-112, 32, 36

α2-macroglobulin, 220
Magnetic force microscopy, 342
Malachite green, 19
Mammalian cells fixation, 68, 71, 76
McDowell's and Trump's fixative, 24, 26, 68, 73, 296
Mercox resin, 226
Mesosomes, 18
Methacrylate resins, 34, 35, 243, 322
Methanol, 296, 311, 312
  use in cryosubstitution, 264, 291
Methylamine tungstate, 220
Methylene blue, 170, 171
Microanalysis, 9, 282, 285–293
  frozen sections for, 260–262
Microbodies, 299, 300, 302
Microfilaments, 12, 20, 177, 218
Microsomes, 38
Microtubules, 12, 16, 17, 18, 20, 22, 177, 218
Microwave fixation, 7
Microwave staining, 167, 180
Milk proteins, 218
Millonig's phosphate buffer, 30
Mitochondria, 38, 39, 68, 71, 73, 75, 103, 129, 218, 265, 306, 313
  fixation, 12, 13, 24, 46
Molecular distillation, 258, 267–270, 272, 291, 310, 313
Molecular dipstick microscopy, 342
Molybdenum, 212, 214, 290
MOPS, 32
Morphometry, 333, 337
Myelin, 18
Myofibers, 280

Negative staining, 179, 209, 218–221, 267, 310, 313, 314
Nitrocellulose, 103, 105
Nitrogen slush, 251, 313
Nuclear emulsions, 324–325
Nuclear envelope, 13, 68, 71, 270, 272
Nuclear pores, 218
Nucleic acids, 342
  fixation, 10, 14, 16, 18, 20
  immunolabeling, 309
  staining, 173
Nucleoli, 175
Nucleoplasm, 9, 13, 46, 68, 71, 73, 74, 272

Optical absorption microscopy, 342
Osmium, 167, 172, 262, 263, 264, 297, 305, 318
  contrast building with, 278, 290
  electron scattering by, 126
  fixation with, 6–12, 21, 22, 30, 33, 42, 44, 46, 175, 180, 237, 243, 267, 268, 270, 272, 312, 313
  post-fixation, 8, 14, 18, 20, 25, 26, 27, 28, 34, 39–41, 46, 172, 237
Osmium/thiocarbohydrazide, 234
Osmium tetroxide: see Osmium

Palladium, 211, 217
Paraffin embedding, 22, 74
Paraformaldehyde, 13, 14, 251, 296, 299
Parlodion, 103, 106; see also Nitrocellulose
Perfusion methods, 23
Periodic acid-Schiff reagent, 19, 74, 171, 302
Permanganates, 12, 179–180
Permount®, 168
Peroxidase methods, 317
  in cytochemistry, 297, 298, 299, 305
Peroxidase-antiperoxidase procedure (PAP), 318
Peroxidase-conjugated antibodies, 309
Phosphate buffer, 12, 21, 23, 25- 28, 30–32, 39, 40, 42, 44–46, 172, 173, 297, 302, 315
Phosphate-buffered formaldehyde, 260
Phosphate-buffered saline (PBS), 7, 30, 39
Phospholipase A2, 299
Phospholipids, 18, 20, 32, 33
  staining, 173
Phosphoproteins, 173
Phosphotungstic acid (PTA), 20, 314
  as a negative stain, 179, 218–219, 220, 221
Phosphotungstic acid/chromic acid, 179
Photography, 125, 184
  copy enhancement, 198–204
  enlargers, 204–206
  film types for, 185–187, 194–198
  paper types, 191–193
  processing, 187–191
Pili, 210, 218, 314, 316
PIPES, 32
Plasma membranes, 179, 303, 305
Plasmalemma, 68, 76, 262
Plasmodesmata, 23, 73, 76
Platinum, 211, 212, 213, 214, 215, 290
Polyacrylamide gel electrophoresis, 267
Polyester resins, 35, 36
Polyethylene glycol, 37, 280
Polyethylene molds, 38
Polyethyleneimine, 20

Polymount®, 169
Polysaccharide staining, 171, 179, 302–305
Propane, 251, 252, 253, 255
Propylene oxide, 35
    dangers in using, 33
    as a dehydrating agent, 32, 33
Protein A, 310, 315, 319
Protein G, 310, 315, 319
Proteins, 342
    fixation, 10, 12, 13, 14, 16, 18, 22, 32, 33,
Protozoa fixation, 21, 68, 71, 76
Pumps: see Vacuum pumps
Pyroantimonate, 299, 306, 307

Radiation damage, 275, 277–278, 281
Remnance, 119
Resin polymerization, 35, 267, 268, 295
    heat-induced, 8, 262, 268, 312
    ultraviolet radiation-induced, 8, 267, 268
Resolution, 115–117, 172
    autoradiographic, 327–330
    computer associated image, 334–335
    in IVEM, 281–282
    in SEM, 230–232
    in TEM, 230
    light-microscopy, 113
Resolution standards, 149
Reticulation, 190, 191
Reynolds' lead citrate, 175, 177, 178, 179, 180
Ribonucleoproteins, 173
Ribosomes, 12, 17, 73, 175
Ruthenium red, 20, 299, 302, 303

Saponin, 309, 311
Sarcoplasmic reticulum, 280
Scanning tunneling microscopy (STM), 114, 341–342
Scanning acoustic microscopy, 342
Scanning electron microscopy
    artifacts, 242–243
    biological applications, 224–228
    coating, 241–242
    computer-assisted imaging, 333, 335
    dehydration agents, 32, 238
    development, 223–224
    electrode guns, 229, 267
    immunocytochemical techniques, 310, 319
    medical applications, 224–228
    photomicrography, 194, 195
    principles, 228–229
    specimen preparation, 237–242, 243
    sputter coating, 216
    x-ray microanalysis, 286, 291

Scanning transmission electron microscopy
    (STEM), 141, 281, 286, 290
Scanning thermal microscopy, 342
Secondary electron imaging (SEI), 228, 231, 233,
    234, 237, 245, 291
SEM stubs, 240, 242
Shadow casting, 209–218
Signal-to-noise ratio (S/N), 229, 278, 326
Silicon monoxide gratings, 149
Silicone, 157, 287, 290
Silver methenamine, 302, 304
Skeletonization, 333, 336
Slam freezing, 251, 261, 263, 267, 271, 272
Slide stripping, 105–111
Sodium in microanalysis, 286, 293
Sorenson's buffer, 30
Specimen holders, 240–241
Spherical aberration, 117, 124–125, 129, 133,
    147, 148, 199, 202, 205, 281
Spurr resin, 32, 36, 37, 68, 91, 174, 179, 267,
    268, 269, 270, 272, 313
Sputter coating, 216–218, 224, 241–242, 353
Sputter ion pumps, 161–162
Stereology, 333, 335, 337–339
Stereoscopy, 280
Stop baths, 189–190
Sucrose, 10, 30, 313
    as a cryoprotectant, 250, 260, 261, 262
Support films, 103–112

Tannic acid, 20
Tantalum, 211
Texas red, 195
Thiocarbohydrazide (TCH), 237, 302
Tissue hardening
    fixation associated, 10–11, 14, 16, 18
Titanium, 253, 290
Toluidine blue O, 67
Tonofilaments, 71, 72
Transmission electron microscopy, 223, 244
    buffers, 24–32
    computer-assisted imaging, 334
    cryotechniques, 249, 250, 260, 262, 265, 270
    dark-field imaging, 180–181
    dehydration, 32–34
    depth of focus, 118–119, 124, 125
    development, 114–117
    electrode guns, 285, 286
    electron scattering, 209
    embedding media, 34–39
    fixatives, 5–24
2,4,6-Trinitrocresol, 20
Tris, 31

Tris-maleate buffer, 302
Triton X-100, 311
Tungsten
    as a filament, 211, 277, 281
    as a source of illumination, 115, 131, 133
    electron gun, 133
    for shadowing, 211

Ultramicrotome, 352
    features, 79–80
    knives, 79–87
    problems with, 97–101
    sectioning, 91–97
Ultramicrotomy, 249
Uranium, 116, 126, 286
    for contrast building, 278, 290
Uranyl acetate, 20, 31, 33, 37, 175, 178, 180,
        263, 264, 272
    as a negative stain, 220

Vacuoles, 76
Vacuum evaporation, 216, 217
Vacuum evaporator, 211–214, 224, 241, 242, 352

Vacuum gauges, 151–155
Vacuum pumps, 151, 155–162, 353
Vacuum seal lubrication, 165
Vacuum systems, 287, 341, 342
Veronal acetate, 31
Vestopal W, 35, 36
Vibration, 279
Vinylcyclohexene dioxide, 37
Viruses, 179, 218, 314
    negative staining, 220
    support films for, 103

Wehnelt assembly, 133, 134, 143, 228
Wetting agents, 221
Wheat germ agglutinin, 305
White radiation, 289

X-ray microanalysis: see Energy-dispersive spec-
    troscopy (EDS)
X-ray fluorescence, 288
X-rays, 141, 148, 229, 230, 236, 322, 323

Zwitterionic buffers, 32